# CONNECTIONIST MODELS AND THEIR IMPLICATIONS: READINGS FROM COGNITIVE SCIENCE

edited by

**David Waltz**
*Thinking Machines Corporation and Brandeis University*

**Jerome A. Feldman**
*University of Rochester*

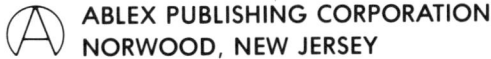

ABLEX PUBLISHING CORPORATION
NORWOOD, NEW JERSEY

Copyright © 1988 by Ablex Publishing Corporation.

All rights reserved. No part of this publication may be reproduced, stored in a retrieval system, or transmitted, in any form or by any means, electronic, mechanical, photocopying, microfilming, recording, or otherwise, without permission of the publisher.

Printed in the United States of America.

**Library of Congress Cataloging-in-Publication Data**

Connectionist models and their implications.

    Includes bibliographies and indexes.
    1. Artificial intelligence—Data processing.   2. Cognition.   3. Machine learning.   I. Waltz, David.   II. Feldman, Jerome A.
Q336.C66    1988              006.3                      87-31881
ISBN 0-89391-456-8

Ablex Publishing Corporation
355 Chestnut St.
Norwood, NJ 07648

# Contents

Preface: Connectionist Models and Their Prospects   **vii**
   *Marvin Minsky*

1. Connectionist Models and Their Implications   **1**
   *David Waltz and Jerome A. Feldman*

2. Connectionist Models and Their Properties   **13**
   *J.A. Feldman and D.H. Ballard*

3. Putting Knowledge in its Place: A Scheme for Programming Parallel Processing Structures on the Fly   **63**
   *James L. McClelland*

4. Positive Feedback in Hierarchical Connectionist Models: Applications in Language Production   **97**
   *Gary S. Dell*

5. A Developmental Neural Model of Visual Word Perception   **119**
   *Richard M. Golden*

6. Computing with Connections in Visual Recognition of Origami Objects   **155**
   *Daniel Sabbah*

7. Massively Parallel Parsing: A Strongly Interactive Model of Natural Language Interpretation   **181**
   *David L. Waltz and Jordon B. Pollack*

8. Feature Discovery by Competitive Learning   **205**
   *David E. Rumelhart and David Zipser*

9. Competitive Learning: From Interactive Activation to Adaptive Resonance   **243**
   *Stephen Grossberg*

10. A Learning Algorithm for Boltzmann Machines   **285**
   *David H. Ackley, Geoffrey E. Hinton, and Terrence J. Sejnowski*

11. Emergence of Grandmother Memory in Feed Forward Networks: Learning with Noise and Forgetfulness   **309**
   *R. Scalettar and A. Zee*

12. An Implementation of Network Learning on the Connection Machine   **329**
   *Charles R. Rosenberg and Guy Blelloch*

13. Connectionist Representation of Concepts  341
    *Jerome A. Feldman*

    Author Index  365

    Subject Index  371

*For Bonnie and Janice*

# Preface:
# Connectionist Models and Their Prospects

MARVIN MINSKY

*M.I.T. Artificial Intelligence Laboratory\**
*& Thinking Machines Corporation*

Why is there so much excitement about connectionist research today? Some researchers simply want machines to do the various sorts of things that are usually called intelligent. Others hope to understand what makes people able to do such things. Still others object to writing programs, year after year after year: how much more pleasant it would be, instead of our having to do all that work, if we could build, once and for all, machines that learn to improve themselves. Whenever we wanted something new, we would simply explain to them what we want—and let them try experiments, or read some books, or go to schools—the sorts of things that people do. Why can't we make machines like us, that grow through learning from experience?

While those researchers all have their different goals, they all seek ways to make learning machines. One approach to pursuing that quest is to start with a top-down strategy: begin at the level of commonsense psychology and try to imagine processes that might explain how someone plays a certain game, solves a certain kind of puzzle, or recognizes a certain kind of object. If you cannot find a simple way to do such complicated things, then do it by constructing a network of many simpler processes. In following this strategy, you keep breaking down those processes into simpler parts until you can actually build them, either in hardware or in software. This type of top-down strategy is typical of the approach to AI called heuristic programming, which has developed productively for several decades. However, over the past few years, it has become noticeably harder to carry out this type of research. This is natural because, after all those decades of development, new workers have much more to learn—and they face a difficulty squared, because so many easy problems already have been solved that the ones that remain are much harder.

Another approach goes the opposite way: to begin with parts that we understand. Then work upwards in complexity, to find out how to interconnect those units to produce the larger scale phenomena that we're trying to explain. In pursuing this bottom-up approach, one can start with almost

---

\* This research was sponsored by the Computer Science Division of the Office of Naval Research.

anything—with small computer programs, elementary logical principles, or simplified models of what brain cells do. This type of bottom-up strategy is typical of the approach to AI called connectionism. This field grew sporadically more than did heuristic programming, in several separate fits and starts. Why did it have to wait so long before it started flourishing? For one thing, heuristics developed so quickly that connectionist networks were swiftly outclassed. Furthermore, connectionist experiments required prodigious amounts of computation that only became available over the past few years. As Rosenberg and Blelloch demonstrate in this book, it has recently become feasible to build computer hardware that is more than adequate enough. But there was yet another, more serious reason why connectionism had to wait so long: we simply did not know enough about how to embody ideas in machines. Effective theories about the subject of knowledge representation did not start to mature until the 1970s. We explain this point in more detail in the Prologue of reference.[1]

Which approach is best to pursue? The answer is simple: we have to use both. In favor of the top-down side, research in AI has told us a little—but only a little—about how to make machines solve problems by using methods that resemble reasoning. In favor of the bottom-up approach, the brain sciences has told us a little—but again, only a little—about what brain cells do. If we knew enough more about brain cells and their connections we could try to work from that to discover how they support our higher level processes. If we understood more about thinking we work down toward finding out how brain cells do it. But right now we're caught in the middle; we know too little at either extreme. The only practical option is to ping-pong between them, searching for materials which with to build a plausible bridge. How can we do that? One way is to focus on inventing various ways to represent knowledge, and then to try to extend those techniques in both directions. On the connectionist side we can try to design neural networks that can learn those representations. Then we can try, on the top-down side, to design higher level systems that can effectively exploit the knowledge thus represented.

The dream of the early connectionists was to start with virtually nothing at all but a loosely connected network of parts that would somehow be able to learn by itself. Our hope was that once some modest goals were achieved, there would then be little more to do except enlarge those miniature nets until they had enough capacity to learn to become intelligent! Several such systems were actually built and they learned to accomplish various things—but none of those systems got very far. What led us to think that they should have done more? There are many different reasons why weighted-connection learning machines seemed promising. For example, connectionist networks

---

[1] Marvin Minsky and Seymour Papert, *Perceptrons,* (2nd edition) MIT Press, 1988.

resemble what we think we see in brains. We knew that they can be designed to recognize many types of patterns. It was obvious that their redundancy could make them resistant to noise and injury. We already knew some simple algorithms that could indeed permit them to learn, and it seemed quite clear that, eventually, they should be able automatically to discover various types of generalizations, without having to be supplied in advance with predefined representation schemes. And because the units of those networks all operate simultaneously, they promise to offer the power and speed of genuine parallel computation.

Given so many advantages, why ever use discrete, symbolic schemes at all? The trouble is that all connectionist schemes are rife with problems of their own. The adventurous essays in this book discuss these problems in various ways, most often in terms of how they bear on particular experiments. Here I'll try to set the stage by focusing on the interplay between these different strategies for synthesizing intelligence.

## LEARNING AND SEARCH

In order for a machine to learn, it needs to have potential ways to represent what it may learn. What representation should we choose? Over the years researchers in AI have made many theories about this, and have done many experiments on knowledge representation schemes, such as those called Semantic Networks, Conceptual Dependency, Frames, Predicate Calculus, Rule-Based Productions, Procedural Representations, and quite a few others. This is not the place to review the features of such representations. Instead we'll contrast all of them with the methods used in connectionist nets. The basic goal in that enterprise is to embody knowledge into the conductivities or weights assigned to connections among a network of nodes. The most common form of such a node consists of a linear part that "adds up evidence" and a nonlinear part that "makes a decision".

In principle we can construct such a network to represent any computable function. Consequently, any of those other types of knowledge representation could be encoded into such a network. In practice, however, the linear, additive aspect of typical connectionist nodes can lead to problems because addition itself is so fundamentally opaque. By that I mean that *once several numbers are added up, one cannot recover, from their sum, the inputs that were thus combined.* There is a spectrum of possible ways to deal with this basic problem of opacity; in fact, this can be seen as breaking up the entire field of research on connectionist nets. The problem of opacity becomes increasingly severe as the density of connections grows. When each node connects only to a relatively few others, then we have structures that resemble what AI researchers call Semantic Networks; the elements of those types of knowledge representations are comparatively *localized*. Some of the models in this book are of this highly localized type. Other models in this book use networks in which a typical node sums a relatively large number of contributions from different sources; these are called *highly distributed*. Most of the points of discussion below are directed toward the latter, so it is particularly important, here, to emphasize that there are important differences between localized and distributed representations.

In any case, once we can represent knowledge in terms of connection weights, it becomes very easy to formulate the problem of learning. This is because we can imagine using any of the well known techniques for "hill-climbing" or gradient ascent. To do this, we merely need to express our evaluation of a network's performance in terms of a single numerical success function. Then problem of learning can be reformulated in terms of searching to find the maximum value of that evaluation functions. The terrain to be searched is simply the vector space of the connection coefficients inside our network. Of course, this problem is simple only in principle, because any strategy based on gradient ascent can fail by getting stuck upon a local, isolated peak whose altitude is relatively insignificant. There simply is no local way to ensure that any such procedure will always reach a global maximum, instead of becoming trapped upon some local feature of topography such as a terrace, ridge, or peak.

Sometimes one can escape from traps by making occasional random jumps—and many people even hold that in this lies the key to creativity. One such strategy is the method called "annealing"; the essay by Ackley, Hinton, and Sejnowski demonstrates that this can work effectively on certain types of problems. But we still have little insight into which classes of problems can be treated that way; indeed, it is easy to construct examples in which annealing leads to *worse* results than would come from complete, exhaustive search. Such problems not only arise in AI: they lie at the heart of evolution itself. For example, it seems almost a truism that "most mutations are deleterious," but it is important to see why this is so. Whenever we see a live animal, we're seeing a system that is highly evolved: in other words, it is virtually certain already to stand on a local peak! Because much of the nearby territory has already been explored, the present location where it stands is likely to be quite close to the best that lies in the structural neighborhood. Therefore, mutations will tend to be bad because they will naturally tend to *undo* the work of selection that was already done in the animal's evolution. To be sure, there is always a chance to find a better place by making very large random jumps. But unless we do this selectively, the results can be worse than exhaustive search. Annealing may seem efficacious at first, when applied to systems in random states—but it won't fare so well when applied to systems that are already in more highly evolved states. For the more we've invested in finding *this* peak, the more will be wasted, of what we have learned, as soon as we jump away from it.

So, contrary to common belief, such methods can only ameliorate, but never eliminate all the types of difficulties that gave rise in the 1950s to the field of heuristic programming. No matter how hard we continue to try to extend the powers of methods based on local search, such methods can take us just so far; that search itself will end up trapped upon some abstract, unknown peak in the strategy space of search machines. Finding a peak is a means, not an end; it rarely is our real goal. Instead of just seeking escapes from traps, we might better use those peaks as clues at a deeper level of analysis. When the problem we're solving is easy enough, it may suffice just to climb its hill. But when our problems are deeper than that, then, in place of simply climbing those hills, a better goal would be, instead, to ask ourselves what causes those hills. Which of those peaks reflect inherent structural aspects of the problems we're trying to solve, and which of them are artificats of the representations we happen to choose?

## INSULATION AND INTERACTION

To see what could cause a local peak, imagine a certain animal in which some new mutated gene produces a substance S that comes to play two different, vital roles at once—one in the heart and one in the brain. If that animal's descendants can now do more by using less, the natural selection will tend to further improve that gene and disperse it among the flock. But consider the cost of that short-term gain. That double-purpose protein will be an obstacle to further improvements in either heart and brain! Any change in S that strengthens one would almost surely weaken the other, because the earlier form of the gene has already evolved to constitute the best available compromise. Now our new, mutated animal is doomed to be very slow to evolve because each further change in S disrupts so many processes. By breaking down the separateness of the mechanisms of the heart and the brain, S constitutes a new constraint that keeps them from learning independently. Each such constraint makes it harder to change—and the short term gain from finding two different uses for S is a long term evolutionary liability. The peak upon which our species is stuck is actually an artifact—produced by causes that interact only through an evolutionary accident that constrains a sum to a smaller result than we could get by separately climbing two different peaks.

```
       A/\               /\B     (P traces hill)
      /   X————P————X     \      (Supply Drawing)
  ___/     _____/      \___
```

How does our evolution ever manage to escape such double-purpose deadlock states? Our very early ancestors evolved a trick that has become an essential part of all of our subsequent evolution. The secret lies in the simple fact that the processes used to copy genes are prone to make *duplicate strings of genes.* Then, whenever a duplicate gene mutates, its unchanged twin can still perform its original function, while the variant gene can drift along a separate evolutionary track. Two versions of the gene for S could thus then manufacture different chemicals, one of which can improve the heart, while the other enhances the brain. We usually think in positive terms about making wholes by combining parts. But when interactions lead to inefficiency, we may need negative connections—call them insulations—to keep things from getting confused.

The Double-Purpose Deadlock

It is not only in connectionist nets that multi-purpose deadlocks can be deleterious. It can happen as well in symbolic realms. Suppose that a certain concept C was learned in a certain context X—but now we want to apply that skill to another, different context Y. If C attempts to continue to learn in both contexts, then the more that X depends on C's original details, the more X may be handicapped when those details are modified to suit the demands of Y. Again we could try to avoid this by providing Y with its own duplicate of C—but if too many copies learn different things, our system will tend to split into incoherent accumulations of non-sharable information. Our decades of experience with serial computers have taught us something about how to avoid massive duplication using inheritance-based "virtual" copies. But we still know very little about how to approximate the virtues of virtual copies in connectionist networks. One way is to use special recording units (called K-lines[2]) to remember various features of a network's activation state. Such records will be useful only to the extent to which the network actually represents features that can help to distinguish between the contexts in question. Consequently, it is very important that our networks be able to represent concepts like the "microfeatures" described in the essay by Waltz and Pollack.

How to combine the advantages of symbolic AI and connectionist nets? Symbolic certainly has developed better methods for controlling search. We

---

[2] Marvin Minsky, *The Society of Mind,* Simon and Schuster, 1987.

should be able to embody these into systems in which some networks supervise the experiments that occur inside some other nets. Symbolic AI researchers have developed many powerful ways to represent knowledge—but have not developed adequate ways to make those systems develop good new representations; here connectionist methods could come to their aid. And today's symbolic systems are particularly poor at discovering, without external help, particular knowledge to fill their rules and frames and scripts and semantic nets. Connectionist methods could help in all these areas—but not until they get better at search. Perhaps one way to build a bridge could be through making systems that augment our schemes for local search by knowledge-based analysis. That could begin by developing multi-section learning schemes in which some sections supervise how other sections work and learn. In chapters 4, 5, and 6, Sabbah, Golden, and Dell all discuss some prospects of, and problems in, assembling hierarchical networks. Eventually, as we learn more about such matters, we shall begin to attack the longer range goal of discovering how to enable such systems to construct productive new reformulations of problems.

## INTERMEDIATE UNITS AND SIGNIFICANCE

The problem of opacity becomes increasingly acute as representations become more distributed. As McClelland points out, the unlocalized ingredients of such representations make it hard to assign credit where credit is due. More generally, diffuseness makes it difficult for external systems to exploit the knowledge enmeshed inside those nets, say, for making new hypotheses. So, while distributed systems may work very well, they eventually should begin to fail when confronted with problems on larger scales. Unless the process of learning somehow leads to internal simplification and localization, we can expect to find such systems reaching limits to their conceptual growths. But I don't intend this to sound so bad since, for all we know at the present time, the scales at which such systems crash may actually be large enough for many of our purposes. Indeed, the thesis we call the Society of Minds holds that the most of the "agents" that grow in our brains do indeed need only to operate on scales so small that each, by itself, might seem as scarcely more than a toy. (See footnote 2.) But when we combine enough of them—but *not* in too highly distributed ways—we can make them do most anything.

In any case, we shouldn't conclude that we always can—or always should—avoid the use of opaque schemes. The circumstances of everyday life constantly compel us to make decisions based, in effect, on "adding up the evidence". We frequently find, when we value our time, that, even if we had the means, it wouldn't pay to analyze. Nor does the Society of Mind approach make everything more tractable; on the contrary, that theory, too, makes us expect to encounter incomprehensible representations at every level of the mind. Rarely can an agency do very much more than exploit what it learns about what other agencies can do. Few parts of our mind are capable of learning how any other parts work. In chapter 13, Feldman mentions various writers who rejoice in the holistic quality of representations from which one cannot extract significant parts and relationships. The trouble is that this limits growth because "the idea of a thing with no parts provides nothing that we can use as pieces of explanation." (See footnote 2.)

I did not mean to insinuate that these problems of opacity are in any way peculiar to connectionist models. We can see a complementary kind of obscurity in those popular symbolic schemes that represent functions in terms of disjunctions composed completely of Boolean terms, each of which has virtually no individual significance. Those canonical forms are capable in principle of representing anything—but only in terms of components so small that they disperse all traces of the representee's character. It would seem that we can lose our game with representations at both extremes—that are either too diffuse or too discrete. We may not recognize the loss for problems that are easy to solve, but trouble will come when we ascend to higher scales. This is because comprehending complex situations usually hinges on being able to find compact summaries or, what is the same thing, expressive analogies. And solving hard problems almost always turns around composing meaningful variations on themes with which we are already familiar. But it is virtually impossible to do this with representations, like logical forms or linear sums, whose elements seem meaningless because they are either too small or too large. Again we must seek some middle ground—and there are plenty of intermediate-level representations described in this book, *viz.,* those microfeatures, or the grandmother cells of Scalettar and Zee, or the groupings produced by competitive learning schemes like those described in the chapters by Grossberg and by Rumelhart and Zipser.

## THE DREAM OF ESCAPING FROM PROGRAMMING

The future work of mind design will not be much like what we do today. Some programmers will continue to use traditional languages and processes. Other programmers will turn toward new kinds of knowledge-based expert systems. But eventually all that will be succeeded by systems that exploit two new kinds of resources. One type will come supplied with hugh preprogrammed reservoirs of commonsense knowledge. The second and complementary type will arrive with a wide variety of powerful learning machines equipped with no knowledge at all.

Then what we know as programming will change its character entirely—to an activity that I envision as very much like sculpturing. To program today, we must describe things very carefully, because nowhere is there any margin for error. But once we have modules that know how to learn, we won't have to specify nearly so much. Instead we'll be able to get what we want by using techniques that I imagine as vaguely analogous to what Feldman and Ballard call "Coarse-Fine Coding": we'll program a vastly grander scale, with learning filling in all small details.

This doesn't mean, I hasten to add, that things will be simpler than they are now; instead we'll make our projects more ambitious. Designing an artificial mind will be much like evolving an animal. Imagine yourself at a terminal, assembling various parts of a brain. You'll be specifying the sorts of things that we've only seen described heretofore in texts about neuroanatomy. *"Here,"* you'll find yourself thinking, *"we'll need two similar networks that can learn to shift time-signals into spatial patterns so that they can be compared by, umm, by a feature extractor sensitive to a context about, umm, this wide."* Then you'll have to sketch the architectures of organs that can learn to supply appropriate inputs to those agencies, and draft the outlines of intermediate organs for learning to suitably encode the outputs to suit the needs of other agencies.[3] A functional sketch of such a design might turn out to involve dozens of different sorts of organs, centers, layers, and pathways. The human brain might have many thousands of such components.

A functional sketch is only the start. Whenever you employ a learning machine, you must specify a great deal more than merely the sources and destination of the data-level information. For every learning organ needs some signals to indicate what it should learn—for comparing results, testing hypotheses, and selecting suitable goals. Each choice of design must somehow determine how long the learner should persist when progress slows. How to decide when enough has been done; which particular learning procedures to employ; when to decide that things have gone badly wrong; how to determine the allocation of hardware, time, memory, and other resources? When enough such things are taken into account, our sculpturing art may look more like a weird form of management skill than anything we would recognize today as programming. For we will have to decide which agencies should provide what *incentives* for which others—and then we'll have to decide who will watch those watchers. As in any society, every such decision about one agency imposes additional constraints and requirements on several others—and then we have to specify how to train *those* other agencies as well.

You might object that all this, too, might only be a transient step. One eventual outcome could be for us to end up having developed enough large, off-the-shelf systems to do most of what we want. Another outcome that would be discovering how to build a single, great net that can learn to do all

---

[3] Section 31.3 of footnote 2 suggests how the genetic systems for human brains might sculpture the outlines of an agency that then would be predestined to learn to recognize the presence of particular human individuals.

those things by itself. But the only example of this that we know was the slow and prodigal evolution that eventually led to our own human brains. We *could* regard this as proving that just such a large-scale learning search is feasible—but only by making ourselves ignore the colossal size of that billion year search, and its cost of the lives of chemical *moles* of very respectable animals. Remember, too, that even after all that time, not all of the problems have been solved. We have to raise one issue more: what will we do when our sculptures don't work? Consider a few of the wonderful bugs that we ourselves remain subject to:

- obsessive preoccupation with inappropriate goals.
- inattention and inability to concentrate.
- bad representations.
- excessively broad or narrow generalizations.
- excessive accumulation of useless information.
- superstition; defective credit assignment schema.
- unrealistic cost/benefit analyses.
- unbalanced, fanatical search strategies.
- formation of defective categorizations.
- inability to deal with exceptions to rules.
- improper staging of development, or living in the past.
- unwillingness to acknowledge loss.
- depression or maniacal optimism.
- excessive confusion from cross-coupling.

To suppress the emergence of serious bugs, our systems will require intricate arrangements of interlocking checks and balances, in which every agency is supervised by several others, each of which must learn when and how to use the resources available to it. What makes our human brains so versatile and reliable? In *The Society of Mind* I argue that it is no accident that our brains contain so many different and specialized brain centers. In most cases we simply don't yet know what those various organs do, but I'm willing to bet that many of them serve mainly to regulate others so as to keep the system as a whole from frequently falling prey to the sorts of bugs we mentioned above. Until we start building brains ourselves, to learn what bugs are probable, it may remain hard for us to guess what much of that hardware is actually for.

There are countless wonders to be discovered today, in these exciting new fields of research. We still can learn a great many things from experiments, on even the simplest nets. We'll learn even more from trying to make theories about what we see. Soon we shall start to prepare for that future art of mind design, by experimenting with societies of nets that embody more subtle strategies. Eventually we'll develop constructive theories about such systems and start to use them to analyze the networks that make up our own human minds.

CHAPTER 1

# Connectionist Models and Their Implications

The use of computational theories and machines to study and emulate intelligent activity is an old enterprise. In the era of electronic digital computers, this effort has taken form and has led to considerable scientific and practical progress. Until quite recently, attempts to model intelligence have followed two apparently incompatible paths (Newell, 1983). One line of research started with the computational properties of neurons and neural assemblies and tried to understand how such systems could compute and learn. The other, more dominant, tradition started from behavioral descriptions of intelligent activity and modeled such activity using computer programs and other formal systems of rules. Despite considerable success, the logical rule approach to artificial intelligent (AI) has encountered difficulties that have led some workers to question the paradigm itself. At the same time, rapid progress in the behavioral and brain sciences has led to a much richer understanding of the structure and function of natural intelligence. One view of connectionist research, such as the papers collected here, is an attempt to synthesize the best features of the neural and logical approaches. The central ideas of neural modeling are cooperative computing, error tolerance, and adaptation. The key notions from logic-based AI are representation and inference. These need not be incompatible; the articles in this book indicate how scientists from a variety of perspectives are beginning to explore the properties of connectionist models.

For most of its history, the heuristic search, logic, and "physical symbol system" (Newell, 1980) paradigms have dominated AI. AI was conceived at about the same time that protocol analysis was in vogue in psychology (Miller & Pribram, 1954); such protocols could be implemented on the then-new von Neumann machines fairly well. Protocol analysis suggested that people operate by trial and error, using word-like objects as primitive units. For the first ten years of its history, AI entertained both symbolic/heuristic search models and neural net/perceptron (Rosenblatt, 1962) models, but by 1967, it had become clear that the neurally-inspired models in use had serious limitations (Minsky & Papert, 1969), and AI interest in them abruptly waned. AI stuck almost exclusively with heuristic search and symbol systems for the next 15 years, using them in a wide variety of AI systems and programming languages, ranging from ATN's, most other natural language parsing systems, and planning based models (e.g., for pragmatics) to Prolog and Planner (Hewitt, 1970).

However, by the early 1980s it was becoming clear that "traditional" symbolic AI models had serious limitations as well:

- Problems in scaling up "expert systems" led to increased interest in learning and "knowledge acquisition," and also led to strong interest in "non-monotonic logics" for solving the "frame problem" (roughly, deciding what to change and what not to change when a robot's model of the world changes) and the "qualification problem" (trying to list all the exceptions that could preclude the applicability of a rule).
- Problems in representing fine shades of meaning led to an abandonment of most of the natural language representation schemes of the 1970s. At the same time, natural language processing systems based on heuristic search have had little success in modeling human linguistic performance. Natural language understanding clearly involves subtle and intricate subconscious processes (e.g., memory, inference, weighing and combining of evidence) that seems very difficult to model with traditional AI methods.
- Computer vision, which never had much stake in the heuristic search model, and had pursued its own path of mathematically grounded and partially neurally inspired models, flourished, and began to exploit new special-purpose parallel hardware (Ballard, Hinton & Sejnowski, 1983). Neuroscience continued a steady advance in knowledge and technology: much finer probes of nervous system activity became available, including non-invasive methods such as those that detect nuclear magnetic resonance (Crick & Asanuma, 1986). Experimental psychology developed refined methodologies and began to study temporal and other interactions more seriously (Posner, 1978).

In this atmosphere, the field of "connectionist modeling" has experienced a strong resurgence that shows little sign of leveling off.

Connectionist systems have stirred a great deal of excitement for a number of reasons:

- They're novel. Connectionism seems to be a good candidate for a major new paradigm in AI where there have only been a handful of paradigms (heuristic search; constraint propagation; blackboard systems; marker passing).
- They have cognitive science potential. While connectionist neural nets are not accurate models of neurons, they do seem brain-like and capable of modeling a substantial range of cognitive phenomena. (cf. Chapters 4, 5, 6, 7)

- Connectionist systems have exhibited non-trivial learning; they are able to self-organize, given only examples as inputs. (cf. Chapters 5, 8, 9, 10, 11, 12)
- Connectionist systems can be made fault-tolerant and error-correcting, degrading gracefully for cases not encountered previously (Waltz & Pollack, 1985). (cf. Chapters 4-12)
- Appropriate and scalable connectionist hardware is rapidly becoming available. This is important, both for actually testing models, and because the kinds of brain and cognitive models that we build are very heavily dependent on available and imaginable hardware (Backus, 1977; Feldman, Fanty, Goddard, & Lynne; Pylyshyn, 1980). (cf. Chapter 13)
- Connectionist *architectures* also scale well, in that modules can be interconnected rather easily. This is because messages passed between modules are generally activation levels, not symbolic messages. No systems large enough to exploit this have yet been built, but the way to proceed is clear.

Nonetheless, there are considerable difficulties still ahead for connectionist models. It is premature to generalize based on our experience with them to date. So far all systems built have either learned relatively small numbers of items (Blelloch & Rosenberg, 1987; Grossberg, 1987; Rumelhart & Zipser, 1985), or they have been toy systems, hand built for some particular task (Cottrell & Small, 1983; Sabbah, 1985; Shastri, 1985; Waltz & Pollack, 1985). The kinds of learning shown to date are hardly general. It seems very unlikely to us that it will be possible for a single, large, randomly wired module to learn *everything*. If we want to build a system out of many modules, we must devise an architecture for the system with input and output specifications for modules and/or a plan for interconnecting the internal nodes of different modules. Finally, connectionist models cannot yet be argued to offer a superset of traditional AI operations: certain operations such as variable binding and recursion cannot yet be performed efficiently in connectionist networks.

**Best Match versus Exact Match**
Despite these very real difficulties, the connectionist paradigm offers some fundamental advantages. In general, it is not possible to specify completely the conditions for any sort of decision; arbitrarily large numbers of rules with arbitrarily long sets of conditions may be required. Connectionist models inherently are able to integrate *all* available evidence, most pieces of which will be irrelevant or only weakly relevant for most decisions. Moreover, one does not have to find logically necessary and sufficient conditions; statistical correlations between actions and the facts of the world can be represented as weighted connections. In basic character, connectionist reasoning is *evidential* rather than *logical*.

Reasoning that is *apparently* logical can emerge from connectionist models in at least two ways. (1) A programmer can encode individual alternatives for conceptual interpretations of inputs as nodes which compete with or support each other; the processing of an input then involves clamping the values of some input nodes, and allowing the whole network to settle. For "regular" inputs, strong pathways, which "collaborate" in reinforcing each other, can give the appearance of rule-like behavior. Given similar inputs, one can expect similar outputs. Most natural language connectionist work has been rule-like in this sense (Cottrell & Small, 1983; Selman & Hirst; Small, 1980; Waltz & Pollack, 1985). (2) Connectionist learning networks and associative memory models can find activation patterns or memories which are close to a current input event or situation. Such systems degrade gracefully as inputs vary slightly or as noise is added to inputs. Thus, they too exhibit rule-like emergent behavior. Unlike expert systems, these "rules" apply to full situations, and chaining is not involved. (See also Grossberg, 1987; Stanfill & Waltz, 1987).

In contrast, traditional AI systems based on logic, unification and exact matching are inevitably brittle (i.e., situations even slightly outside the realm of those encoded in the rules fail completely, and the system exhibits discontinuous behavior). We see no way to repair this property of such systems (See also Nilsson, 1983; Pentland & Fischler, 1983).

**Match with Psychological Results**

Psychological research on categorization (Berlin & Kay, 1969; Lakoff, 1987; Rosch & Mervis, 1975; Smith & Medin, 1981) has shown that category formation cannot be explained in a classical logical model. That is, the conditions of category membership are not merely logical conditions (result of expressions with connectives 'and,' 'or,' and 'not'). Rather, categories are organized around "focus concepts" or prototypes, and exhibit graceful degradation for examples that differ from the category focus along any of a number of possible dimensions (Lakoff, 1987). Connectionist systems seem well-suited for modeling such category structure (though such modeling has not been explored very extensively (Hinton & Anderson, 1981)).

**Massive Parallelism**

"Toy" AI problems and microworlds can shed only limited light on intelligence. Eventually we must run large problems and be able to see the results of experiments in finite time. Small scale experiments (involving fewer than thousands of nodes or concepts) are inadequate to really advance the frontiers.

Fortunately, dramatic strides are being made in massively parallel computer architecture at just the time that connectionist theoretical models are being explored. These fields are not unrelated. Connectionist models

(Fahlman, 1982; Quillian, 1968) served as initial inspiration to designers of new generation hardware (Hillis, 1985), though many parallel architectural ideas were already being explored purely in the pursuit of greater speed. This followed the realization that we were approaching asymptotes for speeds possible with serial uniprocessors. The rate of hardware progress has been very rapid, especially for massively parallel processors that are ideally matched for connectionist models. It seems obvious that developing appropriate hardware will prove to be the easiest part of building full-scale AI systems.

**Integration of Modules**
Connectionist models allow a much easier integration of modules than is possible with symbolic/heuristic search-based systems. Generally, symbolic systems require either a very simple architecture (e.g., natural language's traditional phonetic—syntactic—semantic—pragmatic bottom-up model adopted from classical linguistics) or a sophisticated communications facility (for example, a blackboard (Nii, 1986)) in order to build a system composed of many modules. In the blackboard model, each module must in general have a *generator* for complex messages as well as an *interpreter* for such messages.

In contrast, connectionist models allow an integration of modules by links that can go directly to the nodes (concepts or microfeatures) that co-vary with the activation patterns of other modules, and messages themselves can be extremely simple (e.g., numerical activation levels, or markers (Hendler, 1986)). In some cases, link weights can be generated based on an analysis of the statistical correlations between various concepts or structures; in other cases weights can be generated by learning schemes (Rumelhart, McClelland, & PDP Research Group, 1986). Nonetheless, still there is a potentially large set of cases where weights will have to be generated by hand, or by yet-to-be-discovered learning methods. Clearly, every concept cannot be connected to every other directly. (This would require $n^2$ connections for $n$ concepts, where $n$ is at least $10^6$.) Some solutions have been suggested (e.g., the microfeature ideas in (Waltz & Pollack, 1985) and the three-way connecting nodes of (Shastri & Feldman, 1985)) but none seems easy to program.

**Fault Tolerance**
Since a large number of nodes (or modules) have a bearing on a single connectionist decision then not all of them need to be active in order to make a correct decision; some variation of values can be tolerated. Many models, especially those used in associative memory models, exhibit "attractors," i.e., regions in input space, all of which drive the system to the same state (Hinton & Anderson, 1981; Hopfield, 1982). Distributed connectionist models (e.g., Ackley, Hinton, & Sejnowski, 1985; Blelloch & Rosenberg,

1987; Grossberg, 1987; Rumelhart, Hinton, & Williams, 1986) degrade gracefully when nodes are moved or weights altered.

**Learning**

Learning is one of the most exciting aspects of connectionist models for both the AI and psychology communities. For example, the back propagation error learning (Rumelhart, et al., 1986), Boltzmann machine (Sejnowski & Rosenberg, 1986), and adaptive resonance methods (Grossberg, 1987) have proved quite effective for teaching input/output patterns. However, such learning is not a panacea. Much remains to be done, especially for dealing with learning that takes place through time, with only occasional feedback. For the models listed above, full input/output patterns pairs must be known for all items to be taught; if any of several outputs would be satisfactory for a given input, such systems cannot easily learn this. Moreover, in the real world, one only occasionally gets immediate feedback, and then the feedback is generally just a reward/punishment signal. (See Barto & Anandan, 1984; Sutton & Barto, 1981 for recent progress on this type of problem).

## KEY PROBLEMS FOR CONNECTIONIST LANGUAGE MODELS

**Learning from "Experience"**

As suggested above, learning is both a key achievement of connectionism, and a key open issue for a full cognitive system.

The difficulty for cognitive learning theories of any sort is the observation that perception has to be prior to language. In turn, perception itself seems to require *a priori,* innate organization. Just how large must an innate component be? We believe it will have to account at least for such phenomena as figure/ground organization of scenes, the ability to appropriately segment events, (both to separate them from the experiences that precede and follow them and also to articulate their internal structure); the notion of causality; and general structuring principles for creating memory instances. This suggests to us that a large portion of a learning system must be wired initially, probably into fairly large internally regular modules, which are subject only to rudimentary learning via parameter adjustment.

This conclusion follows from the observation that if brains could completely self-organize, this method, being simpler than present reality, would have been discovered first by evolution. Our guess is that such total self-organization would require far too long, since it requires exploring vast space of weight assignments. Even given extensive *a priori* structure, humans require some twenty years to mature. We think that we cannot avoid programming cognitive architecture.

## Variable Binding

Some operations that programmers have traditionally taken for granted have proven difficult to map onto connectionist networks. One such key operation is variable binding. Assume that we have devised a good schema representation or learning system, and stored a number of schemas: What happens when a new input triggers a schema and we would like to store this experience in long-term memory? It seems that we need to create an instance of the schema with the particular agents, objects, patients, so on, bound to case roles. It is not obvious how this ought to be done in a connectionist model. One can treat the general problem of binding structures as an instance of association and use traditional matrix models (Smolensky, 1987).

Another possibility is to make a copy of the entire schema structure for each new instance, but this seems to lack neurophysiological plausibility. A more appealing direction is suggested both by Minsky (Minsky, 1980; 1987) and Feldman and Shastri (Feldman & Ballard, 1982; Shastri & Feldman, 1985): A very large number of nodes are randomly connected to each other such that nodes that have never been used before form a kind of pool of potential binding units for novel combinations of schemas and role fillers. When a new instance is encountered, all the participants which are active can be bound together using one or more of these previously unutilized binding nodes, and those nodes can then be removed from the "free binders pool."

Variable binding may turn out to be a profound issue in serial and parallel computation. Assigning a value to a global variable is inherently sequential and seems to be an irreducible bottleneck for all logic-based formalisms. Trying several bindings in parallel can quickly lead to combinatorial explosion unless the assigned values can interact. All of the proposed connectionist solutions to the binding problem are also serial and this might be a fundamental limitation. If so, we would have an instance where an arbitrarily parallel processor would be forced to resort to sequential operations.

There are important open questions in any case: For example, are different modules responsible for sentence processing, perceptual processing, short-term memory and long-term memory (Fodor, 1982)? If so, how are these interconnected and "controlled"? If not, how can we account for these different processing modes?

## Timing and Judging When Processing is Complete

Connectionist systems for processing input sentences have assumed that these sequences will be preceded and followed by quiescent periods. The resulting pattern of activations on nodes in the system can then be "read" whenever appropriate, and the time sequence of node actuations interpreted as desired. There is a real difficulty in knowing how and when one should interpret the internal operation of a system. Should we wait until activation levels on nodes have settled, that is, changed less than a certain amount on each cycle? Should we wait for activity to either be completely on or completely off in various nodes? Should we wait a fixed amount of time and then evaluate the network activation pattern? If so, how do we set the clock rate of the relaxation network relative to the rate at which input words arrive? What should be done to the activation pattern of a set of nodes after a sequence has been "understood"? Should the levels be zeroed out? Should they remain active? Under what circumstances and by what methods should items be transferred to (or transformed into) long-term memory? Are the nodes used in understanding the same ones responsible for long-term memory storage or is there some sort of copying or transfer mechanism?

All these questions cry out for crisper answers and principles. For natural language systems, processing must be approximately complete soon after the completion of a sentence so that processing of the next sentence can start, since sentences or clauses can occur with very little separation. This suggests in turn that expectations play a important role in sentence processing and further that really important material ought to appear or be expected well before the end of a sentence if the processing of the next sentence is not to be interfered with. Similar constraints seem to hold for perception and real-time problem solving as well.

**Debugging and Understanding Systems**

While local connectionist models (i.e., ones where nodes correspond to concepts (Feldman, 1982; Waltz & Pollack, 1985) can be understood, at least to some degree, by inspection, it is difficult to tell exactly what systems with distributed knowledge representations know or don't know. Such systems cannot explain what they know, nor can a person look at their structures and tell whether or not they are complete and robust, except in very simple cases (Hinton et al., 1986). The only way to test such systems is by giving them examples and judging them on the basis of their performance. This problem is a quite serious one for systems that are designed to be fault tolerant: debugging a system in which some modules frequently compensate for and cover up the errors of others may prove to be a rather intractable problem.

**Role of Learning**

It may be impossible to use human-like learning methods for connectionist systems (or for any computer-based intelligent system). It may also be undesirable. Unlike people, computers are capable of remembering literally the contents of large text files and complete dictionaries while at the same time they lack perceptual and reasoning facilities. The combination suggests that infant-like learning may not be appropriate for computer-based language systems, even if a brain-like machine can be built. The entire issue of how connectionist models relate to natural and artificial intelligence remains problematical.

**Contents of This Volume**

The articles collected here contain many common themes and many contrasting perspectives. Taken as a whole, they constitute a fair sample of the work on connectionist models in the mid-1980s. The chapters are organized in a structure based on the starting point of the investigation in each case. The chapters by Feldman and Ballard and by McClelland are focussed on computational problems arising in connectionist models and on possible solutions to them. Dell and Golden are among the many psychologists using connectionist models to address the traditional behavioral scientist's task of fitting experimental data. Dell uses an extremely compact ("grandmother cell") representation while Golden employs a highly distributed one and contrasts his model with a compact treatment of the same phenomena. The articles by Sabbah and by Waltz and Pollack are firmly in the tradition of AI performance programs. Sabbah is concerned with vision and Waltz and Pollack with natural language, but both articles stress the evidential nature of connectionist models and how it helps overcome the rigidity of logical formulations.

Only one of the early chapters (Golden's) deals seriously with learning. In contrast, the next five articles take learning as the central issue. The papers by Rumelhart and Zipser and by Grossberg are solidly in the tradition of neural net learning, although the three researchers have broader interests. The adversarial character of articles like Grossberg's is part of the development of the connectionist paradigm. The chapter by Ackley, Hinton and Sejnowski and the one by Scalettar and Zee approach learning in connectionist systems from the entirely different perspective, based originally on physics. Again, there is an interesting contrast in that the first article proposes an extremely diffuse representation while the second suggests the optimality of "grandmother cells" in some situations. The Blelloch and Rosenberg article describes additional experiments with back-propagation learning (Rumelhart & Zipser, 1985) and indicates the problems and potential of engineering connectionist solutions on highly parallel machines.

Massively parallel computational models are sometimes viewed as having profound implications for other cognitive sciences including philosophy. The article by Thagard and the subsequent exchange convey some of the flavor of this type of exploration. The final article suggests that connectionist computational considerations may be able to help resolve core issues in Cognitive Science such as the nature of neural concept representation. Like the rest of the book, it is an early and possibly faulty step.

DAVID WALTZ
*Cambridge, MA*

JEROME FELDMAN
*Rochester, NY*

May 1987

# REFERENCES

Ackley, D.H., Hinton, G.E., & Sejnowski, T.J. (1985). A learning algorithm for boltzmann machines. *Cognitive Science, 9,* 147-169.
Backus, J. (1987, August). Can Programming Be Liberated from the von Neumann style? A Functional Style and Its Algebra of Programs. *Communications of the ACM, 21*(8), 613-641. (1977 ACM Turing Award Lecture).
Ballard, D.H., Hinton, G.E., & Sejnowski, T.J. (1983, November). Parallel visual computation. *Nature, 306*(5938), 21-26.
Barto, A.G., & Anandan, P. (1984, December). *Pattern recognizing stochastic learning automata.* (Tech. Rep. No. 84-30). Amherst, MA: Computer and Information Science, University of Massachusetts at Amherst.
Berlin, B., & Kay, P. (1969). *Basic color terms: Their universality and evolution.* Berkeley and Los Angeles: University of California Press.
Blelloch, G., & Rosenberg, C. (1987). *Network learning on the connection machine.* (Tech. Rep. No. RL87-3). Cambridge, MA: Thinking Machines Corporation.
Cottrell, G.W., & Small, S.L. (1983). A connectionist scheme for modeling word sense disambiguation. *Cognition and Brain Theory, 6,* 89-120.
Crick, F., & Asanuma, C. (1986). Certain aspects of the anatomy and physiology of the cerebral cortex. In D.E. Rumelhart, J.L. McClelland, & the PDP Research Group. (Ed.), *Paralell distributed processing.* Cambridge, MA: MIT Press.
Fahlman, S.E. (1979). *NETL: A system for representing and using real-world knowledge.* Cambridge, MA: MIT Press.
Feldman, J.A. (1982). Dynamic connections in neural networks. *Biological Cybernatics, 47,* 27-39.
Feldman, J.A., & Ballard, D.H. (1982). Connectionist models and their properties. *Cognitive Science, 6,* 205-254.
Feldman, J.A., Fanty, M.A., Goddard, N., & Lynne, K. (1987). Computing with structured connectionist networks. *Communications of the ACM.*
Fodor, J. (1982). *The modularity of mind.* Cambridge, MA: MIT Press.
Grossberg, S. (1987). Competitive learning: From interactive activation to adaptive resonance. *Cognitive Science, 11,* 23-63.
Hendler, J. (1986). *Integrating Marker-passing and Problem Solving.* Doctoral dissertation, Brown University.
Hewitt, C. (1970). PLANNER: A language for manipulating models and proving theorems in a robot. (AI Memo 168) MIT Artificial Intelligence Lab, Cambridge, MA.
Hillis, D. (1985). *The connection machine.* Cambridge, MA: MIT Press.
Hinton, G.E., & Anderson, A. (Eds.). (1981). *Parallel models of associative memory.* Hillsdale, NJ: Erlbaum.
Hinton, G.E., McClelland, J.L., & Rumelhart, D.E. (1986). *Distributed representations: Parallel distributed processing.* Cambridge, MA: MIT Press.
Hopfield, J.J. (1982). Neural networks and physical systems with emergent collective computational abilities. In *Proceedings, National Academy of Sciences of the United States of America* (pp. 2554-2558).
Lakoff, G. (1987). *Women, fire and dangerous things.* Chicago: University of Chicago Press.
Miller, G.A., Gelanter, & Pribram, K. (1954). *Plans and the structure of behavior.* New York: Holt, Rinehart, and Winston.
Minsky, M.L. (1980). K-lines: A theory of memory. *Cognitive Science, 4,* 117-133.
Minsky, M.L. (1987). *The society of mind.* New York: Simon and Schuster.
Minsky, M.L., & Papert, S. (1969). *Perceptrons: An introduction to computational geometry.* Cambridge, MA: MIT Press.

Newell, A. (1983). Intellectual issues in the history of artificial intelligence. In F. Machlup & U. Mansfield, (Eds.), *The study of information: Interdisciplinary messages.* New York: Wiley.

Newell, A. (1980). Physical symbol systems. *Cognitive Science, 4,* 135-183.

Nii, H.P. (1986). Blackboard systems part two: Blackboard application systems. *AI Magazine, 7,* 82-106.

Nilsson, N.J. (1983). Artificial intelligence prepares for 2001. *AI Magazine, 4,* 7.

Pentland, A.P., & Fischler, M.A. (1983). A more rational view of logic or, up against the wall, logic imperialists! *AI Magazine, 4,* 15-18.

Posner, M.I. (1978). *Chronometric explorations of mind.* Hillsdale, NJ: Erlbaum.

Pylyshyn, Z.W. (1980). Computation and cognition: Issues in the foundations of cognitive science. *The Behavioral and Brain Sciences, 3,* 111-169.

Quillian, M.R. (1968). Semantic memory. In M. Minsky, (Ed.), *Semantic information processing.* Cambridge, MA: MIT Press.

Rosch, E., & Mervis, C. (1975). Family resemblances: Studies in the internal structure of categories. *Cognitive Psychology, 7,* 573-605.

Rosenblatt, F. (1962). *Principles of neurodynamics, perceptrons and the theory of brain mechanisms.* Washington, DC: Spartan Books.

Rumelhart, D.E., Hinton, G.E., & Williams, R.J. (1986). Learning internal representations by error propagation. In J.L. McClelland, D.E. Rumelhart, & PDP Research Group. *Parallel distributed processing.* Cambridge, MA: MIT Press.

Rumelhart, D.E., McClelland, J.L., & the PDP Research Group (Eds.). (1986). *Parallel distributed processing: Explorations in the microstructure of cognition.* (Volume 1 and 2). Cambridge, MA: MIT Press.

Rumelhart, D.E., & Zipser, D. (1985). Feature Discovery by Competitive Learning. *Cognitive Science, 9,* 75-112.

Sabbah, D. (1985). Computing with Connections in Visual Recognition of Origami Objects. *Cognitive Science, 9,* 25-50.

Sejnowski, T.J., & Rosenberg, C.R. (1986). *NETtalk: A parallel network that learns to read aloud.* (Tech. Rep. No. JHU/EECS-86-01), Electrical Engineering and Computer Science, The Johns Hopkins University.

Shastri, L. (1985, September). *Evidential reasoning in semantic networks: A formal theory and its parallel implementation.* Doctoral dissertation. Computer Science Department, University of Rochester. (Tech. Rep. No. 166).

Shastri, L., & Feldman, J.A. (1985, August). Evidential Reasoning in Semantic Networks: A Formal Theory. In *Proceedings of the IJCAI,* pp. 465-474. Los Angeles, CA.

Small, S. (180). *Word expert parsing: A theory of distributed word-based natural language understanding.* (Tech. Rep. 954), Department of Computer Science, University of Maryland, Baltimore.

Smith, E.E., & Medin, D. (1981). *Categories and Concepts.* Cambridge, MA: Harvard University Press.

Smolensky, P. (1987). *On variable binding and the representation of symbolic structures in connectionist systems.* (Tech. Rep. No. CU-CS-355-87). Dept. of Computer Science, University of Colorado, Boulder.

Stanfill, C., & Waltz, D.L. (1987). *The memory-based reasoning paradigm.* (Tech. Rep. No. RL87-2), Cambridge, MA: Thinking Machines Corporation.

Sutton, R.S., & Barto, A.G. (1981). Toward a modern theory of adaptive networks: Expectation and prediction. *Psychological Review, 88,* 135-170.

Waltz, D.L., & Pollack, J.B. (1985). Massively parallel parsing: A strongly interactive model of natural language interpretation. *Cognitive Science, 9,* 51-74.

CHAPTER 2

# Connectionist Models and Their Properties

J. A. Feldman and D. H. Ballard

*Computer Science Department*
*University of Rochester*
*Rochester, NY 14627*

Much of the progress in the fields constituting cognitive science has been based upon the use of explicit information processing models, almost exclusively patterned after conventional serial computers. An extension of these ideas to massively parallel, connectionist models appears to offer a number of advantages. After a preliminary discussion, this paper introduces a general connectionist model and considers how it might be used in cognitive science. Among the issues addressed are: stability and noise-sensitivity, distributed decision-making, time and sequence problems, and the representation of complex concepts.

## 1. INTRODUCTION

Much of the progress in the fields constituting cognitive science has been based upon the use of concrete information processing models (IPM), almost exclusively patterned after conventional sequential computers. There are several reasons for trying to extend IPM to cases where the computations are carried out by a parallel computational engine with perhaps billions of active units. As an introduction, we will attempt to motivate the current interest in massively parallel models from four different perspectives: anatomy, computational complexity, technology, and the role of formal languages in science. It is the last of these which is of primary concern here. We will focus upon a particular formalism, connectionist models (CM), which is based explicitly on an abstraction of our current understanding of the information processing properties of neurons.

Animal brains do not compute like a conventional computer. Comparatively slow (millisecond) neural computing elements with complex, parallel connections form a structure which is dramatically different from a high-speed, predominantly serial machine. Much of current research in the neurosciences is concerned with tracing out these connections and with discovering how they transfer information. One purpose of this paper is to suggest how connectionist theories of the brain can be used to produce testable, detailed models of interesting behaviors. The distributed nature of information processing in the brain is not a new discovery. The traditional view (which we shared) is that conventional computers and languages were Turing universal and could be made to simulate any parallelism (or analog values) which might be required. Contemporary computer science has sharpened our notions of what is "computable" to include bounds on time, storage, and other resources. It does not seem unreasonable to require that computational models in cognitive science be at least plausible in their postulated resource requirements.

The critical resource that is most obvious is time. Neurons whose basic computational speed is a few milliseconds must be made to account for complex behaviors which are carried out in a few hundred milliseconds (Posner, 1978). This means that *entire complex behaviors are carried out in less than a hundred time steps*. Current AI and simulation programs require millions of time steps. It may appear that the problem posed here is inherently unsolvable and that there is an error in our formulation. But recent results in computational complexity theory (Ja'Ja', 1980) suggest that networks of active computing elements can carry out at least simple computations in the required time range. In subsequent sections we present fast solutions to a variety of relevant computing problems. These solutions involve using massive numbers of units and connections, and we also address the questions of limitations on these resources.

Another recent development is the feasibility of building parallel computers. There is currently the capability to produce chips with 100,000 gates at a reproduction cost of a few cents each, and the technology to go to 1,000,000 gates/chip appears to be in hand. This has two important consequences for the study of CM. The obvious consequence is that it is now feasible to fabricate massively parallel computers, although no one has yet done so (Fahlman, 1980; Hillis, 1981). The second consequence of this development is the renewed interest in the basic properties of highly parallel computation. A major reason why there aren't yet any of these CM machines is that we do not yet know how to design, assemble, test, or program such engines. An important motivation for the careful study of CM is the hope that we will learn more about how to do parallel computing, but we will say no more about that in this paper.

The most important reason for a serious concern in cognitive science for CM is that they might lead to better science. It is obvious that the choice of technical language that is used for expressing hypotheses has a profound influence on the form in which theories are formulated and experiments undertaken. Artificial intelligence and articulating cognitive sciences have made great progress by employing models based on conventional digital computers as theories of intelligent behavior. But a number of crucial phenomena such as associative memory, priming, perceptual rivalry, and the remarkable recovery ability of animals have not yielded to this treatment. A major goal of this paper is to lay a foundation for the systematic use of massively parallel connectionist models in the cognitive sciences, even where these are not yet reducible to physiology or silicon.

Over the past few years, a number of investigators in different fields have begun to employ highly parallel models (idiosyncratically) in their work. The general idea has been advocated for animal models by Arbib (1979) and for cognitive models by Anderson (Anderson et al., 1977) and Ratcliff (1978). Parallel search of semantic memory and various "spreading activation" theories have become common (though not quite consistent) parts of information processing modeling. In machine perception research, massively parallel, cooperative computational theories have become a dominant paradigm (Marr & Poggio, 1976; Rosenfeld et al., 1976) and many of our examples come from our own work in this area (Ballard, 1981; Sabbah, 1981). Scientists looking at performance errors and other non-repeatable behaviors have not found conventional IPM to be an adequate framework for their efforts. Norman (1981) has recently summarized arguments from cognitive psychology, and Kinsbourne and Hicks (1979) have been led to a similar view from a different perspective. It appears to us that all of these efforts could fit within the CM paradigm outlined here.

One of the most interesting recent studies employing CM techniques is the partial theory of reading developed in (McClelland & Rumelhart, 1981). They were concerned with the word superiority effect and related questions in the perception of printed words, and had a large body of experimental data to explain. One major finding is that the presence of a printed letter in a brief display is easier to determine when the letter is presented in the context of a word than when it is presented alone. The model they developed (cf. Figure 1) explicitly represents three levels of processing: visual features of printed letters, letters, and words. The model assumes that there are positive and negative (circular tipped) connections from visual features to the letters that they can (respectively, cannot) be part of. The connections between letters and words can go in either direction and embody the constraints of English. The model assumes that many units can be simultaneously active, that units form algebraic sums of their inputs and output values proportionally. The activity of a unit is bounded from above and below, has some memory, and decays with time. All of these features, and several more, are captured in the abstract unit described in Section 2.

This idea of simultaneously evaluating many hypotheses (here words) has been successfully used in machine perception for some time (Hanson & Riseman, 1978). What has occurred to us relatively recently is that this is a natural mode of computation for widely interconnected networks of active elements like those envisioned in connectionist models. The generalization of these ideas to the connectionist view of brain and behavior is that all im-

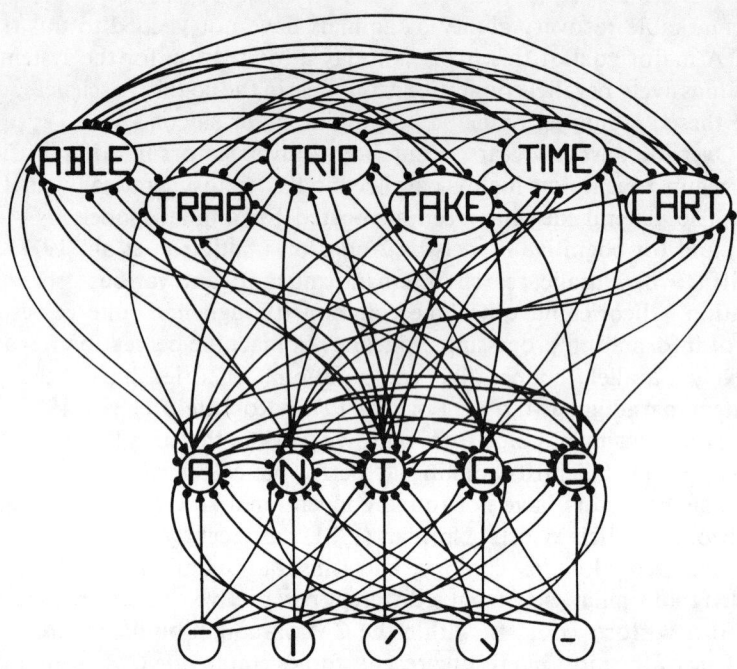

Figure 1. A few of the neighbors of the node for the letter "t" in the first position in a word, and their interconnections (McClelland & Rumelhart, 1981).

portant encodings in the brain are in terms of the relative strengths of synaptic connections. The fundamental premise of connectionism is that individual neurons *do not transmit large amounts of symbolic information.* Instead they compute by being *appropriately connected* to large numbers of similar units. This is in sharp contrast to the conventional computer model of intelligence prevalent in computer science and cognitive psychology.

The fundamental distinction between the conventional and connectionist computing models can be conveyed by the following example. When one sees an apple and says the phrase "wormy apple," some information must be transferred, however indirectly, from the visual system to the speech system. Either a sequence of special *symbols* that denote a wormy apple is transmitted to the speech system, or there are special *connections* to the speech command area for the words. Figure 2 is a graphic presentation of the two alternatives. The path on the right described by double-lined arrows depicts the situation (as in a computer) where the information that a wormy apple has been seen is encoded by the visual system and sent as an abstract message (perhaps frequency-coded) to a general receiver in the speech system which decodes the message and initiates the appropriate speech act. Notice that a complex message would presumably have to be transmitted sequentially on this channel, and that each end would have to

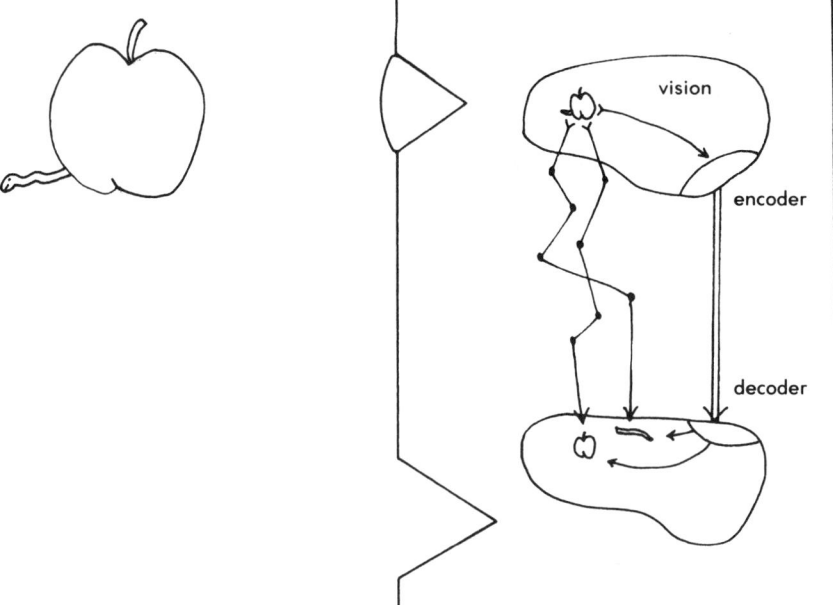

Figure 2. Connectionism vs. symbolic encoding.
⇒ Assumes some general encoding
→ Assumes individual connections

learn the common code for every new concept. No one has yet produced a biologically and computationally plausible realization of this conventional computer model.

The only alternative that we have been able to uncover is described by the path with single-width arrows. This suggests that there are (indirect) links from the units (cells, columns, centers, or what-have-you) that recognize an apple to some units responsible for speaking the word. The connectionist model requires only very simple messages (e.g. stimulus strength) to cross a channel but puts strong demands on the availability of the right connections. Questions concerning the learning and reinforcement of connections are addressed in Feldman, (1981b).

For a number of reasons (including redundancy for reliability), it is highly unlikely that there is exactly one neuron for each concept, but the point of view taken here is that the activity of a small number of neurons (say 10) encodes a concept like apple. An alternative view (Hinton & Anderson, 1981) is that concepts are represented by a "pattern of activity" in a much larger set of neurons (say 1,000) which also represent many other concepts. We have not seen how to carry out a program of specific modeling in terms of these diffuse models. One of the major problems with diffuse models as a parallel computation scheme is cross-talk among concepts. For example, if concepts using units (10, 20, 30, ...) and (5, 15, 25, ...) were simultaneously activated, many other concepts, e.g., (20, 25, 30, 35, ...) would be active as well. In the example of Figure 2, this means that diffuse models would be more like the shared sequential channel. Although a single concept could be transmitted in parallel, complex concepts would have to go one at a time. Simultaneously transmitting multiple concepts that shared units would cause cross-talk. It is still true in our CM that many related units will be triggered by spreading activation, but the representation of each concept is taken to be compact.

Most cognitive scientists believe that the brain appears to be massively parallel and that such structures can compute special functions very well. But massively parallel structures do not seem to be usable for general purpose computing and there is not nearly as much knowledge of how to construct and analyze such models. The common belief (which may well be right) is that there are one or more intermediate levels of computational organization layered on the neuronal structure, and that theories of intelligent behavior should be described in terms of these higher-level languages, such as Production Systems, Predicate Calculus, or LISP. We have not seen a reduction (interpreter, if you will) of any higher formalism which has plausible resource requirements, and this is a problem well worth pursuing.

Our attempts to develop cognitive science models directly in neural terms might fail for one of two reasons. It may be that there really is an interpreted symbol system in animal brains. In this case we would hope that our efforts would break down in a way that could shed light on the nature of this symbol system. The other possibility is that CM techniques are directly applicable but we are unable to figure out how to model some important capacity, e.g., planning. Our program is to continue the CM attack on problems of increasing difficulty (and to induce some of you to join us) until we encounter one that is intractable in our terms. There are a number of problems that are known to be difficult for systems without an interpreted symbolic representation, including complex concepts, learning, and natural language understanding. The current paper is mainly concerned with laying out the formalism and showing how it applies in the easy cases, but we do address the problem of complex concepts in Section 4. We have made some progress on the problem of learning in CM systems (Feldman, 1981b) and are beginning to work seriously on natural language processing and on higher-level vision. Our efforts on planning and long-term memory reorganization have not advanced significantly beyond the discursive presentation in (Feldman, 1980).

We will certainly not get very far in this program without developing some systematic methods of attacking CM tasks and some building-block circuits whose properties we understand. A first step towards a systematic development of CM is to define an abstract computing unit. Our unit is rather more general than previous proposals and is intended to capture the current understanding of the information processing capabilities of neurons. Some useful special cases of our general definition and some properties of very simple networks are developed in Section 2. Among the key ideas are local memory, non-homogeneous and non-linear functions, and the notions of mutual inhibition and stable coalitions.

A major purpose of the rest of the paper is to describe building blocks which we have found useful in constructing CM solutions to various tasks. The constructions are intended to be used to make specific models but the examples in this paper are only suggestive. We present a number of CM solutions to general problems arising in intelligent behavior, but *we are not suggesting that any of these are necessarily employed by nature.* Our notion of an adequate model is one that accounts for *all* of the established relevant findings and this is not a task to be undertaken lightly. We are developing some preliminary sketches (Ballard & Sabbah, 1981; Sabbah, 1981) for a serious model of low and intermediate level vision. As we develop various building blocks and techniques we will also be trying to bury some of the contaminated debris of past neural modeling efforts. Many of our constructions are intended as answers to known hard problems in CM computation. Among the issues addressed are: stability and noise-sensitivity, distributed decision-making, time and sequence problems, and the representation of complex concepts. The crucial questions of learning and change in CM systems are discussed elsewhere (Feldman, 1981b).

## 2. NEURON-LIKE COMPUTING UNITS

As part of our effort to develop a generally useful framework for connectionist theories, we have developed a standard model of the individual unit. It will turn out that a "unit" may be used to model anything from a small part of a neuron to the external functionality of a major subsystem. But the basic notion of unit is meant to loosely correspond to an information processing model of our current understanding of neurons. The particular definitions here were chosen to make it easy to specify detailed examples of relatively complex behaviors. There is no attempt to be minimal or mathematically elegant. The various numerical values appearing in the definitions are arbitrary, but fixed finite bounds play a crucial role in the development. The presentation of the definitions will be in stages, accompanied by examples. A compact technical specification for reference purposes is included as Appendix A. Each unit will be characterized by a small number of discrete states plus:

p—a continuous value in $[-10, 10]$, called *potential* (accuracy of several digits)
v—an *output value,* integers $0 \le v \le 9$
i—a vector of *inputs* $i_1, \ldots, i_n$

## P-Units

For some applications, we will be able to use a particularly simple kind of unit whose output v is proportional to its potential p (rounded) when $p > 0$ and which has only one state. In other words

$$p \leftarrow p + \beta \Sigma w_k i_k \qquad [0 \leq w_k \leq 1]$$
$$v \leftarrow \textit{if } p > \theta \textit{ then } \text{round } (p - \theta) \textit{ else } 0 \qquad [v = 0...9]$$

where $\beta$, $\theta$ are constants and $w_k$ are weights on the input values. The weights are the sole locus of change with experience in the current model. Most often, the potential and output of a unit will be encoding its *confidence,* and we will sometimes use this term. The "←" notation is borrowed from the assignment statement of programming languages. This notation covers both continuous and discrete time formulations and allows us to talk about some issues without any explicit mention of time. Of course, certain other questions will inherently involve time and computer simulation of any network of units will raise delicate questions of discretizing time.

The restriction that output take on small integer values is central to our enterprise. The firing frequencies of neurons range from a few to a few hundred impulses per second. In the 1/10 second needed for basic mental events, there can only be a limited amount of information encoded in frequencies. The ten output values are an attempt to capture this idea. A more accurate rendering of neural events would be to allow 100 discrete values with noise on transmission (cf. Sejnowski, 1977). Transmission time is assumed to be negligible; delay units can be added when transit time needs to be taken into account.

The p-unit is somewhat like classical linear threshold elements (Minsky & Papert, 1972), but there are several differences. The potential, p, is a crude form of memory and is an abstraction of the instantaneous membrane potential that characterizes neurons; it greatly reduces the noise sensitivity of our networks. Without local memory in the unit, one must guarantee that all the inputs required for a computation appear simultaneously at the unit.

One problem with the definition above of a p-unit is that its potential does not decay in the absence of input. This decay is both a physical property of neurons and an important computational feature for our highly parallel models. One computational trick to solve this is to have an inhibitory connection from the unit back to itself. Informally, we identify the negative self feedback with an exponential decay in potential which is mathematically equivalent. With this addition, p-units can be used for many CM tasks of intermediate difficulty. The Interactive Activation models of McClelland and Rumelhart can be described naturally with p-units, and some of our own work (Ballard, 1981) and that of others (Marr

& Poggio, 1976) can be done with p-units. But there are a number of additional features which we have found valuable in more complex modeling tasks.

## Disjunctive Firing Conditions and Conjunctive Connections

It is both computationally efficient and biologically realistic to allow a unit to respond to one of a number of alternative conditions. One way to view this is to imagine the unit having "dendrites" each of which depicts an alternative enabling condition (Figure 3). For example, one could extend the network of Figure 1 to allow for several different type fonts activating the same letter node, with the higher connections unchanged. Biologically, the firing of a neuron depends, in many cases, on local spatio-temporal summation involving only a small part of the neuron's surface. So-called dendritic spikes transmit the activation to the rest of the cell.

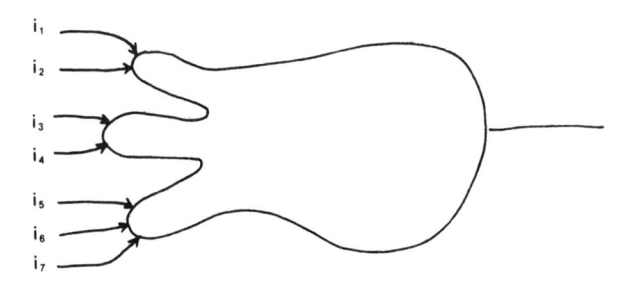

Figure 3. Conjunctive connections and disjunctive input sites.

In terms of our formalism, this could be described in a variety of ways. One of the simplest is to define the potential in terms of the maximum of the separate computations, e.g.,

$$p \leftarrow p + \beta \text{Max}(i_1 + i_2 - \varphi,\ i_3 + i_4 - \varphi,\ i_5 + i_6 - i_7 - \varphi)$$

where $\beta$ is a scale constant as in the p-unit and $\varphi$ is a constant chosen (usually $>10$) to suppress noise and require the presence of multiple active inputs (Sabbah, 1981). The minus sign associated with $i_7$ corresponds to its being an inhibitory input.

It does not seem unreasonable (given current data, Kuffler & Nicholls, 1976) to model the firing rate of some units as the maximum of the rates at its active sites. Units whose potential is changed according to the maximum of a set of algebraic sums will occur frequently in our specific models. One advantage of keeping the processing power of our abstract unit close to that of a neuron is that it helps inform our counting arguments. When we attempt to model a particular function (e.g., stereopsis), we expect to require that the number of units and connections as well as the execution time required by the model are plausible.

The max-of-sum unit is the continuous analog of a logical OR-of-AND (disjunctive normal form) unit and we will sometimes use the latter as an approximate version of the former. The OR-of-AND unit corresponding to Figure 3 is:

$$p \leftarrow p + \alpha \text{ OR } (i_1 \& i_2, i_3 \& i_4, i_5 \& i_6 \& (\text{not } i_7))$$

This formulation stresses the importance that nearby spatial connections *all* be firing before the potential is affected. Hence, in the above example, $i_3$ and $i_4$ make a *conjunctive connection* with the unit. The effect of a conjunctive connection can always be simulated with more units but the number of extra units may be very large.

## Q-Units and Compound Units

Another useful special case arises when one suppresses the numerical potential, p, and relies upon a finite-state set $\{q\}$ for modeling. If we also identify each input of **i** with a separate named input signal, we can get classical finite automata. A simple example would be a unit that could be started or stopped from firing.

One could describe the behavior of this unit by a table, with rows corresponding to states in $\{q\}$ and columns to possible inputs, e.g.,

|   | $i_1$ (start) | $i_2$ (stop) |
|---|---|---|
| **Firing** | Firing | Null |
| **Null** | Firing | Null |

One would also have to specify an output function, giving output values required by the rest of the network, e.g.,

$$v \leftarrow \text{if } q = \text{Firing then 6 else 0.}$$

This could also be added to the table above. An equivalent notation would be transition networks with states as nodes and inputs and outputs on the arcs.

In order to build models of interesting behaviors we will need to employ many of the same techniques used by designers of complex computers and programs. One of the most powerful techniques will be encapsulation and abstraction of a subnetwork by an individual unit. For example, a system that had separate motor abilities for turning left and turning right (e.g., fins) could use two start-stop units to model a turn-unit, as shown in Figure 4.

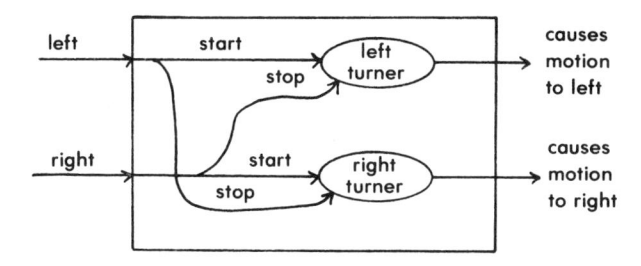

Figure 4. A Turn Unit.

Note that the compound unit here has two distinct outputs, where basic units have only one (which can branch, of course). In general, compound units will differ from basic ones only in that they can have several distinct outputs.

The main point of this example is that the turn-unit can be described abstractly, independent of the details of how it is built. For example, using the tabular conventions described above,

|          | Left      | Right    | Values Output        |
|----------|-----------|----------|----------------------|
| a gauche | a gauche  | a droit  | $v_1 = 7, v_2 = 0$   |
| a droit  | a gauche  | a droit  | $v_1 = 0, v_2 = 8$   |

where the right-going output being larger than the left could mean that we have a right-finned robot. There is a great deal more that must be said about the use of states and symbolic input names, about multiple simultaneous inputs, etc., but the idea of describing the external behavior of a system only in enough detail for the task at hand is of great importance. This is one of the few ways known of coping with the complexity of the magnitude needed for serious modeling of biological functions. It is not strictly necessary that the same formalism be used at each level of functional abstraction and, in the long run, we may need to employ a wide range of models. For example, for certain purposes one might like to expand our units in terms of compartmental models of neurons like those of (Perkel, 1979). The advantage of keeping within the same formalism is that we preserve intuition, mathematics, and the ability to use existing simulation programs. With sufficient care, we can use the units defined above to represent large subsystems without giving up the notion that each unit can stand for an abstract neuron. The crucial point is that a subsystem must be elaborated into its neuron-level units for timing and size calculations, but can (hopefully) be described much more simply when only its effects on other subsystems are of direct concern.

**Units Employing p and q**

It will already have occurred to the reader that a numerical value, like our p, would be useful for modeling the amount of turning to the left or right in the last example. It appears to be generally true that a single numerical value and a small set of discrete states combine to provide a powerful yet tractable modeling unit. This is one reason that the current definitions were chosen. Another reason is that the mixed unit seems to be a particularly convenient way of modeling the information processing behavior of neurons, as generally described. The discrete states enable one to model the effects in neurons of polypeptide modulators, abnormal chemical environments, fatigue, etc. Although these effects are often continuous functions of unit parameters, there are several advantages to using discrete states in our models. Scientists and laymen alike often give distinct names (e.g., cool, warm, hot) to parameter ranges that they want to treat differently. We also can exploit a large literature on understanding loosely-coupled systems as finite-state machines (Sunshine, 1979). It is also traditional to break up a function into separate ranges when it is simpler to describe that way. We have already employed all of these uses of discrete states in our detailed work (Feldman, 1981b; Sabbah, 1981). One example of a unit employing both p and q non-trivially is the following crude neuron model. This model is concerned with saturation and assumes that the output strength, v, is something like average firing frequency. It is not a model of individual action potentials and refractory periods.

We suppose the distinct states of the unit $q \in \{normal, recover\}$. In *normal* state the unit behaves like a p-unit, but while it is *recovering* it ignores inputs. The following table captures almost all of this behavior.

|  |  | $-1 < p < 9$ | $p > 9$ | Output Value |
|---|---|---|---|---|
| (incomplete) | normal | $p \leftarrow p + \Sigma i$ | $p \leftarrow -p/$ recover | $v \leftarrow \alpha p - \theta$ |
|  | recover | normal | <impossible> | $v \leftarrow 0$ |

Here we have the change from one state to the other depending on the value of the potential, p, rather than on specific inputs. The recovering state is also characterized by the potential being set negative. The unspecified issue is what determines the duration of the recovering state—there are several possibilities. One is an explicit dishabituation signal like those in Kandel's experiments (Kandel, 1976). Another would be to have the unit sum inputs in the recovering state as well. The reader might want to consider how to add this to the table.

The third possibility, which we will use frequently, is to assume that the potential, p, decays toward zero (from both directions) unless explicitly changed. This implicit decay $p \leftarrow p_0 e^{-kt}$ can be modeled by self inhibition; the decay constant, k, determines the length of the recovery period.

The general definition of our abstract neural computing unit is just a formalization of the ideas presented above. To the previous notions of p, v, and *i* we formally add

$$\{q\} \text{—a set of } discrete\ states,\ <10$$

and functions from old to new values of these

$$p \leftarrow f(i,p,q)$$
$$q \leftarrow g(i,p,q)$$
$$v \leftarrow h(i,p,q)$$

which we assume, for now, to compute continuously. The form of the f, g, and h functions will vary, but will generally be restricted to conditionals and simple functions. There are both biological and computational reasons for allowing units to respond (for example) logarithmically to their inputs and we have already seen important uses of the maximum function.

The only other notion that we will need is modifiers associated with the inputs of a unit. We elaborate the input vector *i* in terms of received values, weights, and modifiers:

$$\forall j, i_j = r_j \cdot w_j \cdot m_j \qquad\qquad j = 1,\ldots,n$$

where $r_j$ is the *value* received from a predecessor $[r=0\ldots9]$; $w_j$ is a changeable *weight*, unsigned $[0 \leq w_j \leq 1]$ (accuracy of several digits); and $m_j$ is a synapto-synaptic *modifier* which is either 0 or 1.

The weights are the only thing in the system which can change with experience. They are unsigned because we do not want a connection to change from excitatory to inhibitory. The modifier or gate simplifies many of our detailed models. Learning and change will not be treated technically in this paper, but the definitions are included in the Appendix for completeness (Feldman, 1981b).

We conclude this section with some preliminary examples of networks of our units, illustrating the key idea of mutual (lateral) inhibition (Fig. 5). Mutual inhibition is widespread in nature and has been one of the basic computational schemes used in modeling. We will present two examples of how it works to help aid in intuition as well as to illustrate the notation. The basic situation is symmetric configurations of p-units which mutually inhibit one another. Time is broken into discrete intervals for these examples. The examples are too simple to be realistic, but do contain ideas which we will employ repeatedly.

**Two P-Units Symmetrically Connected**

$$\text{Suppose } w_1 = 1, w_2 = -.5$$
$$p(t+1) = p(t) + r_1 - (.5)r_2 \qquad r_j = \text{received}$$
$$v = \text{round}(p) \ [0 \ldots 9]$$

Referring to Figure 5a, suppose the initial input to the unit A.1 is 6, then 2 per time step, and the initial input to B.1 is 5, then 2 per time step. At each time step, each unit changes its potential by adding the external value ($r_1$) and substracting half the output value of its rival. This system will stabilize to the side of the larger of two instantaneous inputs.

**Two Symmetric Coalitions of 2-Units**

$$w_1 = 1$$
$$w_2 = .5$$
$$w_3 = -.5$$
$$p(t+1) = p(t) + r_1 + .5(r_2 - r_3)$$
$$v = \text{round}(p)$$

A,C start at 6; B,D at 5;
A,B,C,D have no external input for $t > 1$

The connections for this system are shown in Figure 5b. This system converges faster than the previous example. The idea here is that units A and C form a "coalition" with mutually reinforcing connections. The competing units are A vs. B and C vs. D. The last example is the smallest network depicting what we believe to be the basic mode of operation in connectionist systems. The faster convergence is not an artifact; the *positive feedback* among members of a coalition will generally lead to faster convergence than in separate competitions. It is the amount of positive feedback rather than just the size of the coalition that determines the rate of convergence (Feldman & Ballard, 1982). In terms of Figure 1, this could represent the behavior of the rival letters A and T in conjunction with the rival words ABLE and TRAP, in the absence of other active nodes.

Competing coalitions of units will be the organizing principle behind most of our models. Consider the two alternative readings of the Necker

Figure 5a.

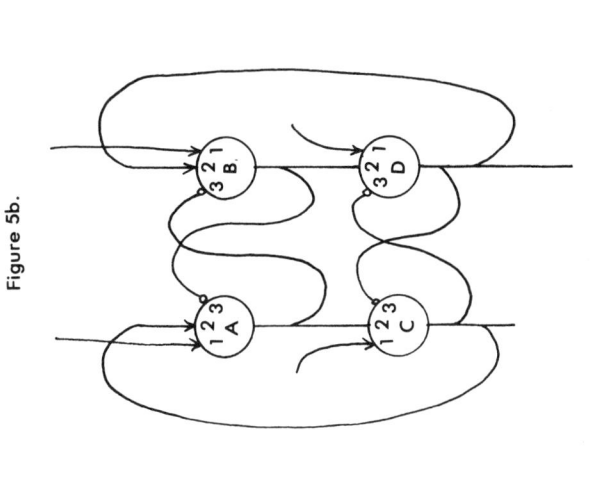

Figure 5b.

Suppose A₁ received an input of 6 units, then 2 per time step
Suppose B₁ received an input of 5 units, then 2 per time step

| t | P(A) | P(B) |
|---|------|------|
| 1 | 6    | 5    |
| 2 | 5.5  | 4    |
| 3 | 5.5  | 3.5  |
| 4 | 6    | 3    |
| 5 | 6.5  | 2    |
| 6 | 7.5  | 1    |
| 7 | 9.5  | 0    |
| 8 | Sat  | 0    |

| t | P(A) | P(B) | P(C) | P(D) |
|---|------|------|------|------|
| 1 | 6    | 5    | 6    | 5    |
| 2 | 6.5  | 4.5  | 6.5  | 4.5  |
| 3 | 7.5  | 3.5  | 7.5  | 3.5  |
| 4 | 9.5  | 1.5  | 9.5  | 1.5  |
|   | Sat  | 0    | Sat  | 0    |

Figure 5 (a) and (b). Small Symmetric Networks.

cube shown in Figure 6. At each level of visual processing, there are mutually contradictory units representing alternative possibilities. The dashed lines denote the boundaries of coalitions which embody the alternative interpretations of the image. A number of interesting phenomena (e.g., priming, perceptual rivalry, filling, subjective contour) find natural expression in this formalism. We are engaged in an ongoing effort (Ballard, 1981; Sabbah, 1981) to model as much of visual processing as possible within the connectionist framework. The next section describes in some detail a variety of simple networks which we have found to be useful in this effort.

## 3. NETWORKS OF UNITS

The main restriction imposed by the connectionist paradigm is that no symbolic information is passed from unit to unit. This restriction makes it difficult to employ standard computational devices like parameterized functions. In this section, we present connectionist solutions to a variety of computational problems. The sections address two principal issues. One is: Can the networks be connected up in a way that is sufficient to represent the problem at hand? The other is: Given these connections, how can the networks exhibit appropriate dynamic behavior, such as making a decision at an appropriate time?

**Using a Unit to Represent a Value**

One key to many of our constructions is the dedication of a separate unit to each value of each parameter of interest, which we term the unit/value principle. We will show how to compute using unit/value networks and present arguments that the number of units required is not unreasonable. In this representation the output of a unit may be thought of as a confidence measure. Suppose a network of depth units encodes the distance of some object from the retina. Then if the unit representing depth = 2 saturates, the network is expressing confidence that the distance is two units. Similarly, the "G-hidden" node in Figure 6 expresses confidence in its assertion. There is much neurophysiological evidence to suggest unit/value organizations in less abstract cortical maps. Examples are edge sensitive units (Hubel & Wiesel, 1979) and perceptual color units (Zeki, 1980), which are relatively insensitive to illumination spectra. Experiments with cortical motor control in the monkey and cat (Wurtz & Albano, 1980) suggest a unit/value organization. Our hypothesis is that the unit/value organization is widespread, and is a fundamental design principle.

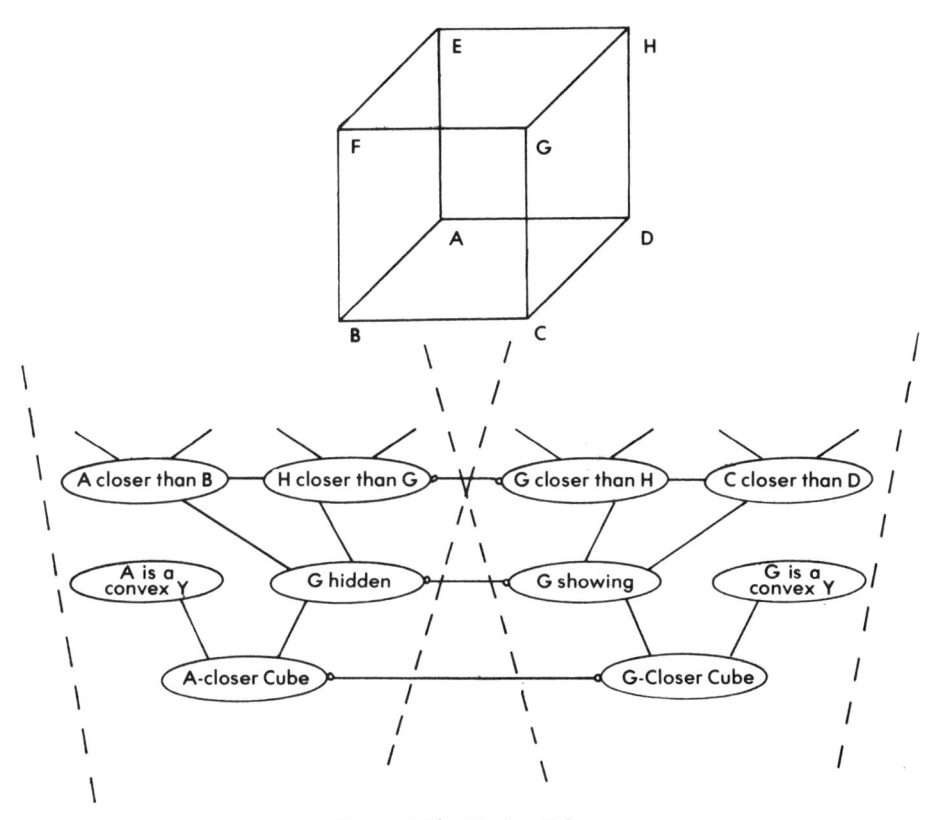

Figure 6. The Necker Cube.

Although many physical neurons do seem to follow the unit/value rule and respond according to the reliability of a particular configuration, there are also other neurons whose output represents the range of some parameter, and apparently some units whose firing frequency reflects both range and strength information (Scientific American, 1979). Both of the latter types can be accommodated within our definition of a unit, but we will employ only unit/value networks in the remainder of this paper.

In the unit/value representation, much computation is done by table look-up. As a simple example, let us consider the multiplication of two variables, i.e., $z = xy$. In the unit/value formalism there will be units for *every* value of x and y that is important. Appropriate pairs of these will make a conjunctive connection with another unit cell representing a specific value for the product. Figure 7 shows this for a small set of units representing values for x and y. Notice that the confidence (expressed as output value) that a particular product is an answer can be a linear function of the maximum of the sums of the confidences of its two inputs. A major problem with function tables (and with CM in general) is the potential combinatorial explosion in the number of units required for a computation. A naive approach would demand $N^2$ units to represent all products of numbers from 1 to N. The network of Figure 7 requires many fewer units because each product is represented only once, another advantage of conjunctive connections. We could use even fewer units by exploiting positional notation and replacing each output connection with a conjunction of outputs from units representing multiples of 1, 10, 100, etc. The question of efficient ways of building connection networks is treated in detail in Section 4 (cf. also Hinton, 1981a; 1981b).

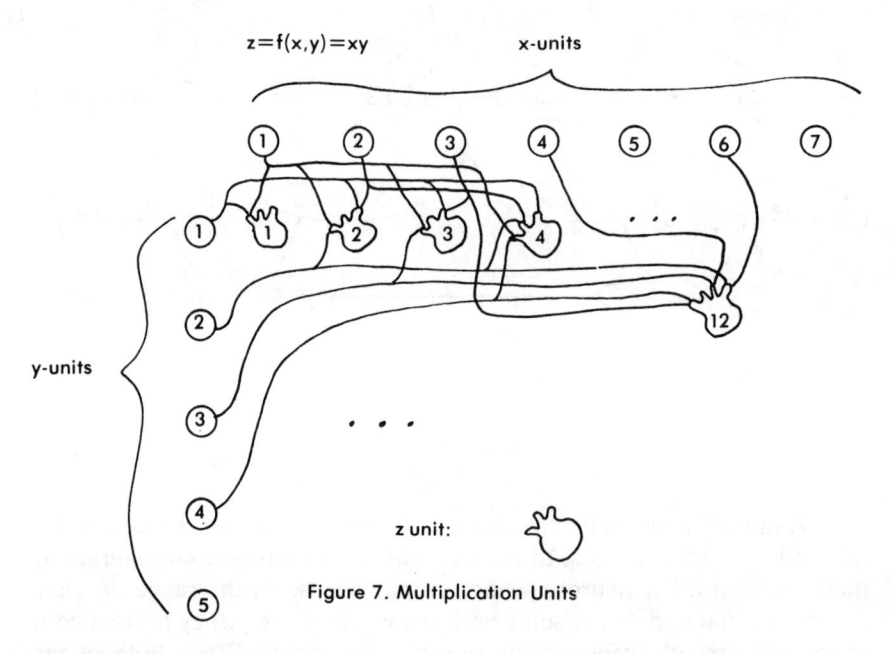

Figure 7. Multiplication Units

**Modifiers and Mappings**

The idea of function tables (Fig. 7) can be extended through the use of *variable mappings*. In our definition of the computational unit, we included a binary modifier, m, as an option on every connection. As the definition specifies, if the modifier associated with a connection is zero, the value v sent along that connection is ignored. Thus the modifier denotes inhibition, or blocking. There is considerable evidence in nature for synapses on synapses (Kandel, 1976) and the modifiers add greatly to the computational simplicity of our networks. Let us start with an initial informal example of the use of modifiers and mappings. Suppose that one has a model of grass as green except in California where it is brown (golden), as shown in Figure 8.

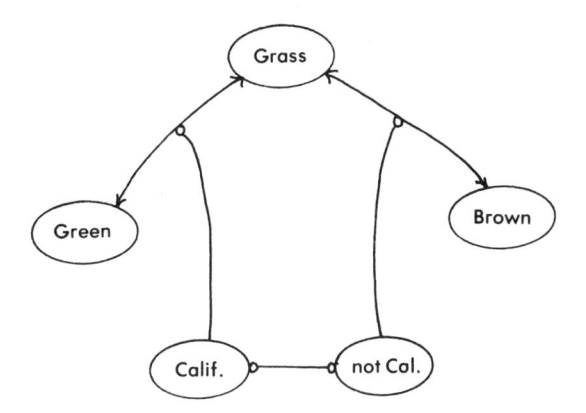

Figure 8. Grass is Green connection modified by California.

Here we can see that grass and green are potential members of a coalition (can reinforce one another) except when the link is blocked. This use is similar to the cancellation link of (Fahlman, 1979) and gives a crude idea of how context can effect perception in our models. Note that in Figure 8 we are using a shorthand notation. A modifier touching a double-ended arrow actually blocks two connections. (Sometimes we also omit the arrowheads when connection is double-ended.)

Mappings can also be used to select among a number of possible values. Consider the example of the relation between depth, physical size, and retinal size of a circle. (For now, assume that the circle is centered on and orthogonal to the line of sight, that the focus is fixed, etc.) Then there is a fixed relation between the size of retinal image and the size of the physical circle for any given depth. That is, each depth specifies a *mapping* from retinal to physical size (see Fig. 9). Here we suppose the scales for depth and the two sizes are chosen so that unit depth means the same numerical size. If we knew the depth of the object (by touch, context, or magic) we would know its physical size. The network above allows retinal size 2 to reinforce physical size 2 when depth = 1 but inhibits this connection for all other depths. Similarly, at depth 3, we should interpret retinal size 2 as physical size 8, and inhibit other interpretations. Several remarks are in order. First, notice that this network implements a function phys = f(ret, dep) that maps from retinal size and depth to physical size, providing an example of how to replace functions with parameters by mappings. For the simple case of looking at one object perpendicular to the line of sight, there will be one consistent coalition of units which will be stable. The work does something more, and this is crucial to our enterprise; the network can represent the consistency relation R among the three quantities: depth, retinal size, and physical size. It embodies not only the function f, but its two inverse functions as well (dep = $f_1$(ret,phys), and ret = $f_2$(phys,dep)). (The network as shown does not include the links for $f_1$ and $f_2$, but these are similar to those for f.) Most of Section 5 is devoted to laying out networks that embody theories of particular visual consistency relations.

The idea of modifiers is, in a sense, complementary to that of conjunctive connections. For example, the network of Figure 9 could be transformed into the following network (Fig. 10). In this network the variables for physical size, depth, and retinal size are all given equal weight. For example, physical size = 4 and depth = 1 make a *conjunctive connection* with retinal size = 4. Each of the value units in a competing row could be connected to all of its competitors by inhibitory links and this would tend to make the network activate only one value in each category. The general issue of rivalry and coalitions will be discussed in the next two sub-sections.

When should a relation be implemented with modifiers and when should it be implemented with conjunctive connections? A simple, non-rigorous answer to this question can be obtained by examining the size of two sets of units: (1) the number of units that would have to be inhibited by modifiers; and (2) the number of units that would have to be reinforced with conjunctive connections. If (1) is larger than (2), then one should choose modifiers; otherwise choose conjunctive connections. Sometimes the choice is obvious: to implement the brown Californian grass example of Figure 8 with conjunctive connections, one would have to reinforce all units representing places that had green grass! Clearly in this case it is easier to handle the exception with modifiers. On the other hand, the depth relation R(phy,dep,ret) is more cheaply implemented with conjunctive connections. Since our modifiers are strictly binary, conjunctive connections have the additional advantage of continuous modulation.

To see how the conjunctive connection strategy works in general, suppose a constraint relation to be satisfied involves a variable x, e.g., f(x,y,z,w) = 0. For a particular value of x, there will be triples of values of y, z, and w that satisfy the relation f. Each of these triples should make a conjunctive connection with the unit representing the x-value. There could also be 3-input conjunctions at each value of y,z,w. Each of these four different kinds of conjunctive connections corresponds to an interpretation of the *relation* f(x,y,z,w) = 0 as a *function,* i.e., x = $f_1$(y,z,w), y = $f_2$(x,z,w), z = $f_3$(x,y,w), or w = $f_4$(x,y,z). Of course, these functions need not be single-valued. This network connection pattern could be extended to more than four variables, but high numbers of variables would tend to increase its sensitivity to noisy inputs. Hinton has suggested a special notation for the situation where a network exactly captures a consistency relation. The mutually consistent values are all shown to be centrally linked (Fig. 11). This notation provides an ele-

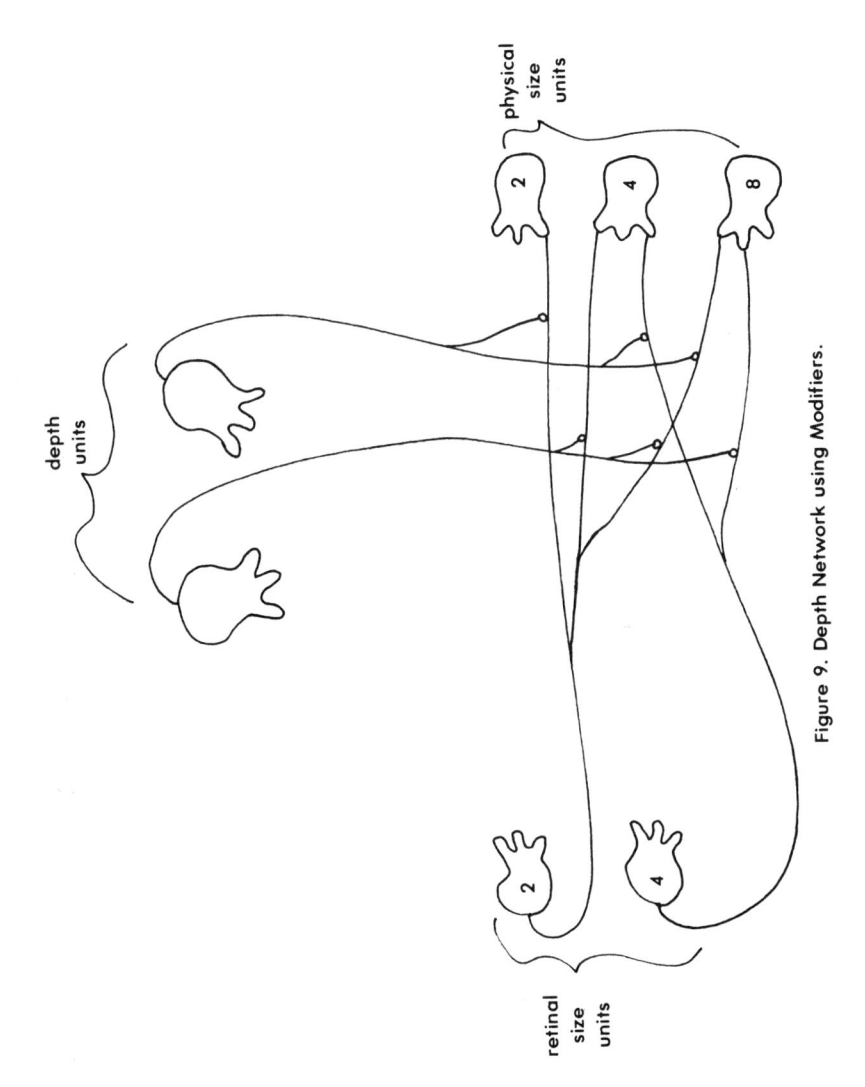

Figure 9. Depth Network using Modifiers.

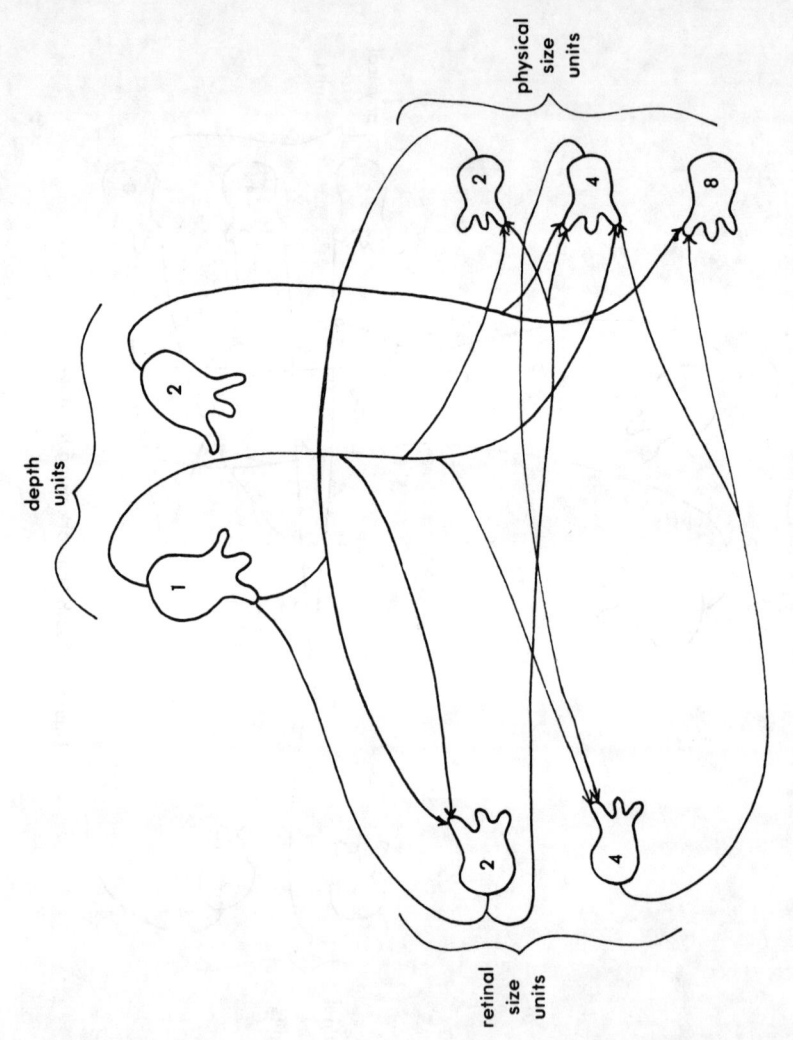

Figure 10. Depth Network using Conjunctive Connections.

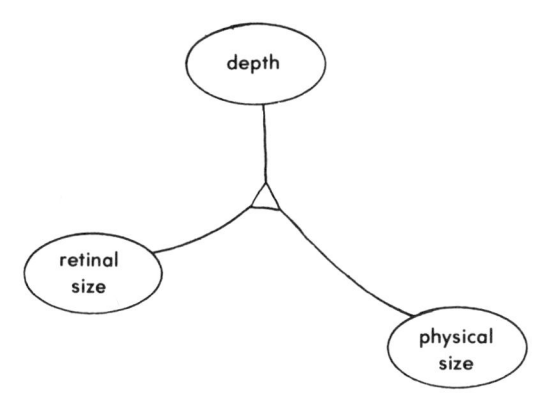

Figure 11. Notation for consistency relations.

gant way of presenting the interactions among networks, but must be used with care. Writing down a triangle diagram does not insure that the underlying mappings can be made consistent or computationally well-behaved.

**Winner-Take-All Networks and Regulated Networks**

A very general problem that arises in any distributed computing situation is how to get the entire system to make a decision (or perform a coherent action, etc.). Biologically necessary examples of this behavior abound; ranging from turning left or right, through fight-or-flight responses, to interpretations of ambiguous words and images. Decision-making is a particularly important issue for the current model because of its restrictions on information flow and because of the almost linear nature of the p-units used in many of our specific examples. Decision-making introduces the notions of *stable states* and *convergence* of networks.

One way to deal with the issue of coherent decisions in a connectionist framework is to introduce *winner-take-all* (WTA) networks, which have the property that only the unit with the highest potential (among a set of contenders) will have output above zero after some setting time (Fig. 12). There are a number of ways to construct WTA networks from the units described above. For our purposes it is enough to consider one example of a WTA network which will operate in one time step for a set of contenders each of whom can read the potential of all of the others. Each unit in the network computes its new potential according to the rule:

$$p \leftarrow \mathit{if}\ p > \max(i_j, .1)\ \mathit{then}\ p\ \mathit{else}\ 0.$$

That is, each unit sets itself to zero if it knows of a higher input. This is fast and simple, but probably a little too complex to be plausible as the behavior of a single neuron. There is a standard trick (apparently widely used by nature) to convert this into a more plausible scheme. Replace each unit above with two units; one computes the maximum of the competitor's inputs and inhibits the other. The circuit above can be strengthened by adding a reverse inhibitory link, or one could use a modifier on the output, etc. Obviously one could have a WTA layer that got inputs from some set of competitors and settled to a winner when triggered to do so by some downstream network. This is an exact analogy of strobing an output buffer in a conventional computer.

One problem with previous neural modeling attempts is that the circuits proposed were often unnaturally delicate (unstable). Small changes in parameter values would cause the networks to oscillate or converge to incorrect answers. We will have to be careful not to fall into this trap, but would like to avoid detailed analysis of each particular model for delicacy in this paper. What appears to be required are some building blocks and combination rules that preserve the desired properties. For example, the WTA subnetworks of the last example will not oscillate in the absence of oscillating inputs. This is also true of any symmetric mutually inhibitory subnetwork. This is intuitively clear and could be proven rigorously under a variety of assumptions (cf. Grossberg, 1980). If every unit receives inhibition proportional to the activity (potential) of each of its rivals, the instantaneous leader will receive less inhibition and thus not lose its lead unless the inputs change significantly.

Another useful principle is the employment of lower-bound and upper-bound cells to keep the total activity of a network within bounds (Fig. 13). Suppose that we add two extra units, LB and UB, to a network which has coordinated output. The LB cell compares the total (sum) activity of the units of the network with a lower bound and sends positive activation uniformly to all members if the sum is too low. The UB cell inhibits all units equally if the sum of activity is too high. Notice that LB and UB can be parameters set from outside the network. Under a wide range of conditions (but not all), the LB-UB augmented network can be designed to preserve order relationships among the outputs $v_j$ of the original network while keeping the sum between LB and UB.

We will often assume that LB-UB pairs are used to keep the sum of outputs from a network within a given range. This same mechanism also goes far towards eliminating the twin perils of uniform saturation and uniform silence which can easily arise in mutual inhibition networks. Thus we will often be able to reason about the computation of a network assuming that it stays active and bounded.

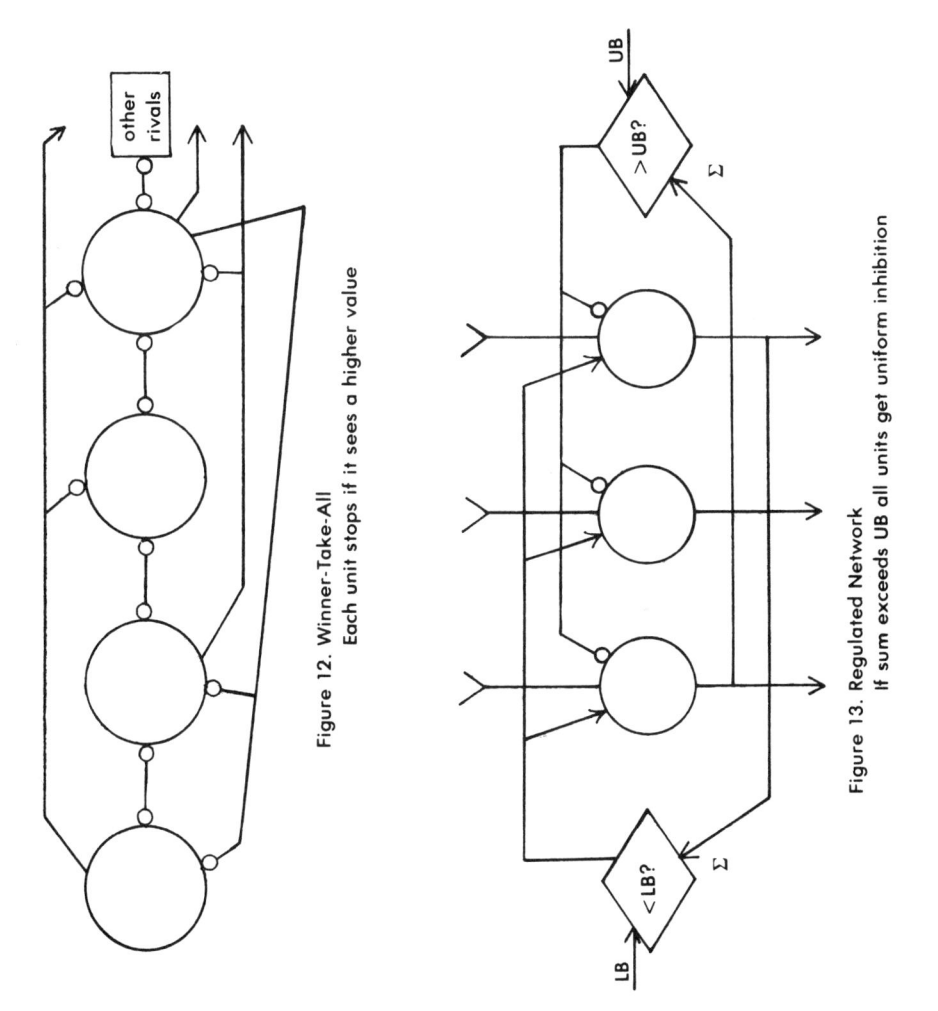

Figure 12. Winner-Take-All
Each unit stops if it sees a higher value

Figure 13. Regulated Network
If sum exceeds UB all units get uniform inhibition

## Stable Coalitions

For a massively parallel system to actually make a decision (or do something), there will have to be states in which some activity strongly dominates. Such stable, connected, high confidence units are termed *stable coalitions*. A stable coalition is our architecturally-biased term for the psychological notions of percept, action, etc. We have shown some simple instances of stable coalitions, in Figure 5b and the WTA network. In the depth networks of Figures 9 and 10, a stable coalition would be three units representing consistent values of retinal size, depth, and physical size. But the general idea is that a very large complex subsystem must stabilize, e.g., to a fixed interpretation of visual input, as in Figure 1. The way we believe this to happen is through mutually reinforcing coalitions which dominate all rival activity when the decision is required. The simplest case of this is Figure 5b, where the two units A and B form a coalition which suppresses C and D. Formally, *a coalition will be called stable when the output of all its members is nondecreasing.* Notice that a coalition is not a particular anatomical structure, but an instantaneously mutually reinforcing set of units, in the spirit of Hebb's cell assemblies (Jusczyk & Klein, 1980).

What can we say about the conditions under which coalitions will become and remain stable? We will begin informally with an almost trivial condition. Consider a set of units $\{a,b,...\}$ which we wish to examine as a possible coalition, $\pi$. For now, we assume that the units in $\pi$ are all p-units and are in the non-saturated range and have no decay. Thus for each u in $\pi$,

$$p(u) \leftarrow p(u) + \text{Exc} - \text{Inh},$$

where Exc is the weighted sum of excitatory inputs and Inh is the weighted sum of inhibitory inputs. Now suppose that $\text{Exc}|\pi$, the excitation from the coalition $\pi$ only, were greater than INH, the largest possible inhibition receivable by u, for each unit u in $\pi$, i.e.,

$$(\text{SC}) \quad \forall u \in \pi \; ; \; \text{Exc}|\pi > \text{INH}$$

Then it follows that

$$\forall u \in \pi \; ; \; p(u) \leftarrow p(u) + \delta \text{ where } \delta > 0.$$

That is, the potential of every unit in the coalition will increase. This is not only true instantaneously, but remains true as long as nothing external changes (we are ignoring state change, saturation, and decay). This is because $\text{Exc}|\pi$ continues to increase as the potential of the members of $\pi$ increases. Taking saturation into account adds no new problems; if all of the units in $\pi$ are saturated, the change, $\delta$, will be zero, but the coalition will remain stable.

The condition that the excitation from other coalition members alone, $\text{Exc}|\pi$, be greater than any possible inhibition INH for each unit may appear to be too strong to be useful. It is certainly true that coalitions can be stable without condition (SC) being met. The condition (SC) is useful for model building because it may be relatively easy to establish. Notice that INH is directly computable from the description of the unit; it is the largest negative weighted sum possible. If inhibition in our networks is mutual, the upper-bound possible after a fixed time $\tau$, $\text{INH}\tau$, will depend on the current value of potential in each unit u. The simplest case of this is when two units are "deadly rivals"—each gets all its inhibition from the other. In such cases, it may well be feasible to show that after some time $\tau$, the stable coalition condition will hold (in the absence of decay, fatigue, and changes external to the network). Often, it will be enough to show that the coalition has a stable "frontier," the set of units with outputs to some system under investigation.

There are a number of interesting properties of the stable coalition principle. First notice that it does not prohibit multiple stable coalitions nor single coalitions which contain units which mutually inhibit one another (although excessive mutual inhibition is precluded). If the units in the coalition had non-zero decay, the coalition excitation $\text{Exc}|\pi$ would have to exceed both INH and decay for the coalition to be stable. We suppose that a stable coalition yields control when its input elements change (fatigue and explicit resets are also feasible). To model coalitions with changeable inputs, we add boundary elements, which also had external "Input" and thus whose condition for being part of a stable coalition, $\pi$, would be:

$$\text{Exc}|\pi + \text{Input} > \text{INH}.$$

This kind of unit could disrupt the coalition if its Input went too low. The mathematical analysis of CM networks and stable coalitions continues to be a problem of interest. We have achieved some understanding of special cases (Feldman & Ballard, 1982) and these results have been useful in designing CM too complex to analyze in closed form.

## 4. CONSERVING CONNECTIONS

It is currently estimated that there are about $10^{11}$ neurons and $10^{15}$ connections in the human brain and that each neuron receives input from about $10^3$–$10^4$ other neurons. These numbers are quite large, but not so large as to present no problems for connectionist theories. It is also important to remember that neurons are not switching devices; the same signal is propagated along all of the outgoing branches. For example, suppose some model called for a separate, dedicated path between all possible pairs of units in two layers in size N. It is easy to show that this requires $N^2$ intermediate sites. This means, for example, that there are not enough neurons in the brain to provide such a cross-bar switch for substructures of a million elements each. Similarly, there are not enough neurons to provide one to represent each complex object at every position, orientation, and scale of visual space. Although the development of connectionist models is in its perinatal period, we have been able to accumulate a number of ideas on how some of the required computations can be carried out without excessive resource requirements. Five of the most important of these are described below: (1) functional decomposition; (2) limited precision computation; (3) coarse and coarse-fine coding; (4) tuning; and (5) spatial coherence.

**Functional Decomposition**

When the number of variables in the function becomes large, the fan-in or number of input connections could become unrealistically large. For example, with the function $t = f(u,v,w,x,y,z)$ implemented with 100 values of t, when each of its arguments can have 100 distinct values, would require an average number of inputs per unit of $10^{12}/10^2$, or $10^{10}$. However, there are simple ways of trading units for connections. One is to replicate the number of units with each value. This is a good solution when the inputs can be partitioned in some natural way as in the vision examples in the next section. A more powerful technique is to use intermediate units when the computation can be decomposed in some way. For example, if $f(u,v,w,x,y,z) = g(u,v)$ o $h(w,x,y,z)$, where o is some composition, then separate networks of value units for f(g,h), g(u,v), and h(w,x,y,z) can be used. The outputs from the g and h units can be combined in conjunctive connections according to the composition operator o in a third network representing f. An example is the case of word recognition. Letter-feature units would have to connect to vastly more word units without the imposition of the intermediate level of letter units. The letter units limit the ways letter-feature units can appear in a word.

## Limited Precision Computation

In the multiplication example $z = xy$, the number of z units required is proportional to $N_x N_y$, even when redundant value units are eliminated, and in general the number of units could grow exponentially with the number of arguments. However, there are several refinements which can drastically reduce the number of required units. One way to do this is to fix the number of units at the *precision* required for the computation. Figure 14 shows the network of Figure 7 modified when less computational accuracy is required.

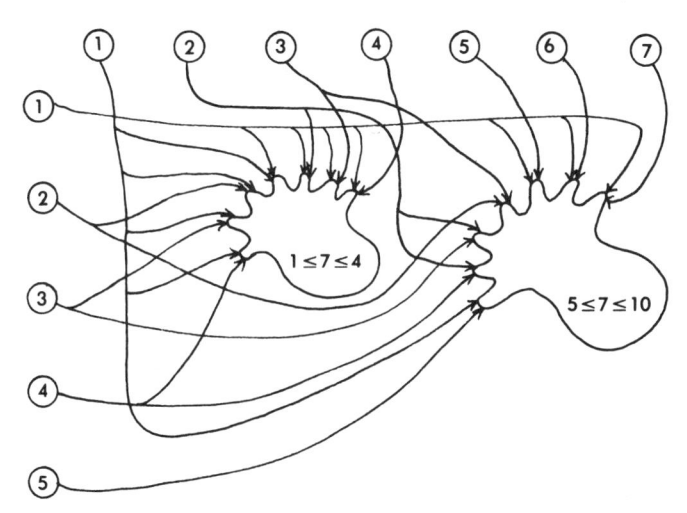

Figure 14. Modified Multiplication Table using Less Units.

This is the same principle that is incorporated in integer calculations in a sequential computer: computations are rounded to within the machine's accuracy. Accuracy is related to the number of bits and the number representation. The main difference is that since the sequential computer is general purpose, the number representations are conservative, involving large number of bits. The neural units need only represent sufficient accuracy for the problem at hand. This will generally vary from network to network, and may involve very inhomogeneous, special purpose number representations.

## Coarse and Coarse-Fine Coding

Coarse coding is a general technical device for reducing the number of units needed to represent a range of values with some fixed precision, due to Hinton (1980). As Figure 15a suggests, one can represent a more precise value as the simultaneous activation of several (here 3) overlapping coarse-valued units. In general, D simultaneous activations of coarse cells of diameter D precise units suffice. For a parameter space of dimension k, a range of F values can be captured by only $F^k/D^{k-1}$ units rather than $F^k$ in the naive method. The coarse coding trick and the related coarse-fine trick to be described next both depend on the input at any given time being sparse relative to the set of all values expressible by the network.

The coarse-fine coding technique is useful when the space of values to be represented has a natural structure which can be exploited. Suppose a set of units represents a vector parameter v which can be thought of as partitioned into two components (r,s). Suppose further that the number of units required to represent the subspace r is $N_r$ and that required to represent s is $N_s$. Then the number of units required to represent v is $N_r N_s$. It is easy to construct examples in vision where the product $N_r N_s$ is too close to the upper bound of $10^{11}$ units to be realistic. Consider the case of trihedral (v) vertices, an important visual cue. Three angles and two position coordinates are necessary to uniquely define every possible trihedral vertex. (Two angles define the types of vertex (arrow, y-joint); the third specifies the rotation of the joint in space.) If we use 5 degree angle sensitivity and $10^5$ spatial sample points, the number of units is given by $N_r \approx 3.6 \times 10^5$ and $N_s = 10^5$ so that $N_r N_s \approx 3.6 \times 10^{10}$. How can we achieve the required representation accuracy with less units?

In many instances, one can take advantage of the fact that the *actual occurrence* of parameters is sparse. In terms of trihedral vertices, one assumes that in an image, such vertices will rarely occur in tight spatial clusters. (If they do, they cannot be resolved as individuals simultaneously.) Given that simultaneous proximal values of parameters are unlikely, they can be represented accurately for other computations, without excessive cost.

The solution is to decompose the space v into two subspaces, r and s, each with unilaterally reduced resolution.

Instead of $N_r N_s$ units, we represent v with two spaces, one with $N_{r'} N_s$ units where $N_{r'} << N_r$, and another with $N_r N_{s'}$ units where $N_{s'} << N_s$.

To illustrate this technique with the example of trihedral vertices we choose

$$N_{s'} = 0.01 N_s \text{ and } N_{r'} = 0.01 N_r.$$

Thus the dimensions of the two sets of units are:

$$N_{s'} N_r = 3.6 \times 10^8$$

and

$$N_s N_{r'} = 3.6 \times 10^8.$$

The choices result in one set of units which accurately represent the angle measurements and fire for a specific trihedral vertex anywhere in a fairly broad visual region, and another set of units which fire only if a general trihedral vertex is present at the precise position. The coarse-fine technique can be viewed as replacing the square coarse-valued covering in Figure 15a with rectangular (multi-dimensional) coverings, like those shown in Figure 16. In terms of our value units, the coarse-fine representation of trihedral vertices is shown in Figure 15b.

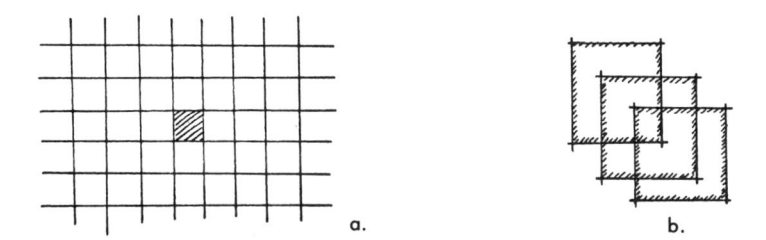

Figure 15a. Coarse coding example. In a two-dimensional measurement space, the presence of a measurement can be encoded by making a single unit in the fine resolution space have a high confidence value. The same measurement can be encoded by making overlapping coarse units in three distinct coarse arrays have high confidence values.

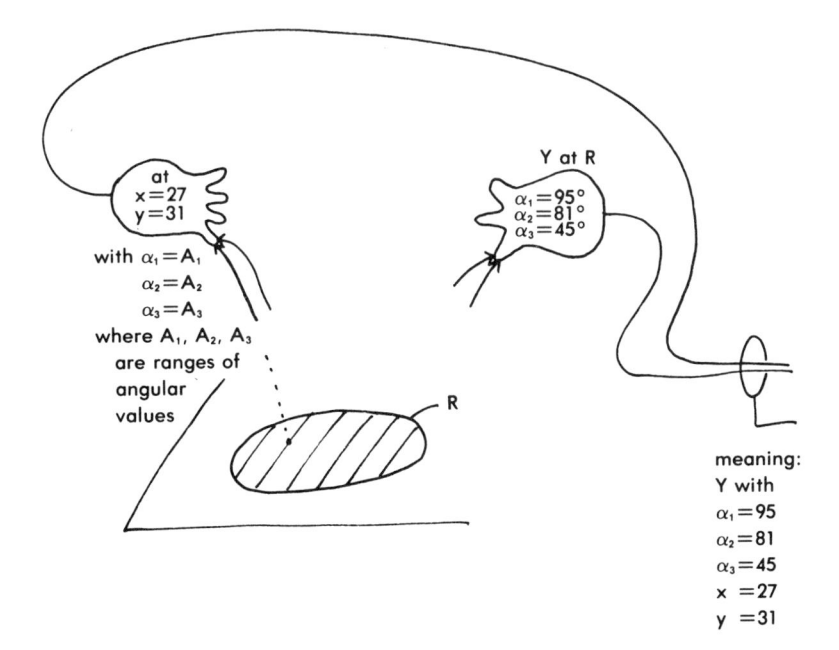

meaning:
Y with
$\alpha_1 = 95$
$\alpha_2 = 81$
$\alpha_3 = 45$
x = 27
y = 31

Figure 15b. Coarse angle—fine position and coarse position—fine angle units combine to yield precise values of all five parameters.

If the trihedral angle enters into another relation, say R(v,α), where both its angle and position are required accurately, one conjunctively connects pairs of appropriate units from each of the reduced resolution spaces to appropriate R-units. The conjunctive connection represents the *intersection* of each of its components' *fields*. Essentially the same mechanism will suffice for conjoining (e.g.) accurate color with coarse velocity information.

An important limitation of these techniques, however, is that the input must be sparse. If inputs are too closely spaced, "ghost" firings will occur. In Figure 16, two sets of overlapping fields are shown, each with unilaterally reduced resolution. Actual input at points A and B will produce an erroneous indication of an input at C, in addition to the correct signals. The sparseness requirement has been shown to be satisfied in a number of experiments with visual data (Ballard & Kimball, 1981a, 1981b; Ballard & Sabbah, 1981).

The resolution device involves a units/connections tradeoff, but in general, the tradeoff is attractive. To see this, consider a unit that receives input from a network representing a vector parameter v. If n is the number of places where the output is used, and conjunctive connections are used to conjoin the D firing units, then Dn synapses are required. Thus if A is the number of non-coarse coded units to achieve a given acuity, then coarse coding is attractive when $A/D^{k-1} >$ Dn, assuming connections and units are equally scarce. This result is optimistic in that, when other uses of conjunctive connections are taken into account, the number of conjunctive units could be unrealistically large.

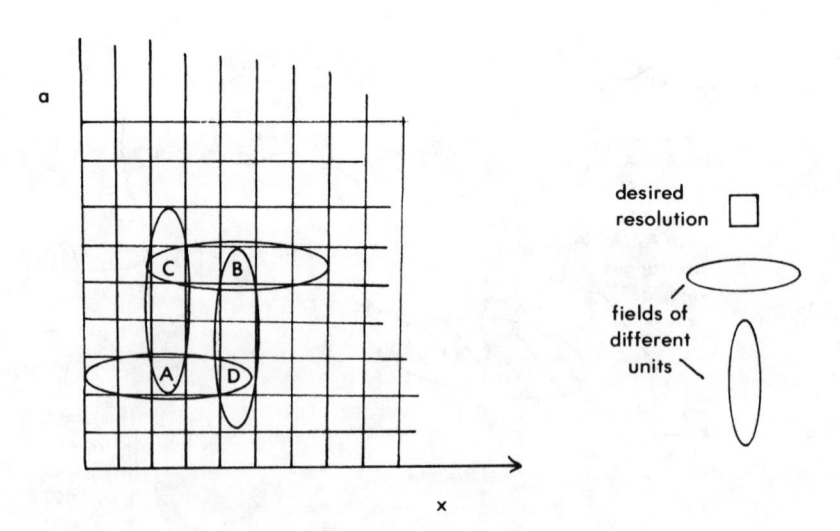

Figure 16. Inputs at A & B cause ghosts at C & D.

## Tuning

The idea of tuning further exploits networks composed of coarsely- and finely-grained units. Suppose there are n fine resolution units of a feature A and n fine resolutions for a feature B. To have explicit units for feature values AB, $n^2$ units would be required. This is an untenable solution for large feature spaces (the number of units grows exponentially with the number of features), so alternatives must be sought. One solution to this problem is to vary the grain of the AB units so that they are only coarsely represented. This solution has its attendant disadvantages in that separate stimuli within the limits of the coarse resolution grain cannot be distinguished. Also, a set of weak stimuli can be misinterpreted. A better solution is to have a coarse unit that would respond only to a single saturated unit within its input range. In that way a collection of weak inputs is not misinterpreted.

This situation can be achieved by having the units in each finely-tuned network that are in the field of a coarse unit laterally inhibit each other, e.g., in the WTA network of Figure 5a. The outputs of these individual feature units then form disjunctive connections with appropriate coarse resolution multiple feature units. If m is the grain of the coarse resolution units along with each feature dimension, the number of disjunctions per coarse unit is $(n/m)^2$. The result of this connection strategy is that a coarse unit responds with a strength that varies as the strengths of the largest maximum in the subnetwork of each of the finely-tuned units that correspond to its field. The response of a coarse-tuned unit is the maximum of the sums of the conjunctive inputs from the finely tuned units which connect to it. In terms of Figure 15, a tuned coarse-angle cell would respond only to one high-confidence pair of angles in its range, and not to several weak ones (which couldn't correctly appear all at one position). This is a better property than just having unstructured coarse units and it will be exploited in the next section, when we deal with perceiving complex objects.

## Spatial Coherence

The most serious problem which requires conserving connections is the representation of complex concepts. The obvious way of representing concepts (sets of properties) is to dedicate a separate unit to each conjunction of features. In fact, it first appears that one would need a separate unit for each combination at each location in the visual field. We will present here a simple way around the problem of separate units for each location and deal with the more general problem in the next section.

The basic problem can be readily seen in the example of Figure 17. Suppose there were one unit each for finally recognizing concepts like colored circles and squares. Now consider the case when a red circle (at $x = 7$) and a blue square (at $x = 11$) simultaneously appear in the visual field. If the various "colored figure" units simply summed their inputs, the incorrect "blue circle" unit would see two active inputs, just like the correct "red circle" and "blue square" units. This problem is known as cross-talk, and is always a potential hazard in CM networks. The solution presented in Figure 17 is quite general. Each unit is assumed to have a separate conjunctive connection site for each position of the visual field. In our example, the correct units get dual inputs to a single site (and are activated) while the partially matched units receive separated inputs and are not activated. Only sets of properties which are spatially coherent can serve to activate concept units. This example was meant to show how spatial coherence could be used with conjunctive connections to eliminate cross-talk. There are a number of additional ways of using spatial coherence, each of which involves different tradeoffs. These are discussed in the next section, which considers some sample applications in more detail.

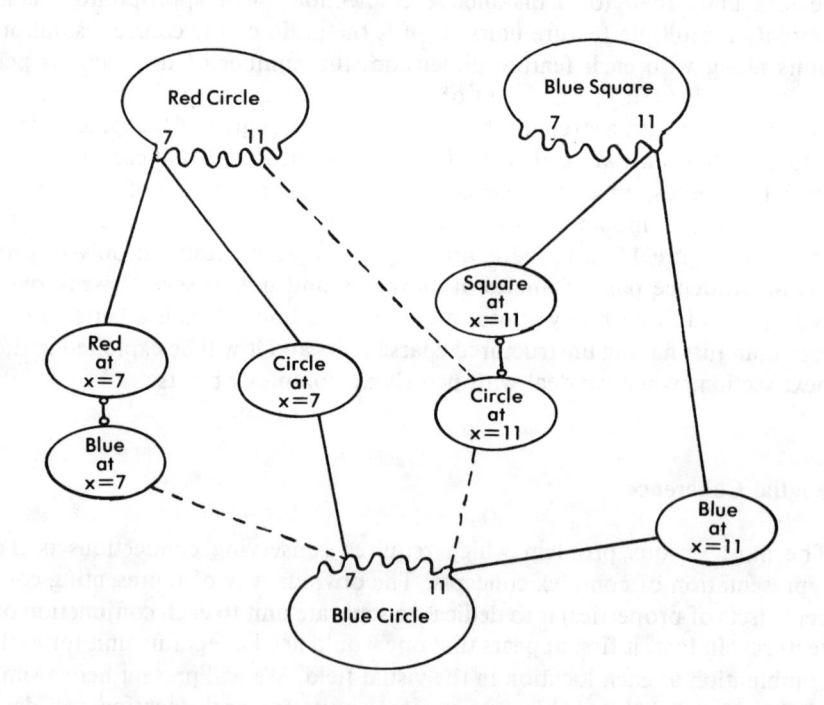

Figure 17. Spatial coherence on inputs can represent complex concepts without cross-talk. Solid lines show active inputs and dashed lines (some of the) inactive inputs.

## 5. APPLICATIONS

This section illustrates the power of the CM paradigm via two groups of examples. The first shows how the various techniques for conserving connections can be used in an idealized form of perception of a complex object. Here the point is that an object has multiple features which are computed in parallel via the transform methodology. The second group of examples starts with a relatively simple problem, that of vergence eye movements, to illustrate motor control using value units. In this example, control is immediate; a visual signal produces an instantaneous output (within the settling time constants of the units). Extensions of this idea use space as a buffer for time. For motor output, space allows the incorporation of more complex motor commands. For speech input, spatial buffering allows for phoneme recognition based on *subsequent* information.

These examples were chosen to show that CM can provide a unified representation for both perception and motor control. This is important since an animal is hardly ever passively responding to its environment. Instead, it seems involved in what Arbib has called a perception-action cycle (Arbib, 1979). Perceptions result in actions which in turn cause new perceptions, and so on. Massive parallelism changes the way the perception-action cycle is viewed. In the traditional view, one would convert the input to a language which uses variables, and then use these variables to direct motor commands. CM suggests that we think of accomplishing the same actions via a transformation: sensory input is transformed (connected to) to abstract representational units, which in turn are transformed (connected to) to motor units. This will obviously work for reflex actions. The examples are intended to sugest how more flexible command and control structures can also be represented by systems of value units.

**Object Recognition**

The examples of Figures 1 and 6 are representative of the problem of gestalt perception: that of seeing parts of an image as a single percept (object). An "object" is indicated by the "simultaneous" appearance of a number of "visual features" in the correct relative spatial positions. In any realistic case, this will involve a variety of features at several different levels of abstraction and complex interaction among them. A comprehensive model of this process would be a prototype theory of visual perception and is well beyond the scope of this paper. What we will do here is consider the prerequisite task of constructing CM solutions to the problems of detecting non-punctate visual features and of forming sets of the features which could help characterize a percept. We will refer throughout to the prototype problem of detecting Fred's frisbee, which is known to be round, baby-blue, and moving fairly fast. The development suppresses many important issues such as hierarchical descriptions, perspective, occlusion, and the integration of separate fixations, not to mention learning. A brief discussion of how these might be tackled follows the technical material.

The first problem is to develop a general CM technique for detecting features and properties of images, given that these features are not usually detectable at a single point in some retinotopic map. The basic idea is to find parameters which characterize the feature in question and connect each retinotopic detector to the parameter values consistent with its detectand.

Consider the problem of detecting lines in an image from short edge segments. Different lines can be represented by units having different discrete parameter values, e.g. in the line equation $p = x\cos\theta + y\sin\theta$, the parameters are $p$ and $\theta$. Thus edge units at $(x,y,\alpha)$ could be connected to appropriate line units. Note that this example is analogous to the word recognition example (Fig. 1). Edges are analogous to letters and lines to words. As in the words-letter example, "top-down" connections allow the existence of a line to raise the confidence of a local edge. In our line detection example, lines in the image are high potential (confidence) units in a slope-intercept $(\theta,p)$ parameter space. High confidence edge units produce high confidence line units by virtue of the network connectivity. This general way of describing this relationship between parts of an image (e.g., edges) and the associated parameters (e.g., $p,\theta$ for a line) is a connectionist interpretation of the *Hough transform* (Duda & Hart, 1972). Since each parameter value is determined by a large number of inputs, the method is inherently noise-resistant and was invented for this purpose. A Hough transform network for circles (like Fred's frisbee) would involve one parameter for size plus two for spatial location, and exactly this method has been used for tumor detection in chest radiographs (Kimme et al., 1975). Notice that the circle parameter space is itself retinotopic in that the centers of circles have specified locations; this will be important in registering multiple features.

The Hough transform is a formalism for specifying excitatory links between units. The general requirements are that part of an image representation can be represented by a parameter vector **a** in an image space A and a feature can be represented by a vector **b** which is an element of a feature space B. *Physical constraints* $f(\mathbf{a},\mathbf{b}) = 0$ relate **a** and **b**. The space A represents spatially indexed units, and each individual element $\mathbf{a}_k$ is only consistent with certain elements in the space B, owing to the constraint imposed by the relation f. Thus for each $\mathbf{a}_k$ it is impossible to compute the set

$$B_k = \{\mathbf{b} \mid \mathbf{a}_k \text{ and } f(\mathbf{a}_k,\mathbf{b}) \leq \delta_b\}$$

where $B_k$ is the set of units in the feature space network B that the $\mathbf{a}_k$ unit must connect to, and the constant $\delta_b$ is related to the quantization in the space **B**. Let H(**b**) be the number of active connections the value unit **b** receives from units in A. H(**b**) is the number of image measurements which are consistent with the parameter value **b**. The potential of units in B is given by $p(\mathbf{b}) \leftarrow H(\mathbf{b})/\Sigma_b H(\mathbf{b})$. The value p(**b**) can stand for the confidence that segment with feature value **b** is present in the image. If the measurement represented by **a** is realized as groups of units, e.g., $\mathbf{a} = (\mathbf{a}_1, \mathbf{a}_2)$, then conjunctive connections are required to implement the constraint relation.

Implementing these networks often results in a set of *very sparsely distributed* high-confidence feature space units. In implementations of the line detection example, only approximately 1% of the units have maximum confidence values. This figure is also typical of other modalities. In general, each $\mathbf{a}_k$ and the relationship f will not determine a single unit in $\mathbf{B}_k$ as in the line detection example, but there still will be isolated high-confidence units. Figure 1 shows why this is the case: different $\mathbf{a}_k$ letter-feature units connect to common units in the letter space B.

We have found that parameter spaces combine with the growing body of knowledge on specific physical constraints to provide a powerful and robust model for the simultaneous computation of invariant object properties such as reflectance, curvature, and relative motion (Ballard, 1981).

Of course segmentation must involve ways of associating peaks in several different feature spaces and methods for doing this are discussed presently, but the cornerstone of the techniques are high-confidence units in the individual-modality feature spaces. In extending the single feature case to multiple features, the most serious problem is the immense size of the cross product of the spatial dimensions with those of interesting features such as color, velocity, and texture. Thus to explain how image-like input such as color and optical flow are related to abstract objects such as "a blue, fast-moving thing," it becomes necessary to use all the techniques of the previous sections.

Even if we assume that there is a special unit for recognizing images of Fred's frisbee, it cannot be the case that there is a separate one of these units for each point in the visual field. One weak solution to this kind of problem was given in Figure 17 of the last section. There could conceivably be a separate 3-way conjunctive connection on the Fred's frisbee unit for each position in space. Activation of one conjunct would require the simultaneous activation of circle, baby-blue, and fairly-fast in the same part of the visual field. The solution style with separate conjunctions for every point in space becomes increasingly implausible as we consider more complex objects with hierarchical and multiple descriptions. The spatially registered conjunctions would have to be preserved throughout the structure.

The problem of going from a set of descriptors (features) to the object which is the best match to the set is known in artificial intelligence as the *indexing problem*. The feature set is viewed as an index (as in a data base). There have been several proposed parallel hierarchical network solutions to the indexing problem (Fahlman, 1979; Hillis, 1981) and these can be mapped into CM terms. But these designs assume that the network is presented with sets of descriptors which are already partitioned; precisely the vision problem we are trying to solve. There are three additional mechanisms that seem to be necessary, two of which have already been discussed. Coarse coding and tuning (as discussed in Section 4) make it much less costly to represent conjunctions. In addition, some general concepts (e.g., blue frisbee) might be indexed more efficiently through less precise units. The new idea is an extension of spatial coherence that exploits the fact that the networks respond to activity that occurs together in time. If there were a way to focus the activity of the network on one area at a time, only properties detected in that area would compete to index objects.

The obvious way to focus attention on one area of the visual field is with eye movements, but there is evidence that focus can also be done within a fixation. The general idea of internal spatial focus is shown in Figure 18. In this network, the general "baby-blue" unit is configured to have separate conjunctive inputs for each point in space, like the blue-square units of Figure 17. The difference is that the second input to the conjunction comes from a "focus" unit, and this makes a much more general network. The idea of making a unit (e.g., baby blue) more responsive to inputs from a given spatial position can be implemented in different ways. The conjunctive connection at the $x=7$ lobe of the baby-blue unit is the most direct way. But treating this conjunct as a strict AND would mean that all spatial units would have to be active when there was no focus. An alter-

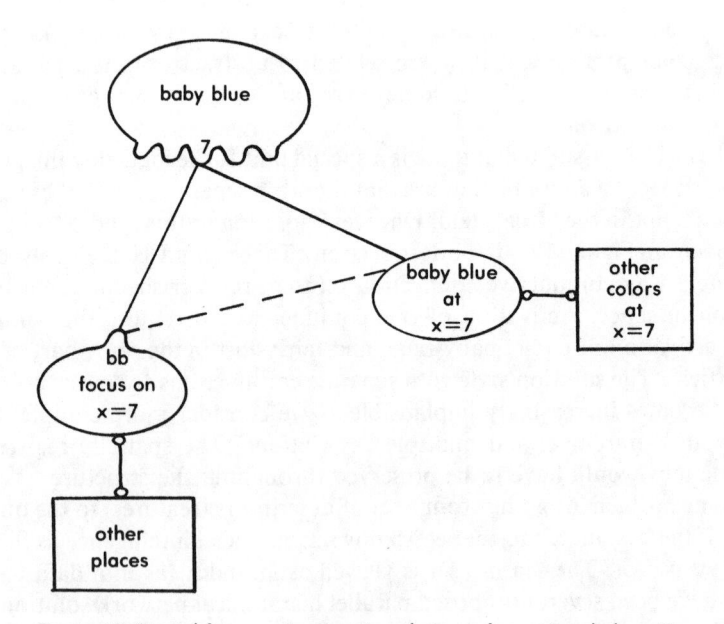

Figure 18. Spatial focus unit can gate only input from attended positions.

native would be to have the "focus on 7" unit boost the output of the "baby blue at 7" unit (and all of its rivals) as shown by the dashed line; this would eliminate the need for separate spatial conjunctions on the baby-blue unit, but would alter the potential of all the units at the position being attended. The trade-offs become even trickier when goal-directed input is taken into account, but both methods have the same effect on indexing. If the system has its attention directed only to $x = 7$, then the only feature units activated at all will be those whose local representatives are dominant (in their WTA) at $x = 7$. In such a case, there would be a time when the only concept units active in the entire network would be those for $x = 7$. This does not "solve" the problem of identifying objects in a visual scene, but it does suggest that sequentially focusing attention on separate places can help significantly. There is considerable reason to suppose (Posner, 1978; Triesman, 1980) that people do this even in tasks without eye movement.

There are other ways of looking at the network of Figure 18. Suppose the system had reason to focus on some particular property (e.g., baby-blue). If we make bi-directional the links from "focus on $x = 7$" to "baby-blue" and "baby-blue at 7," a nice possibility arises. The "focus on 7" unit could have a conjunctive connection for each separate property at its position. If, for example, baby-blue was chosen for focus and was the dominant color at $x = 7$, then the "focus on $x = 7$" unit would dominate its rivals. This suggests another way in which the recognition of complex objects could be helped by spatial focus. Figure 19 depicts the fairly general situation.

In Figure 19, the units representing baby-blue, circular, and fairly-fast are assumed to be for the entire visual field and moderately precise. The dotted arrows to the "Fred's frisbee" node suggest that there might be more levels of description in a realistic system. The spatial focus links involving baby-blue are the same as in Figure 18, and are replicated for the other two properties. Notice that the position-specific sensing units do not have their potentials affected by spatial focus units, so that the sensed data can remain intact. The network of Figure 19 can be used in several ways.

If attention has been focused on $x = 7$ for any reason, the various space-independent units whose representatives are most active at $x = 7$ will become most active, presumably leading to the activation (recognition) of Fred's frisbee. If a top-down goal of looking for Fred's frisbee (or even just something baby-blue) is active, then the "focus on $x = 7$" will tend to defeat its WTA rivals, leading to the same result. A third possibility is a little more complicated, but quite powerful. Suppose that a given image, even in context, activates too many property units so that no objects are effectively indexed. One strategy would be to systematically scan each area of the visual field, eliminating confounding activity from other areas. But it is also possible to be more efficient. If some property unit (say baby-blue) were strongly activated, the network could focus attention on all the positions with that property. In this case it is like putting a baby-blue filter in front of the scene, and should often lead to better convergence in the networks for shape, speed, etc.

One should compare the network of Figure 17 with Figures 18 and 19. In the former, parallel co-existing concepts are possible if we assume delicate arrangements of conjunctive connections. The latter networks are more robust but use sequentiality to eliminate cross-talk.

## Time and Sequence

Connectionist models do not initially appear to be well-suited to representing changes with time. The network for computing some function can be made quite fast, but it will be fixed in functionality. There are two quite different aspects of time variability of connectionist structures. One is time-varying responses, i.e., long-term modification of the networks (through changing weights) and short-term changes in the behavior of a fixed network with time. The second aspect is sequence: the problem of analyzing inherently sequential input (such as speech) or producing inherently sequential output (such as motor commands) with parallel models. The problem of change will be deferred to (Feldman, 1981b). The problem of sequence is discussed here.

There are a number of biologically suggested mechanisms for changing the weight ($w_j$) of synaptic connections, but none of them are nearly rapid enough to account for our ability to hear, read, or speak. The ability to perceive a time-varying signal like speech or to integrate the images from successive fixations must be achieved (according to our dogma) by some dynamic (electrical) activity in the networks. As usual, we will present computational solutions to the problems of sequence that appear to be consistent with known structural and performance constraints. These are, again, too crude to be taken literally but do suggest that connectionist models can describe the phenomena.

*Motor Control of the Eye.* To see how the transform notion of distributed units might work for motor control, we present a simplistic model of vergence eye movements. (The same idea may be valid for fixations, but control probably takes place at higher levels of abstraction.) In this model retinotopic (spatial) units are connected directly to muscle control units. Each retinotopic unit can if saturated cause the appropriate contraction so that the new eye position is centered on that unit. When several retinotopic units saturate, each enables a muscle control unit independently and the muscle itself contracts an average amount.

Figure 20 shows the idea for a one-dimensional retina. For example, with units at positions 2, 4, 5, and 6 saturated, the net result is that the muscle is centered at 17/4 or 4.25. (This idea can be extended to the case where

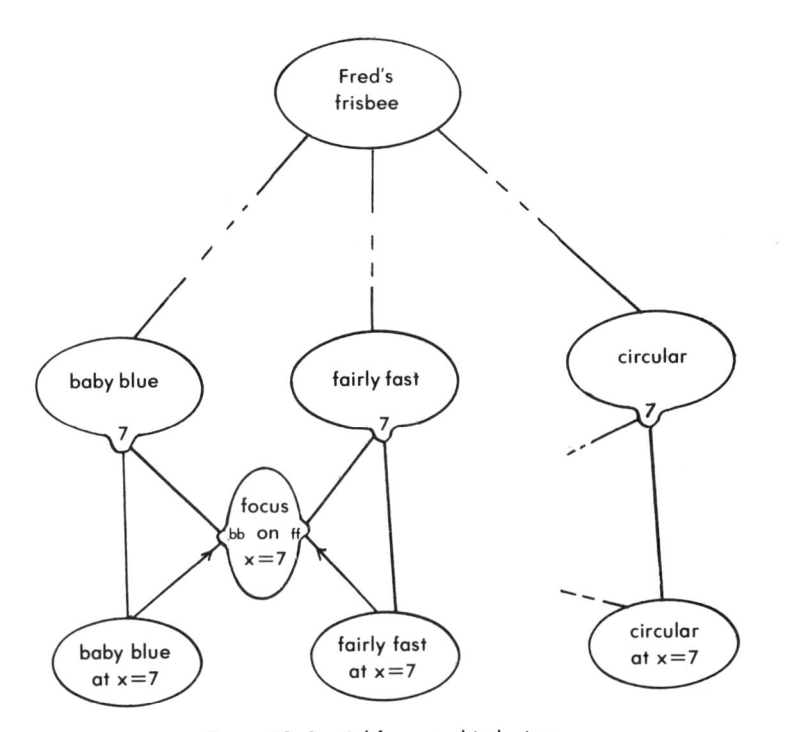

Figure 19. Spatial focus and indexing.

current eye coordinate

retinal
spatial units
[C(x) in Fig. 19]

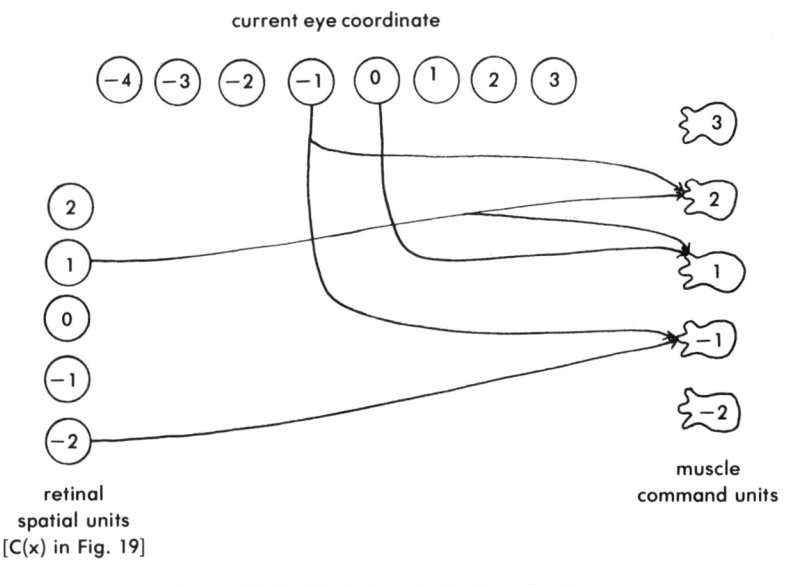

muscle
command units

Figure 20. Distributed Control of Eye Fixations

53

the retinotopic units have overlapping fields.) This kind of organization could be extended to more complex movement models such as that of the organization of the superior colliculus in the monkey (Wurtz & Albano, 1980).

Notice that each retinotopic unit is capable of enabling different muscle control units. The appropriate one is determined by the enabled x-origin unit which inhibits commands to the inappropriate control units via modifiers.

One problem with this simple network arises when disparate groups of retinotopic units are saturated. The present configuration can send the eye to an average position if the features are truly identical. The network can be modified with additional connections so that only a single connected component of saturated units is enabled by using additional object primitives. A version of this WTA motor control idea has already been used in a computer model of the frog tectum (Didday, 1976).

There are still many details to be worked out before this could be considered a realistic model of vergence control, but it does illustrate the basic idea: local spatially separate sensors have *distinct, active* connections which could be averaged at the muscle for fine motor control or be fed to some intermediate network for the control of more complex behaviors.

*Converting Space to Time.* Consider the problem of controlling a simple physical motion, such as throwing a ball. It is not hard to imagine that in a skilled motor performance unit-groups fire each other in a fixed succession, leading to the motor sequence. The computational problem is that there is a unique set of effector units (say at the spinal level) that must receive input from each group at the right time. Figure 21a depicts a simple case in which there are two effector units ($e_1$, $e_2$) that must be activated alternatively. The circles marked 1–4 represent units (or groups of units) which activate their successor and inhibit their predecessor (cf. Delcomyn, 1980). The main point is that a succession of outputs to a single effector set can be modeled as a sequence of time-exclusive groups representing instantaneous coordinated signals. Moving from one time step to the next could be controlled by pure timing for ballistic movements, or by a proprioceptive feedback signal. There is, of course, an enormous amount more than this to motor control, and realistic models would have to model force control, ballistic movements, gravity compensation, etc.

The second part of Figure 21 depicts a somewhat fanciful notion of how a variety of output sequences could share a collection of lower level response units. The network shown has a single "Dixie" unit which can start a sequence and which joins in conjunctive connections with each note to specify its successor. At each time step, a WTA network decides what note gets sounded. One can imagine adding the rhythm network and transposition networks to other keys and to other modalities of output.

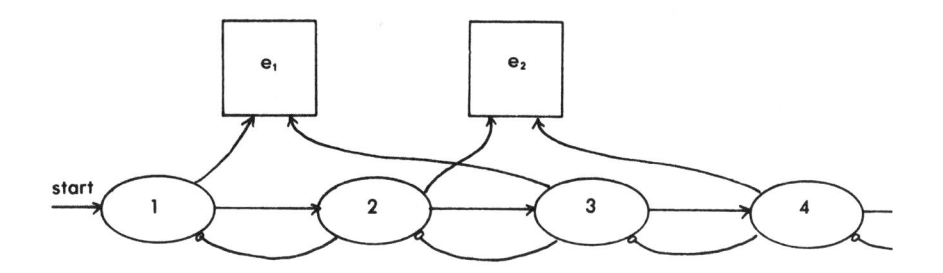

a. Sequence and Suppression

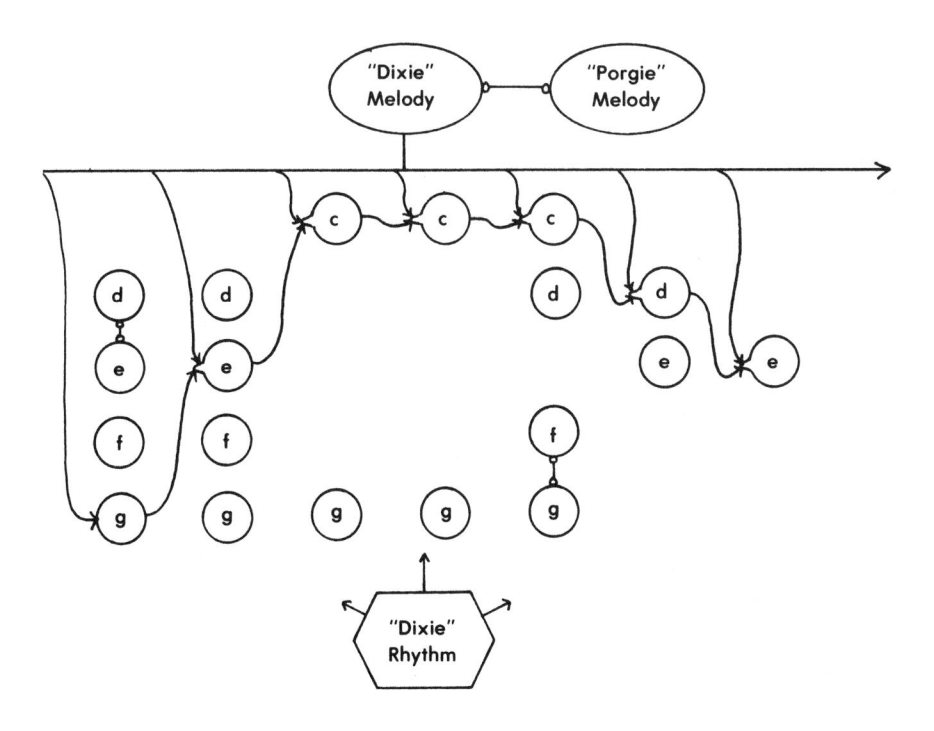

b. Whistling Dixie

Figure 21. Mapping Space to Time.

*Converting Time to Space.* The sequencer model for skilled movements was greatly simplified by the assumption that the sequence of activities was pre-wired. How could one (still crudely, of course) model a situation like speech perception where there is a largely unpredictable time-varying computation to be carried out? One solution is to combine the sequencer model of Figure 21 with a simple vision-like scheme. We assume that speech is recognized by being sequenced into a buffer of about the length of a phrase and then is relaxed against context in the way described above for vision. For simplicity, assume that there are two identical buffers, each having a pervasive modifier ($m_j$) innervation so that either one can be switched into or out of its connections. We are particularly concerned with the process of going from a sequence of potential phonetic features into an interpreted phrase. Figure 22 gives an idea of how this might happen.

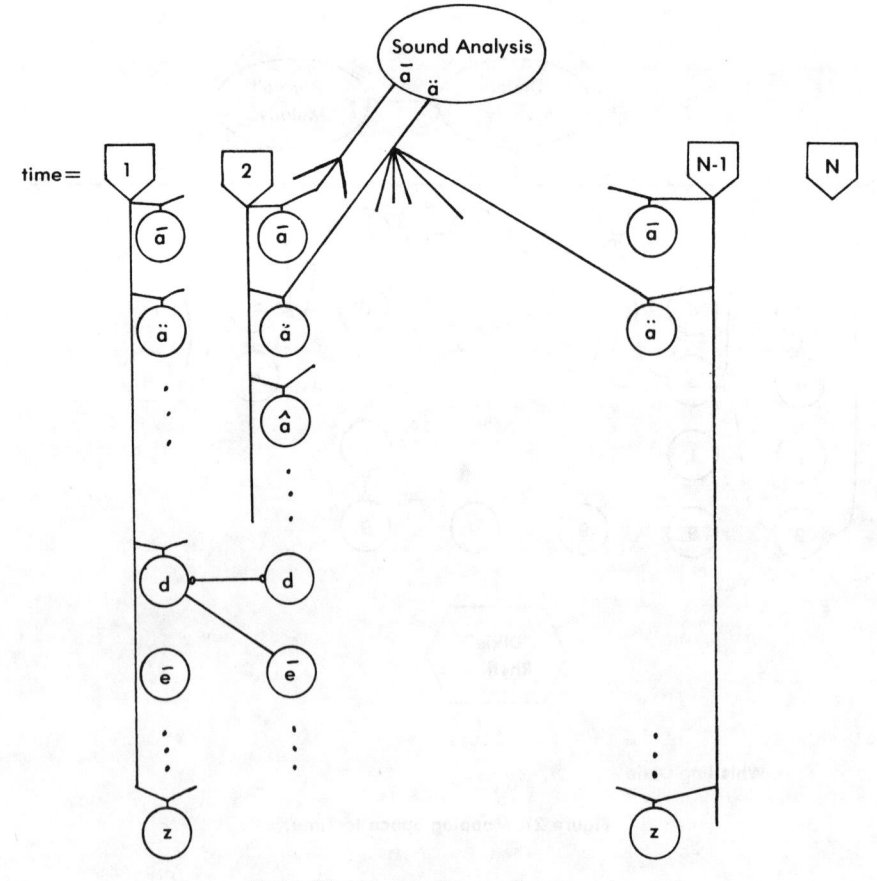

Figure 22. Mapping Time to Space.

Assume that there is a separate unit for each potential feature for each time step up to the length of the buffer. The network which analyzes sound is connected identically to each column, but conjunction allows only the connections to the active column to transmit values. Under ideal circumstances, at each time step exactly one feature unit would be active. A phrase would then be laid out on the buffer like an image on the "mind's eye," and the analogous kind of relaxation cones (cf. Figure 1, 6) involving morphemes, words, etc., could be brought to bear. The more realistic case where sounds are locally ambiguous presents no additional problems. We assume that, at each time step, the various competing features get varying activation. Diphone constraints could be captured by ( + or − ) links to the next column as suggested by Figure 22. The result is a multiple possibility relaxation problem—again exactly like that in visual perception. The fact that each potential feature could be assigned a row of units is essential to this solution; we do not know how to make an analogous model for a sequence of sounds which cannot be clearly categorized and combined. Recall that the purpose of this example is to indicate how time-varying input could be treated in connectionist models. The problem of actually laying out detailed models for language skills is enormous and our example may or may not be useful in its current form. Some of the considerations that arise in distributed modeling of language skills are presented in Arbib and Caplan, (1979).

## CONCLUSIONS

The CM paradigm advanced in this paper has been applied successfully only to relatively low-level tasks. There is no reason, as yet, to be confident that an intermediate symbolic representation will not be required for modeling higher cognitive processes. There is, however, the beginning of a collection of efforts which can be interpreted as attempting CM approaches to higher level tasks. These include work which explicitly uses parallelism in planning (Stefik, 1981) and deduction, and work which incorporates more connectionist architectural notions of value units (Forbus, 1981) and coarse coding (Garvey, 1981).

We have now completed six years of intensive effort on the development of connectionist models and their application to the description of complex tasks. While we have only touched the surface, the results to date are very encouraging. Somewhat to our surprise, we have yet to encounter a challenge to the basic formulation. Our attempts to model in detail particular computations (Ballard & Sabbah, 1981; Sabbah, 1981) have led to a number of new insights (for us, at least) into these specific tasks. Attempts like this one to formulate and solve general computational problems in realistic connectionist terms have proven to be difficult, but less so than we would have guessed. There appear to be a number of interesting technical problems within the theory and a wide range of questions about brains and behavior which might benefit from an approach along the lines suggested in this paper.

## APPENDIX: SUMMARY OF DEFINITIONS AND NOTATION

A **unit** is a computational entity comprising:

$\{q\}$—a set of *discrete states*, $<10$
p—a continuous value in $[-10,10]$, called *potential* (accuracy of several digits)
v—an *output value*, integers $0 \leq v \leq 9$
$i$—a vector of *inputs* $i_1,\ldots,i_n$

and functions from old to new values of these

$$p \leftarrow f(i,p,q)$$
$$q \leftarrow g(i,p,q)$$
$$v \leftarrow h(i,p,q)$$

which we assume to compute continuously. The form of the f, g, and h functions will vary, but will generally be restricted to conditionals and simple functions.

### P-Units

For some applications, we will use a particularly simple kind of unit whose output v is proportional to its potential p (rounded) (when $p > 0$) and which has only one state. In other words

$$p \leftarrow p + \beta \Sigma w_k i_k \qquad [0 \leq w_k \leq 1]$$
$$v \leftarrow \textit{if } p > \theta \textit{ then } \text{round } (p - \theta) \textit{ else } 0 \qquad [v = 0\ldots 9]$$

where $\beta, \theta$ are constants and $w_k$ are weights on the input values.

### Conjunctive Connections

In terms of our formalism, this could be described in a variety of ways. One of the simplest is to define the potential in terms of the maximum, e.g.,

$$p \leftarrow p + \beta \text{Max}(i_1 + i_2 - \varphi, i_3 + i_4 - \varphi, i_5 + i_6 - i_7 - \varphi)$$

where $\beta$ is a scale constant as in the p-unit and $\varphi$ is a constant chosen (usually >10) to suppress noise and require the presence of multiple active inputs. The minus sign associated with $i_7$ corresponds to its being an inhibitory input. The max-of-sum unit is the continuous analog of a logical OR-of-AND (disjunctive normal form) unit and we will sometimes use the latter as an approximate version of the former. The OR-of-AND unit corresponding to the above is:

$$p \leftarrow p + \alpha \text{ OR } (i_1 \& i_2,\ i_3 \& i_4,\ i_5 \& i_6 \& (\text{not } i_7))$$

**Winner-take-all (WTA) networks** have the property that only the unit with the highest potential (among a set of contenders) will have output above zero after some settling time.

A **coalition** will be called stable when the output of all of its members is non-decreasing.

## Change

For our purposes, it is useful to have all the adaptability of networks be confined to changes in weights. While there is known to be some growth of new connections in adults, it does not appear to be fast or extensive enough to play a major role in learning. For technical reasons, we consider very local growth or decay of connections to be changes in existing connection patterns. Obviously, models concerned with developing systems would need a richer notion of change in connectionist networks (cf. von der Malsburg & Willshaw, 1977). We provide each unit with a memory vector $\mu$ which can be updated:

$$\mu \leftarrow c(i, p, q, x, w, \mu)$$

where $\mu$ is the intermediate-term memory vector, $w$ is the weight vector, $i$, $p$, and $q$ are as always, and x is an additional single integer input ($0 \leq x \leq 9$) which captures the notion of the importance and value of the current behavior. Instantaneous establishment of long-term memory imprinting would be equivalent to having $\mu = w$. The assumption is that the consolidation of long-term changes is a separate process.

We postulate that important, favorable or unfavorable, behaviors can give rise to faster learning. The rationale for this is given in (Feldman, 1980; 1981a), which also lays out informally our views on how short- and long-term learning could occur in connectionist networks. A detailed technical discussion of this material, along the lines of this paper, is presented in (Feldman, 1981b). Obviously enough, a plausible model of learning and memory is a prerequisite for any serious scientific use of connectionism.

# REFERENCES

Anderson, J. A., Silverstein, J. W., Ritz, S. A., & Jones, R. S. Distinctive features, categorical perception, and probability learning: Some applications of a neural model. *Psychological Review,* September 1977, *84*(5), 413-451.

Arbib, M. A. *Perceptual structures and distributed motor control.* COINS (Tech. Rep. 79-11). University of Massachusetts, Computer and Information Science, and Center for Systems Neuroscience, June 1979.

Arbib, M. A., & Caplan, D. Neurolinguistics must be computational. *The Brain and Behavioral Sciences,* 1979, *2,* 449-483.

Ballard, D. H. Parameter networks: Towards a theory of low-level vision. *Proceedings of the 7th IJCAI,* Vancouver, BC, August 1981.

Ballard, D. H., & Kimball, O. A. *Rigid body motion from depth and optical flow* (Tech. Rep. 70). New York: University of Rochester, Computer Science Department, in press, 1981. (a)

Ballard, D. H., & Kimball, O. A. *Shape and light source direction from shading* (Tech. Rep.). Rochester, NY: University of Rochester, Computer Science Department, in press, 1981. (b)

Ballard, D. H., & Sabbah, D. On shapes. *Proceedings of the 7th IJCAI,* Vancouver, BC, August 1981.

Collins, A. M., & Loftus, E. F. A spreading-activation theory of semantic processing. *Psychological Review,* November 1975, *82,* 407-429.

Delcomyn, F. Neural basis of rhythmic behavior in animals. *Science,* October 1980, *210,* 492-498.

Dell, G. S., & Reich, P. A. Toward a unified model of slips of the tongue. In V. A. Fromkin (Ed.), *Errors in Linguistic Performance: Slips of the Tongue, Ear, Pen, and Hand.* New York: Academic Press, 1980.

Didday, R. L. A model of visuomotor mechanisms in the frog optic tectum. *Mathematical Bioscience,* 1976, *30,* 169-180.

Duda, R. O., & Hart, P. E. Use of the Hough transform to detect lines and curves in pictures. *Communications of the ACM 15*(1), January 1972, 11-15.

Edelman, G., & Mountcastle, B. *The Mindful Brain.* Boston, MA: MIT Press, 1978.

Fahlman, S. E. *NETL, A System for Representing and Using Real Knowledge.* Boston, MA: MIT Press, 1979.

Fahlman, S. E. The Hashnet interconnection scheme. Computer Science Department, Carnegie-Mellon University, June 1980.

Feldman, J. A. *A distributed information processing model of visual memory* (Tech. Rep. 52). Rochester, NY: University of Rochester, Computer Science Department, 1980.

Feldman, J. A. A connectionist model of visual memory. In G. E. Hinton & J.A. Anderson (Eds.), *Parallel Models of Associative Memory.* Hillsdale, NJ: Lawrence Erlbaum Associates, 1981. (a)

Feldman, J. A. *Memory and change in connection networks* (Tech. Rep. 96). Rochester, NY: University of Rochester, Computer Science Department, October 1981. (b)

Feldman, J. A. *Four frames suffice* (Tech. Rep. 99). Rochester, NY: University of Rochester, Computer Science Department, in press, 1982.

Feldman, J. A., & Ballard, D. H. *Computing with connections* (Tech. Rep. 72). Rochester, NY: University of Rochester, Computer Science Department, 1981; to appear in book by A. Rosenfeld & J. Beck (Eds.), 1982.

Forbus, K. D. Qualitative reasoning about physical processes. *Proceedings of the 7th IJCAI,* Vancouver, BC, August 1981, 326-330.

Freuder, E. C. Synthesizing constraint expressions. *Communications of the ACM,* November 1978, *21*(11), 958-965.

Garvey, T. D., Lowrance, J. D., & Fischler, M. A. An inference technique for integrating knowledge from disparate sources. *Proceedings of the 7th IJCAI,* Vancouver, BC, August 1981, 319-325.

Grossberg, S. Biological competition: Decision rules, pattern formation, and oscillations. *Proc. National Academy of Science USA,* April 1980, *77*(4), 2238-2342.

Hanson, A. R., & Riseman, E. M., (Eds.). *Computer Vision Systems.* New York: Academic Press, 1978.

Hillis, W. D. The connection machine (Computer architecture for the new wave). AI Memo 646, M.I.T., September 1981.

Hinton, G. E. Relaxation and its role in vision. (Ph.D. thesis, University of Edinburgh, December 1977.)

Hinton, G. E. Draft of Technical Report. La Jolla, CA: University of California at San Diego, 1980.

Hinton, G. E. The role of spatial working memory in shape perception. *Proceeding of the Cognitive Science Conference,* Berkeley, CA, August 1981. (a) 56-60.

Hinton, G. E. The role of spatial working memory in shape perception. *Proceedings of the Cognitive Science Conference,* Berkeley, CA, August 1981. (a) 56-60.

Hinton, G. E., & Anderson, J. A. (Eds.). *Parallel Models of Associative Memory.* Hillsdale, NJ: Lawrence Erlbaum Associates, 1981.

Horn, B. K. P., & Schunck, B. G. Determining Optical Flow. AI Memo 572, AI Lab, MIT, April 1980.

Hubel, D. H., & Wiesel, T. N. Brain mechanisms of vision. *Scientific American,* September 1979, 150-162.

Ja'Ja', J., & Simon, J. Parallel algorithms in graph theory: Planarity testing. CS 80-14, Computer Science Department, Pennsylvania State University, June 1980.

Jusczyk, P. W., & Klein, R. M. (Eds.). *The Nature of Thought: Essays in Honor of D. O. Hebb.* Hillsdale, NJ: Lawrence Erlbaum Associates, 1980.

Kandel, E. R. *The Cellular Basis of Behavior.* San Francisco, CA: Freeman, 1976.

Kimme, C., Sklansky, J., & Ballard, D. Finding circles by an array of accumulators. *Communications of the ACM,* February 1975.

Kinsbourne, M., & Hicks, R. E. Functional cerebral space: A model for overflow, transfer and interference effects in human performance: A tutorial review. In J. Requin (Ed.), *Attention and Performance 7.* Hillsdale, NJ: Lawrence Erlbaum Associates, 1979.

Kosslyn, S. M. *Images and Mind.* Cambridge, MA: Harvard University Press, 1980.

Kuffler, S. W., & Nicholls, J. G. *From Neuron to Brain: A Cellular Approach to the Function of the Nervous System.* Sunderland, MA: Sinauer Associates, Inc., Publishers, 1976.

Marr, D. C., & Poggio, T. Cooperative computation of stereo disparity. *Science,* 1976, *194,* 283-287.

McClelland, J. L., & Rumelhart, D. E. An interactive activation model of the effect of context in perception: Part 1. *Psychological Review,* 1981.

Minsky, M., & Papert, S. *Perceptrons.* Cambridge, MA: The MIT Press, 1972.

Norman, D. A. A psychologist views human processing: Human errors and other phenomena suggest processing mechanisms. *Proceedings of the 7th IJCAI,* Vancouver, BC, August 1981, 1097-1101.

Perkel, D. H., & Mulloney, B. Calibrating compartmental models of neurons. *American Journal of Physiology* 1979, *235*(1), R93-R98.

Posner, M. I. *Chronometric Explorations of Mind.* Hillsdale, NJ: Lawrence Erlbaum Associates, 1978.

Prager, J. M. Extracting and labeling boundary segments in natural scenes. *IEEE Trans. PAMI,* January 1980, *2*(1), 16-27.

Ratcliff, R. A theory of memory retrieval. *Psychological Review,* March 1978, *85*(2), 59-108.

Rosenfeld, A., Hummel, R. A., & Zucker, S. W. Scene labelling by relaxation operations. *IEEE Trans. SMC 6*, 1976.

Sabbah, D. Design of a highly parallel visual recognition system. *Proceedings of the 7th IJCAI*, Vancouver, BC, August, 1981.

Scientific American. *The Brain*. San Francisco, CA.: W. H. Freeman and Company, 1979.

Sejnowski, T. J. Strong covariance with nonlinearly interacting neurons. *Journal of Mathematical Biology*, 1977, *4*(4), 303-321.

Smith, E. E., Shoben, E. J., & Rips, L. J. Structure and process in semantic memory: A featural model for semantic decisions. *Psychological Review*, 1974, *81*(3), 214-241.

Stefik, M. Planning with Constraints (MOLGEN: Part 1). *Artificial Intelligence*, *16*(2), 1981.

Stent, G. S. A physiological mechanism for Hebb's postulate of learning. *Proc. National Academy of Science USA*, April 1973, *70*(4), 997-1001.

Sunshine, C. A. Formal techniques for protocol specification and verification. *IEEE Computer*, August 1979.

Torioka, T. Pattern separability in a random neural net with inhibitory connections. *Biological Cybernetics*, 1979, *34*, 53-62.

Triesman, A. M., & Gelade, G. A feature-integration theory of attention. *Cognitive Psychology*, 1980, *12*, 97-136.

Ullman, S. Relaxation and constrained optimization by local processes. *Computer Graphics and Image Processing*, 1979, *10*, 115-125.

von der Malsburg, Ch., & Willshaw, D. J. How to label nerve cells so that they can interconnect in an ordered fashion. *Proc. National Academy of Science USA*, November 1977, *74*(11), 5176-5178.

Wickelgren, W. A. Chunking and consolidation: A theoretical synthesis of semantic networks, configuring in conditioning, S-R versus cognitive learning, normal forgetting, the amnesic syndrome, and the hippocampal arousal system. *Psychologial Review*, 1979, *86*(1), 44-60.

Wurtz, R. H., & Albano, J. E. Visual-motor function of the primate superior colliculus. *Annual Review of Neurscience*, 1980, *3*, 189-226.

Zeki, S. The representation of colours in the cerebral cortex. *Nature*, April 1980, *284*, 412-418.

CHAPTER 3

# Putting Knowledge in its Place: A Scheme for Programming Parallel Processing Structures on the Fly

JAMES L. MCCLELLAND
*Carnegie Mellon University*

This paper introduces a mechanism called CID, the Connection Information Distributor. CID extends connectionism by providing a way to program networks of simple processing elements on line, in response to processing demands. Without CID, simultaneous processing of several patterns has only been possible by prewiring multiple copies of the network needed to process one pattern at a time. With CID, programmable processing structures can be loaded with connection information stored centrally, as needed. To illustrate some of the characteristics of the scheme, a CID version of the interactive activation model of word recognition is described. The model has a single permanent representation of the connection information required for word perception, but it allows several words to be processed simultaneously in separate programmable networks. Multiword processing is not perfect, however. The model produces the same kinds of intrusion errors that human subjects make in processing brief presentations of word-pairs, such as SAND LANE (SAND is often misreported as LAND or SANE). The resource requirements of the mechanism, in terms of nodes and connections, are found to be quite moderate, primarily because networks that are programmed in-response to task demands can be much smaller than networks that have knowledge of large numbers of patterns built in.

Connectionism (Feldman, 1981; Feldman & Ballard, 1982) is the idea that the computations performed by a processing system are controlled by the

---

This research was supported by NSF grant BNS79-24062, ONR contract N00014-82-C-0374, and a Grant from the Systems Development Foundation. The author is the recipient of a Research Scientist Career Development Award 5-K01 MH00385 from the National Institute of Mental Health. This report was prepared while the author was a visitor at Bolt Beranek and Newman, Inc. The author is grateful to Alan Collins, John Frederiksen, Geoff Hinton, David Rumelhart, William Salter, and David Zipser for useful discussions, and encouragement.

connections among a large number of simple processing units. The processing units themselves do very simple things—generally, they simply update the strength of the signal they send based on a simple function of signals they receive from other processing elements. The intelligence of the system —what it knows and what it can do with its knowledge—is determined by the interconnections among the elements.

In designing connectionist mechanisms, one is often tempted to hardwire the knowledge about the objects to be processed directly into the connections between the processing units that process the objects. However, as I shall argue in this paper, there are limitations to this approach. To overcome these limitations, I will propose a way of using connectionist hardware to make *programmable* connectionist mechanisms, in which the connections needed to meet the current processing demands can be set up on line, as the processing demands arise.

The article begins with an example of what a hard-wire connectionist mechanism might look like, and uses the example to illustrate the limitations of the approach. The subsequent sections develop the programmable alternative.

## A HARD-WIRED CONNECTIONIST PROCESSING MECHANISM

An example of a model which might be hard wired into connectionist processing structures is the interactive-activation model of word recognition (McClelland & Rumelhart, 1981; Rumelhart & McClelland, 1981, 1982; Figure 1). The model consists of a large number of basic elements or *nodes*. These serve as detectors for visual features, letters and words. Each node corresponds to the assertion that the item the node represents is present in an input pattern being processed by the network, and the *activation* of the node is monotonically related to the strength of this assertion. Nodes are grouped into several levels, with the feature level consisting of the feature nodes, the letter level consisting of the letter nodes, and the word level consisting of the word nodes. Since the inputs presented to the model contain four letters, there are separate sets of feature and letter nodes for each of the four letter positions. Since the model is intended to process only one word at a time, there is only one set of word nodes, with one node assigned to each four-letter word in English.

Processing in the interactive activation model takes place through excitatory and inhibitory interactions between the nodes. Nodes on different levels that are mutually consistent are excitatory. For example, the node for the letter T in the first position excites (and is excited by) the nodes for features of the letter T in the first position, and the nodes for words begin-

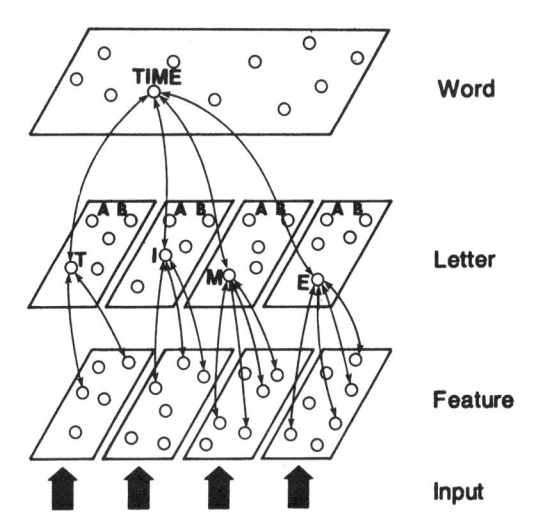

Figure 1. A sketch of some of the nodes in the interactive activation model of word perception (McClelland & Rumelhart, 1981; Rumelhart & McClelland, 1982), illustrating a small fraction of the excitatory and inhibitory influences between a few of the nodes. Nodes within the same rectangle represent mutually exclusive alternatives, and they are all mutually inhibitory; nodes on different levels that are mutually consistent are mutually excitatory. Some of the mutual excitatory interactions are represented by bidirectional arrows.

ning with T, such as TIME, or TAPE. In addition, nodes that represent mutually exclusive interpretations of the input in the same position and at the same level are mutually inhibitory. For example, the node for the T in the first position inhibits all of the other first-position letter nodes. At the word level, the word nodes can be thought of as representing alternative interpretations of the whole string, so they are all mutually inhibitory.

One way to view the interactive activation model is as an abstract description of the time course of information accumulation regarding potential hypotheses, without specification of the actual mechanism whereby these interactions take place. On this view, each node represents a hypothesis, and each excitatory or inhibitory connection represents a weighted contingency between hypotheses. However, it is very easy to visualize a connectionist implementation of the model, in which the nodes are physical processing units and the interactions between them are mediated by physical connections between the processing units. Indeed, Figure 1, taken literally, depicts just such an implementation.

This sort of implementation seems very appealing for a number of reasons. The neural hardware of the brain is, after all, apparently well suited to connectionist mechanisms; and the idea that a processing system might be embodied directly in the brain (perhaps with some redundancy of units and connections) gives us the sense that we have begun to reduce cognition to a level where we might begin to understand its physical basis. Further, systems like Competitive Learning (Grossberg, 1976; Rumelhart & Zipser, 1985; von der Malsberg, 1973) have been proposed which could actually provide mechanisms whereby an unspecialized pool of processing units could learn to behave like the interactive activation model.

There is, however, one difficulty with adopting a hard-wired connectionist implementation of the interactive activation model or any other model involving interactions among large numbers of simultaneous, mutually constraining hypotheses. The difficulty springs from the fact that the knowledge which guides processing is hard-wired into the connections between the processing units in which the processing takes place. This is true, whether the connectionist model is of the *local* variety, in which each hypothesis is represented by a single node, as in the interactive activation model; or of the *distributed* variety, in which each hypothesis is represented by a pattern of activation over a population of nodes, as in the models of Anderson (1983) or Hinton (1981b). In both cases, the knowledge that guides processing is contained in the connections between the processing elements.

Why is this a problem? The reason is that parallel processing of multiple items is purchased at the price of duplication of the knowledge—the connection information—that guides processing. A hard-wired version of the word perception model would be able to process all four letters in a word at one time only because the connections specifying which features make up each letter would be reduplicated in the connections between each of the four banks of feature nodes and the corresponding bank of letter nodes. Only one word could be processed at a time, since there would be only one bank of word nodes. If we wished to process more than one word at a time we could only do so if we were willing to reduplicate the entire model, adding an additional four feature and letter banks and an additional word bank for each additional word we wished to process at the same time.

Even if we were willing to suppose that all of this hardware ought to be dedicated to reading English words, we would have a problem with learning. It would be nice if experience with a pattern when it occurred in one part of the display could result in learning which could be transferred to other parts of the display. But learning in connectionist models amounts to changing the strengths of connections between the nodes, based on what tends to go with what. We would, therefore, need some way of disseminating the changes mandated by events occurring in one bank of detectors to the other banks.

One obvious solution to these problems is just to "go sequential." Rather than reduplicate knowledge, we could put it just in one place—in a single, central connectionist structure—and map inputs into it one at a time. This would solve the learning problem, since patterns arising in different locations would always be processed in the same central set of connections. However, going sequential eliminates the whole point of interactive activation. A basic tenet of the interactive activation model was that processing occurred in parallel, across all of the letters in a word. This fundamental assumption allowed the model to exploit mutual constraints among the letters. It is precisely the possibility of such mutual constraints between larger, higher level units which makes the approach appear appealing in such domains as speech perception (Elman & McClelland, 1984), sentence analysis (Waltz & Pollack, 1985), and language production (Dell, 1985). It becomes cumbersome, if not impossible, to exploit the mutual constraint between items when they are processed sequentially.

Neither duplication of connection information nor sequential processing seem entirely satisfactory. Indeed, it has seemed to me that connectionism would be an unduly limiting computational framework if we were forced to limit the possibility of exploiting mutual constraints in our models to cases where we are willing to postulate duplication of connection information. Putting the point another way, if we could find a way of permitting parallel processing while still retaining the benefits of a single central representation for learning, we would have achieved an important increase in the computational power of connectionist mechanisms.

The rest of this paper describes a solution to this problem. There are four principle sections. The first describes the basic idea behind the approach, and builds up to an implementation of the interactive activation model of word perception which allows multiple words to be processed at the same time, even though it has a single central representation of the connection information specifying which letters go together to make each word. The second section describes a computer simulation of this model, and shows how the approach can account for some interesting data reported recently by Mozer (1983) on the kinds of errors human subjects make in processing two words at the same time. The third section considers the computational resource requirements of the connection information distribution scheme. In it I indicate that the scheme requires much less hardware than it would seem to require at first glance. The discussion section examines the essential properties of CID, and considers how it might be extended beyond the applications implemented here.

The next two sections focus almost exclusively on the identification of words, based on letter information provided by assumed lower levels of processing. Obviously, the word level is but one layer in a very rich language processing system. I chose to focus on this level because it is concrete, accessible, and familiar (at least to me), and because there is interesting evidence that bears on the model at this particular level. The principles embodied in the mechanisms I describe are obviously applicable at other levels, and to other processing tasks besides language processing. Of course, some new problems do arise at other levels. I will say a bit about extending the connection information distribution scheme to handle some of these problems in the discussion section below.

## CID—A CONNECTION INFORMATION DISTRIBUTOR

In the system I propose, information processing takes place in a set of programmable node networks. Each network is a processing system consisting of processing units very similar to those that might be encountered in a direct connectionist implementation of the interactive activation model. Activations in these units stand for hypotheses about what is present where in the input, and information processing unfolds through their excitatory and inhibitory interactions. However, the units are not dedicated permanently to stand for particular hypotheses, and the knowledge that determines the pattern of excitatory and inhibitory interactions is not hard-wired into the connections between them. Rather, the connections in the node network are programmable by inputs from a central network in which the knowledge that guides processing is stored.

The first part of this section describes an individual programmable network. Later parts describe the structures needed to program such networks in response to ongoing processing demands.

### A Programmable Network

Figure 2 presents a very simple hard-wired network. The task of this section is to see how we could replace this hard-wired network with one that could be programmed to do the same work. The network shown in the figure is a very simple interactive activation system, consisting only of a letter and a word level. The figure is different from Figure 1 in organization, in order to highlight the excitatory connections between the units and lay them out in a way which will be convenient as we proceed.

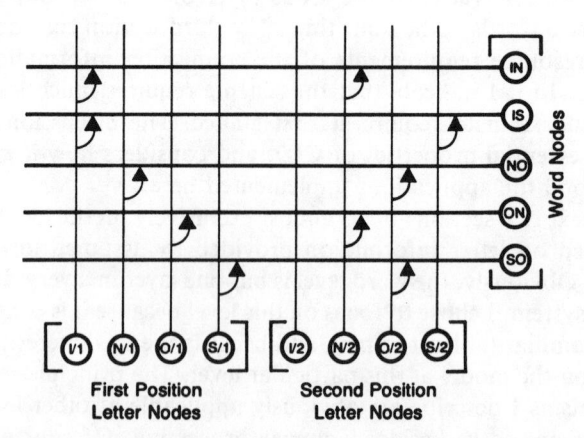

Figure 2. An extremely simple connection mechanism, capable of processing one two-letter string made up of the letters I, N, O, and S. The model knows only the five words that can be made of two of these letters, namely IN, IS, NO, ON, and SO. No top-down connections are included in this simple model. Nodes bracketed together are mutually inhibitory.

In this simple network, there are detectors only for the letters I, N, O and S in each of two letter positions. At the word level, there is a detector for each of the English words that can be made out of two of these letters. For simplicity, this model contains only letter-to-word connections; another matrix would be needed to capture word to letter feedback. Units which are in mutual competition are included in the same square brackets. This is just a shorthand for the bidirectional inhibitory connections, which could also be represented in another connection matrix.

In this diagram, letter units are shown having output lines which ascend from them. Word units are shown having input lines which run from left to right. Where the output line of each letter node crosses the input line of each word node, there is the possibility of a connection between them.

The knowledge built into the system which lets it act as a processor for the words IN, IS, NO, ON, and SO is contained in the excitatory connections between the letter and word nodes. These are represented by the filled triangles in the figure.

Now, we are ready to see how we could build a programmable network, one which we could *instruct* to behave like the hard-wired network shown in Figure 2. Suppose that instead of fixed connections from specific letter nodes to particular word nodes, there is a *potential* connection at the junction between the output line from each letter-level node and the input line to each word-level node. Then, all we would need to do to "program" the network to process the words IN, IS, NO, ON, and SO correctly would be to send in signals from outside turning on the connections which are hard-wired in Figure 2. This proposal is illustrated in Figure 3.

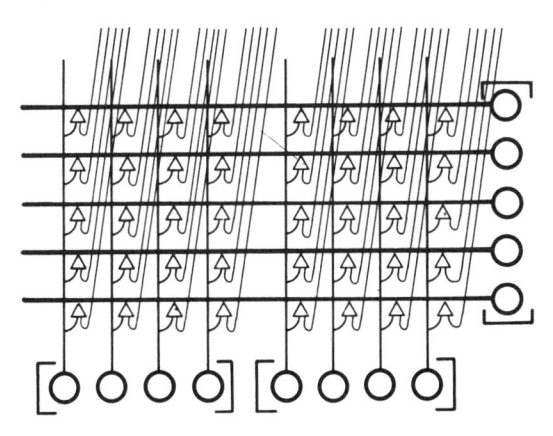

Figure 3. A programmable version of the simplified activation model shown in Figure 2. Each triangle represents a *programmable connection* that can be turned on by a signal coming from the central knowledge store, shown here as lying outside the figure to the upper right. If the triangular connections pass the product of the two signals arriving at their base along to the receiving node, the lines coming into the matrix from above can be thought of as programming the network.

***Multiplicative Interactions Yield Programmable Connections.*** At first glance, the notion of sending instructions to connections may seem to be adding a new kind of complexity to the basic processing elements out of which connectionist mechanisms are built. Actually, though, all we really need to do is to let each connection be a special kind of unit that can multiply two signals before passing along the result.

This point may be appreciated by considering the following equation. For the standard connections used in most connectionist models, the time-varying signal from some node $i$ to some node $j$ is multiplied by the fixed weight or connection stength $w_{ij}$ to determine the value of the input from $i$ to $j$:

$$\text{input}_{ij}(t) = \text{signal}_i(t) * w_{ij}.$$

All we are assuming now is that the signal from node $i$ is multiplied by a second time-varying signal, for example the signal arising from some other node $k$, instead of the fixed connection strength $w_{ij}$:

$$\text{input}_{ij}(t) = \text{signal}_i(t) * \text{signal}_k(t).$$

We can think of the signal from node $k$ as *setting the strength* of the connection between $i$ and $j$. When the value of the second signal at the connection from $i$ to $j$ is greater than 0, we will say that the connection from $i$ to $j$ is *active*.

***History, Implementation, and Function of Programmable Connections.*** The idea of using a second signal to modulate connections has been used in other connectionist models. Hinton (1981a) used such a scheme to map inputs from local (retinocentric) feature detectors onto central (object-centered) feature detectors in a viewpoint-dependent way. My use of multiplicative connections here was inspired by Hinton's. Feldman and Ballard (1982) have also suggested the idea of making connections contingent on the activation of particular nodes. The general notion of using one set of signals to structure the way a network processes another set of signals has previously been proposed by Sejnowski (1981) and Hinton (1981b).

At a neurophysiological level, multiplicative or quasi-multiplicative interactions between signals can be implemented in various ways. Neurons can implement multiplication-like interactions by allowing one signal to bring the unit's activation near threshold, thereby strongly increasing the extent to which another signal can make the unit fire (Sejnowski, 1981). There are other possibilities as well. A number of authors (e.g., Poggio & Torre, 1978) have suggested ways in which multiplication-like interactions could take place in subneuronal structures. Such interactions could also take place at individual synapses, though there is little evidence of this kind of interaction in cortex. For a fuller discussion of these issues, see Shepherd (1979) or Crick and Asanuma (in press).

For our purposes, the implementation is less important than the function. In essence, what connections do is specify *contingencies* between *hypotheses*. A positive weight on the connection from unit $i$ to unit $j$ is like the instruction "if $i$ is active, excite $j$." Fixed connections establish such contingencies in a fixed, permanent way. Programmable connections allow us to specify what contingencies should be in force, in a way which is itself contingent on other signals.

Let's see what we have achieved so far. By using multiplicative interactions between signals, in place of fixed connections, we now have a way of setting from outside a network the functional connections or contingencies between the units inside the network. This means that we can dynamically program processing modules in response to expectations, task demands, etc. The little module shown in Figure 3 could be used for a variety of different processing tasks, if different connection patterns were sent into it at different times. For example, if we sent in different signals from outside, we could reprogram the module so that the word level nodes would now respond to the two-letter words in some other language. In conjunction with reprogramming the connections from feature level nodes to the letter nodes, we could even assign the network to processing words in a language with a different alphabet, or to processing completely different kinds of patterns.

Programmable networks like the one shown in Figure 3 will be called programmable modules. The input nodes will be called programmable letter nodes, and the output nodes will be called programmable word nodes. Though the nodes could be used for other things besides letters and words, these are the roles they will play in the present model.

## Overview of the CID Mechanism

We are now ready to move up to a model containing a number of programmable modules along with the structures required to program them. The system is called a Connection Information Distributor, or CID for short. The basic parts of the model are shown and labeled in Figure 4; they are shown again, with some of the interconnections, in Figure 5.

Basically, CID consists of a central knowledge store, a set of programmable modules, and connections between them. The structure is set up in such a way that all of the connection information that is specific to recognition of words is stored in the central knowledge store. Incoming lines from the programmable modules allow information in each module to access the central knowledge, and output lines from the central knowledge store to the programmable modules allows connection activation information to be distributed back to the programmable modules.

The two programmable modules are just copies of the module shown in Figure 3. It is assumed that lower-level mechanisms, outside of the model itself, are responsible for aligning inputs with the two modules, so that when two words are presented, the left word activates appropriate programmable letter nodes in the left module, and the right one activates appropriate programmable letter nodes in the right module.

*The Central Knowledge Store.* The knowledge store in CID is shown at the top of Figure 4. This is the part of the model that contains the word-level

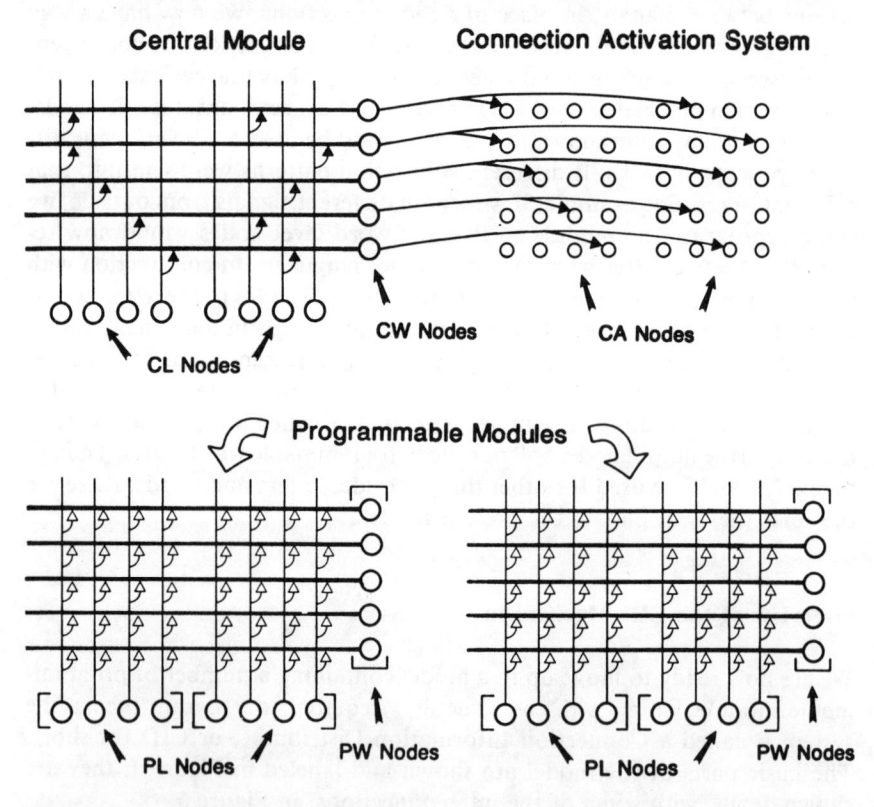

Figure 4. A simplified example of a Connection Information Distributor (CID), sufficient for simultaneous bottom-up processing of two two-letter words. The programmable modules consist of the programmable letter (PL) and programmable word (PW) nodes, and programmable connections between them (open triangles). The central module consists of a set of central letter (CL) nodes, a set of central word (CW) nodes, and hard-wired connections between them (filled triangles). The connection activation system includes the central word nodes, a set of connection activator (CA) nodes, and hard-wired connections between them. Connections between the central knowledge system (central module plus connection activation system) and the programmable modules are shown in Figure 5.

knowledge needed to program the programmable modules. It consists of two parts. One part is called *the central module,* and the other part is called the *connection activation system.*

The central module consists of central letter nodes, central word nodes, and connections between the central letter and the central word nodes. The letter nodes in the local modules project to the letter nodes in the central module, so that whenever a particular letter node is active in either programmable module, the corresponding central letter node is also (Figure 5). Note that the correspondence of local and central letter nodes is quite independent of what letters these nodes stand for.

The central letter nodes are connected to the central word nodes via connections of the standard connectionist variety. These connections allow patterns of letter-level activation to produce corresponding word level activations, just as in the original interactive activation model. However, it should be noted that the central word node activations are based on a superposition of the inputs to each of the two programmable modules. Thus, the activations in the central letter nodes do not specify which module the letters come from, though relative position within each module is encoded. Thus, activations in the central module do not distinguish between the input IN SO and the input SO IN or even SN IO. In short, it cannot correctly determine which aspects of its inputs belong together.

The second part of the central knowledge system, the connection activation system, also consists of two sets of nodes and their interconnections. One of these sets of nodes is the central word nodes—they belong both to the central module and to the connection activation system. The other set is the connection activator (CA) nodes. The purpose of the connection activation system is to translate activations of central word nodes into activations of connections appropriate for processing the corresponding words in the local modules. The CA nodes serve as a central map of the connections in each of the programmable modules, and provide a way to distribute connection information to all of the programmable modules at the same time. (The CA nodes are not strictly necessary computationally, but they serve to maintain the conceptual distinction between that part of the model that contains the knowledge about words, and the parts that simply distribute that knowledge to the local modules). There is one CA node corresponding to the connection between a particular programmable letter node and a particular programmable word node. I have arranged the CA nodes in Figure 4 to bring out this correspondence. Each CA node projects to the corresponding connection in both programmable modules. I have illustrated the projections of two of the CA nodes in Figure 5. For example, the top-left CA node corresponds to the connection between the left-most programmable letter node and the top programmable word node. This CA node projects to its corresponding connection in each of the programmable modules, and provides one of that connection's two inputs. So, when a particular CA node is

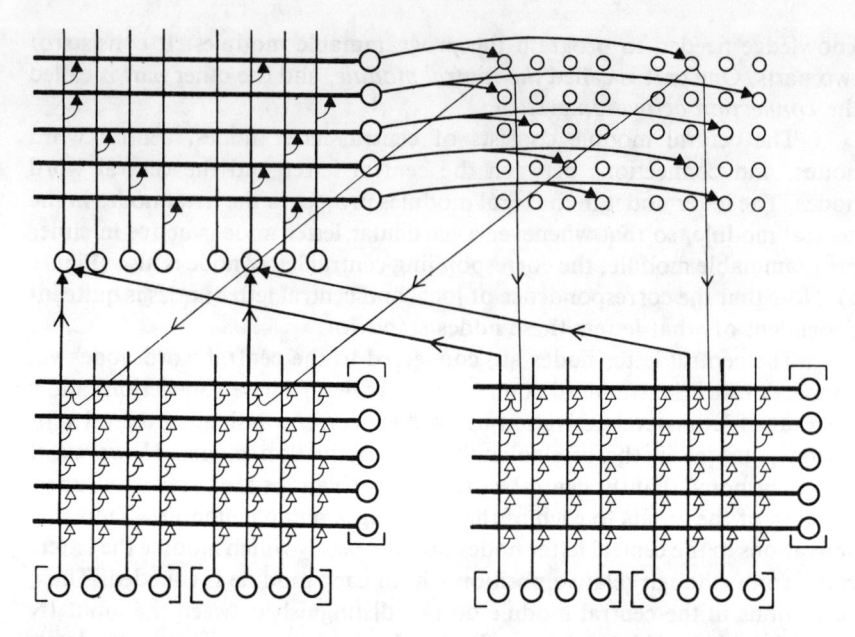

Figure 5. *Each* CA node projects to the corresponding connection in both programmable modules, and *each* central letter node receives projections from the corresponding programmable letter node in both programmable modules. The inputs to two central letter nodes, and the outputs from two CA nodes are shown.

active, it activates the corresponding connection in *all* of the programmable modules. In this way it acts as a sort of master switch.

At a functional level, we can see each CA node as standing for a contingency between two activations. Thus, if we index the programmable letter nodes by subscript $i$, and the programmable word nodes by $j$, the $ij$'th CA node stands for the contingency, "if letter node $i$ is active, excite word node $j$." Thus, we can think of the CA nodes as Contingency Activation, as much as Connection Activation nodes. When we activate a CA node (to a certain degree) we are implementing the contingency it represents (with a corresponding strength) in both of the programmable modules at once.

The central word nodes, of course, are responsible for activating the CA nodes. There are excitatory connections from each word node to each of the CA nodes for the connections needed to process the word. For example, the central word node for IN activates two CA nodes. One is the CA node for the connection between the left-most programmable letter node and the top-most programmable word node. The other is the CA node for the connection from the sixth programmable letter node from the left to the same programmable word node. These connections effectively assign the top programmable word node to be the detector for IN (assuming, of course, that lower levels of processing have been arranged so that I in the first position and N in the second position activate the appropriate programmable letter nodes).

In summary, CID consists of a) the two programmable modules; b) the central knowledge store, including the central module and the connection activation system; c) converging inputs to the central knowledge store from the programmable modules; and d) diverging outputs from the central knowledge store back to the programmable modules.

We can now see how this mechanism allows the programmable modules to be programmed dynamically in response to current inputs. When an input causes activations in some of the programmable letter nodes in one of the programmable modules (say the programmable letter node for I in the first position and N in the second position of the left programmable module), these activations are passed to the corresponding central letter nodes. From these they activate the central word node for IN. Central word nodes for patterns which overlap partially with the input (such as IS and ON) also receive excitation, but only in proportion to their overlap with the input. The central word nodes pass activation to the CA nodes, and these in turn pass activation back to the connections in both programmable modules. Connections are only turned on to the extent that they are consistent with the input. When different patterns are presented to each programmable module, connections appropriate for both patterns are turned on, thereby programming both programmable modules to process either pattern. Central word nodes—and therefore connections—are also turned on for any words that appear in the superimposed input from the two programmable modules. However, the results of processing in each programmable module still depend on the activations of the programmable letter nodes. Thus, the appropriate programmable word node will tend to be the most active in each local module. Although the words are not kept straight at the central level, they are kept straight—though with some tendencies to error—down below. We will examine this matter more closely below.

## A COMPUTER SIMULATION OF CID

To examine the behavior of the CID scheme in more detail and to compare it to the original interactive activation model, I created a computer simulation. The structure I simulated was scaled up from the version in Figures 4 and 5 so that it would be able to process two strings of four letters each. Only three or four different letter alternatives were allowed in each position within each string. These were B, L, P and S in the first position, A, E, I and O in the second position, N, R, and V in the third position, and D, E, and T in the fourth position. The lexicon used in the simulation consisted of the 32 words shown in Table I.

TABLE I
The 32 Words Used in the Simulations

| BAND | BARE | BEND | BIND |
|------|------|------|------|
| BIRD | BOND | BONE | BORE |
| LAND | LANE | LARD | LEND |
| LINE | LINT | LIVE | LONE |
| LORD | LOVE | PANE | PANT |
| PART | PINE | PINT | POND |
| PORE | PORT | SAND | SANE |
| SAVE | SEND | SORE | SORT |

Like the smaller-scale version shown in the figures, the model consisted of two programmable modules, one for each of the two letter strings, and a central knowledge store consisting of the central module and the connection activation system. Each programmable module had 16 programmable letter nodes and 32 programmable word nodes. The programmable letter nodes were grouped into four groups of four, with each group to be used for letters in one display location. The members of each group had mutual, hard-wired, inhibitory connections. Similarly, all of the programmable word nodes in each module were mutually inhibitory. Each programmable module contained $16*32 = 512$ programmable connections, and there were 512 CA nodes, one for each programmable connection. The central module contained 16 letter and 32 word nodes, like the programmable modules. There were no inhibitory connections either between the central word nodes or between the central letter nodes. The connections between the central letter nodes and the central word nodes, and connections from the central word nodes to the appropriate CA nodes, were hard-wired with the connection information needed to make the central letter nodes activate the right central word nodes and to make the central word nodes activate the right CA nodes.

Inputs to the simulation model were simply specifications of bottom-up activations to the programmable letter nodes in either or both programmable modules. Inputs were presented when all the nodes in the model were at their resting activation values, and turned off after some fixed number of time cycles.

### Details of Interaction Dynamics

The programmable letter and word nodes have the same dynamic properties as the letter and word nodes in the original word perception model (McClelland & Rumelhart, 1981). Time is divided into a sequence of discrete processing steps. On each processing step, each programmable node adds up all of its excitatory and inhibitory inputs from all other nodes and from the external input. Then the activation value of each node is updated. If the net input is excitatory, it will tend to increase the activation of the node; if the net input is inhibitory, it will tend to reduce the activation of the node. The effect is graded and gradual, and activation values are always kept between a maximum and minimum value. There is also a tendency for activations to decay back toward resting level, which is set at an activation of $-.05$ for all nodes. Only positive activation values are transmitted to other nodes, so that nodes with activations below 0 are effectively out of the computation. The model has a global letter-to-word excitation constant, called alpha, and a global word-to-word inhibition constant, called gamma, as well as a global decay called beta. The values of alpha and gamma determine the strength of bottom-up excitation relative to within level inhibition. The values used for these three parameters were taken from the original model.

The only difference between the CID version of the model and the original is in the strengths of excitatory connections between nodes. In CID, these strengths vary as a function of the current input, while in the original model they were fixed. Highly simplified activation rules are used to capture the essence of the connection activation process via the central letter, central word, and CA nodes. The activation of a particular central letter node is simply the number of input nodes projecting to it which have activations greater than 0. Thus, the activation of a particular central letter node just gives a count of the corresponding programmable letter nodes that are active. The activation of a central word node is just the sum of the active central letter nodes which have hard-wired connections to the central letter node. The activation of a CA node is just the activation of the central word node that projects to it, and this value is transmitted unaltered to the corresponding programmable connection in each programmable module.

The net effect of these assumptions is to make the activation of the connections coming into a particular programmable word node proportional to the number of active nodes for the letters of the word, summed over both modules. Active letter nodes count only if they stand for letters in appropriate positions, though, within the programmable module of origin.

**Output**

So far, we have said nothing about how the activations which arise in the programmable modules might give rise to overt responses. Following the original interactive activation model, I assume there is a readout mechanism of unspecified implementation which translates activations at either the letter or the word level into overt responses. The readout mechanism can be directed to the word or the letter level of either module, and at the latter it can be directed to a particular letter-position within a module. In cases where more than one stimulus is to be identified on the same trial, the readout of each of the items is independent.

The relation between activation and response probability is based on the choice model of Luce (1963). The probability of choosing a particular response depends on the strength of the node corresponding to that response, divided by the sum of the strengths of all the relevant alternatives (e.g., nodes for words in the same position). The exact relation of strength and activation is described in McClelland and Rumelhart (1981).

The import of these assumptions is that the probability of a particular response is solely a function of the activations of nodes relevant to the response. All interactions between display items are thus attributed to the node and connection activation mechanisms, and not to the readout mechanisms themselves.

## RESULTS OF THE SIMULATIONS

Two principle findings emerged from working with the simulation model. First, when processing a single word, the CID scheme causes the model to behave as though it were sharply tuned to its inputs, thereby eliminating the need for bottom-up inhibition. Second, when processing two words at a time, the connection activation scheme causes the model to make errors similar to those made my human subjects viewing two-word displays. These errors arise as a result of the essential characteristics of the CID scheme.

**One Word at a Time: The Poor get Poorer**

In the original model, bottom-up inhibition from the letter level to the word level was used to sharpen the net bottom-up input to word nodes. For example, consider a display containing the word SAND. Due to bottom-up inhibition, nodes for words matching only three of the four letters shown (e.g., LAND) would receive less than 3/4 as much net bottom-up excitation as the node for the word SAND itself.

The CID version of the model closely emulates this feature of the original, even though it lacks these bottom-up inhibitory connections. In CID, the activation of the *connections* coming into a word node varies with the number of letters of the word that are present in the input. At the same time, the number of inputs to these same connections from the programmable letter nodes also varies with the number of letters in the input that match the word. The result is that in the CID version of the model, the amount of bottom-up activation a programmable word node receives varies as the *square* of the number of letters in common with the input. Poorer matches get penalized twice.

In working with the original model, Rumelhart and I picked values for the bottom-up excitation and inhibition parameters by trial and error, as we settled on an overall set of parameters that fit the results of a large number of experiments. The values we hit upon put the strength of bottom-up inhibition at 4/7 the strength of bottom-up excitation. For words that share two, three or all four letters in common with the input, this ratio produces almost exactly the same relative amounts of net bottom-up activation as is produced by the CID mechanism (Table II). Words with less than two letters in common received net bottom-up inhibition in the old version, whereas in the CID version they simply receive little or no excitation. In both cases, their activation stays below zero due to competition, and thus they have no effect in either case on the behavior of the model.

TABLE II
One Word at a Time:
Bottom-Up Activations of Several Word Nodes in the Original and CID Versions of the Interactive Activation Model

| Node | Letters Shared w/input | Input: SAND | | | |
| --- | --- | --- | --- | --- | --- |
| | | Original | | CID Version | |
| | | Relative Activation | Ratio | Relative Activation | Ratio |
| SAND | 4 | 4 | — | 4*4 | — |
| LAND | 3 | 3-4/7 | .61 | 3*3 | .56 |
| LANE | 2 | 2-8/7 | .21 | 2*2 | .25 |

Note: Ratio is the net bottom-up activation of the node, divided by the net bottom-up activation of the node for SAND.

This analysis shows that the CID version of the model can mimic the original, and even provides an unexpected explanation for the particular value of bottom-up inhibition that turned out to work best in our earlier simulations. As long as the bottom-up input to the letter level was unambiguous, the correspondence of the CID version and a no-feedback version of the original model is extremely close.

When the bottom-up input to the letter level was ambiguous, however, there was a slight difference in the performance of the two versions of the model. This actually revealed a drawback of bottom-up inhibition that is avoided in CID. Consider the input to a word node from the letter nodes in a particular letter position. In the original model, if three or more letter candidates were active, two of them would always produce enough bottom-up inhibition to more than outweigh the excitatory effect any one of them might have on the word. For example, if E, F, and C are equally active in the second letter position, F and C together would inhibit the detectors for words with E in second position more than E will excite them. Thus, if three letters are active in all four letter positions, no word would ever receive a net excitatory input. This problem does not arise in CID, because there is no bottom-up inhibition. Thus, I found that the CID version could pull a word out of a highly degraded display in which several letters were equally compatible with the feature information presented, while the original model could not. It thus appears that CID gives us the benefits of bottom-up inhibition, without the costs.

## Two Words at a Time: Interference and Crosstalk

So far we have seen how CID retains and even improves on some of the important aspects of the behavior of the original interactive activation model. Now, I will show how CID captures important aspects of the data obtained in experiments in which subjects are shown two words at a time. Here CID's structure becomes essential, since simultaneous processing of two patterns introduces considerations which do not arise in the processing of single items.

When letters are presented to both modules, *all* of the letters are combined to turn on connections which are distributed to *both* of the programmable modules. The result is that the connections appropriate for the word presented in one module are turned on in the other module as well. This biases the resulting activations in each module. The programmable word node for the word presented to a particular module will generally receive the most activation. However, the activation of programmable word nodes for words containing letters presented to the other module is enhanced. This increases the probability that incorrect responses to one of the words will contain letters presented in the other.

At first this aspect of the model disturbed me, for I had hoped to build a parallel processor that was less subject to crosstalk between simultaneously presented items. However, it turns out that human subjects make the same kinds of errors that CID makes. Thus, though CID may not be immune to crosstalk, its limitations in this regard seem to be shared by human subjects. I'll first consider some data on human performance, and then examine in detail why CID behaves the same way.

The data come from a recent experiment by Mozer (1983). In his paradigm, a pair of words (e.g., SAND LANE) is displayed, one to the left and one to the right of fixation. The display is followed by a patterned mask which occupies the same locations as the letters in the words that were presented. In addition, the mask display contains a row of underbars to indicate which of the two words the subject is to report. Subjects were told to say the word they thought they saw in the cued location or to say "blank" in case they had no idea.

In his first experiment, Mozer presented pairs of words that shared two letters in common. The pairs of words had the further property that either letter which differed between the two words could be transposed to the corresponding location in the other and the result would still be a word. In our example SAND-LANE, SAND and LANE have two letters in common, and either the L or the E from LANE can be moved into the corresponding position in SAND, and the result would still be a word (LAND and SANE). Of course, it was also always true with these stimuli that the result would be a word if both letters "migrated." The duration of the two-word display was adjusted after each counterbalanced block of trials in an attempt to home in on a duration at which the subject would get approximately 70% of the whole-word responses correct. Thus, the overall error rate was fixed by design, though the pattern of errors was not.

The principal results of Mozer's experiment are shown in Table III. Of the trials when subjects made errors, nearly half involved what Mozer called "migration errors"—errors in which a letter in the context word showed up in the report of the target. To demonstrate that these errors were truly due to the presentation of these letters in the context, Mozer showed that these same error responses occurred much less frequently when the context stimulus did not contain these letters. Such "control" errors are referred to in the table as pseudo-migration errors.

TABLE III
Method and Results of Mozer (1983), Experiment 1

| Method | |
|---|---|
| Example Display | SAND    LANE |
| Target Cue | ↑↑↑↑ |
| Results | |
| Response Type | % of total |
| Correct response (SAND) | 69.0 |
| Single migration (SANE or LAND) | 13.3 |
| Double migration (LANE) | 0.5 |
| Other | 17.2 |
| Total | 100.0 |
| Pseudo-migration rate* | 5.3 |

*Psuedo-migration rate is the percentage of reports of the given single migration responses (SANE, LAND) when a context word which does not contain these letters is presented. In this example, the context string might have been BANK.

As I already suggested, migration errors of the type Mozer reported are a natural consequence of the CID Mechanism. Since the letters from both words are superimposed as they project onto the central module, the connections for words whose letters are present (in the correct letter position) in either of the two input strings are strongly activated in both programmable modules. The result is that programmable nodes for words containing letters from the context are more easily activated than they would be in the absence of the input presented to the other module.

## TABLE IV
### Two Words at a Time: Crosstalk
### Relative Bottom-Up Activations Produced by SAND
### Presented either Alone or with LANE as Context

| Programmable Word Node | Alone | | with LANE | |
|---|---|---|---|---|
| | Activation | Ratio | Activation | Ratio |
| SAND | 4*4 | — | 4*6 | — |
| LAND | 3*3 | .56 | 3*6 | .75 |
| BAND | 3*3 | .56 | 3*5 | .62 |
| SEND | 3*3 | .56 | 3*4 | .50 |
| LANE | 2*2 | .16 | 2*6 | .50 |

Note: Ratio refers to the bottom-up activation of the node, divided by bottom-up activation of SAND.

Table IV compares relative programmable word node activations for various words, for two different cases: In one case, the word SAND is presented alone; in the other, it is presented in the context of the word LANE. When SAND is presented alone, all words which share three letters with it receive $(3/4)^2$ or 9/16's as much bottom up activation as the node for SAND itself—we already explored this property of the CID model in the previous section. When SAND is presented with LANE, however, words fitting the pattern (L or S)-(A)-(N)-(D or E) all have their connections activated to an equal degree, because of the pooling of the input to the connection activation apparatus from both modules. These words are, of course, SAND and LANE themselves, and the single migration error words LAND and SANE. Indeed, over both letter strings, there are 6 occurrences of the letters of each of these words (the A and the N each occur twice). The result is that the excitatory input to the programmable word nodes in the left module for LAND and SANE is 3/4 of that for SAND, as opposed to 9/16. Other words having three letters in common with the target have their connections less activated. Their bottom-up activation is either 5/8 or 1/2 that of SAND, depending on whether two of the letters they have in common with the target are shared with the context (as in BAND) or not (as in SEND). Thus, we expect LAND and SANE to be reported more often than other words sharing three letters in common with SAND.

The reader might imagine that the effect would be rather weak. The difference between 3/4 and 5/8 or 1/2 does not seem strikingly large. However, a raw comparison of the relative bottom-up activation does not take into account the effects of within-level inhibition. Within-level inhibition greatly amplifies small differences in bottom-up activation. This is especially true when two or more nodes are working together at the same level of activation. In this case, the nodes for LAND and SANE act together. Neither can beat out the other, and both "gang up" on those receiving slightly less bottom-up activation, thereby pushing these other alternatives out. This "gang effect" was observed in the original version of the model—see McClelland and Rumelhart (1981), for details. This behavior of the model is illustrated in Figure 6. Through the mutual inhibition mechanism, SAND and LANE come to dominate over other words that share three letters in common with the target. Some of these, in turn, dominate words that share but two letters in common with the target, including, for example, LANE, even though the connections for LANE are strongly activated. This result of the simulation accords with the experimental result that double or "whole-word" migrations are quite rare in Mozer's experiment, as shown in Table III.

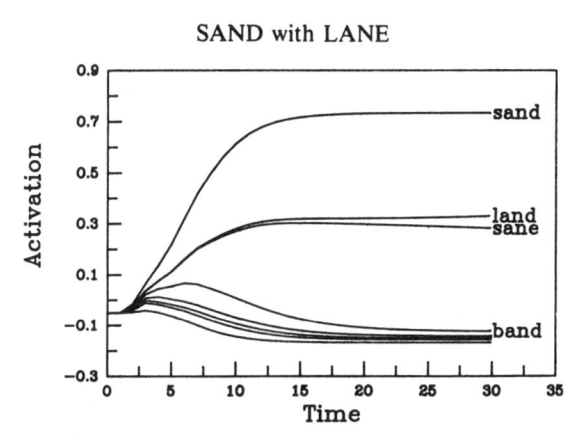

SAND with LANE

Figure 6. Activation curves for various programmable word nodes in the module to which SAND is shown, when the input to the other module is LANE. The x axis represents time cycles from the onset of the two-word display.

Mozer (1983) reported several additional findings that are consistent with the CID model. First, he showed that letter migrations are more likely to be "copies" than "exchanges." That is, when subjects (in a second experiment) were asked to report both words in the display, the probability of a copy or duplication error, in which both reported items contained an L or an S (as in LAND-LANE or SAND-SANE), was much greater than the probability of an exchange, in which the S and L exchanged places (to make LAND-SANE). The data are not consistent with models which attribute migration errors to a mechanism which conserves the number of occurrences of each letter, such as the feature integration model of Treisman and Gelade (1980). However, if as I assume for the CID model the subject selects the best response independently for each of the two inputs, then we expect the probability of an exchange to be equal to the probability of two independent errors occurring at the same time. That is, the probability of an exchange should be equal to the probability of the left letter turning up on the right times the probability of the right letter turning up on the left. The expected probability of an exchange based on these considerations is .006 for this experiment, within experimental error of the value of .008 actually obtained.

Second, Mozer showed that migrations are more common when target and context share letters in common than when they do not. Thus, the probability of saying LAND was much higher when SAND was flanked by LANE than when SAND was flanked by LOVE. The relevant data are displayed in Table V.

TABLE V
Percent Correct Migration and Other
Responses from Mozer (1983), Experiment 3

| | Context Type | |
|---|---|---|
| | Common Letters (SAND-LANE) (%) | No Common Letters (SAND-LOVE) (%) |
| Correct response (SAND) | 64 | 74 |
| Single migration (SANE or LAND) | 11 | 6 |
| Other | 25 | 19 |

Note: Pseudo migration rate (probability of reporting LAND or SANE when context contained no letters which could form a word with the target) was 3%. In this example (Target word LAND), the context string might have been COMB.

At first sight it might appear that the CID model would not expect this difference. When SAND is presented with LANE, connections for all the (S/L)(A)(N)(D/E) words receive 6 units of activation because of the repetition of the A and N in both letter-strings. When SAND is shown in the context of LOVE, connections for these same words all receive 4 units. The ratios of bottom-up activation for correct and single-migration words are the same in both cases. However, once again, this ratio is not the whole story. The absolute *magnitude* of bottom-up activation is greater in the case where there are letters in common than in the case where there are not. The higher the overall magnitude of bottom-up activation, the less difference the ratio makes, due to the tendency of node activations to saturate at high activation levels. When bottom-up activation is reduced overall, sharper differences in the pattern of activation emerge. The result is that there is far more activation of migration words when the two words shown have letters in common than when they do not. When there is less overlap between target and context, the tendency of the correct answer to dominate the pattern of activation is sharply increased. This is illustrated in Figure 7, which shows much less activation for migration error words than Figure 6.

In summary, the CID version of the interactive activation model appears to provide fairly accurate accounts of the intriguing perceptual interactions reported by Mozer. Mozer's basic finding, that letters in one display position tend to show up in reports of the contents of the other position, is a

SAND with LOVE

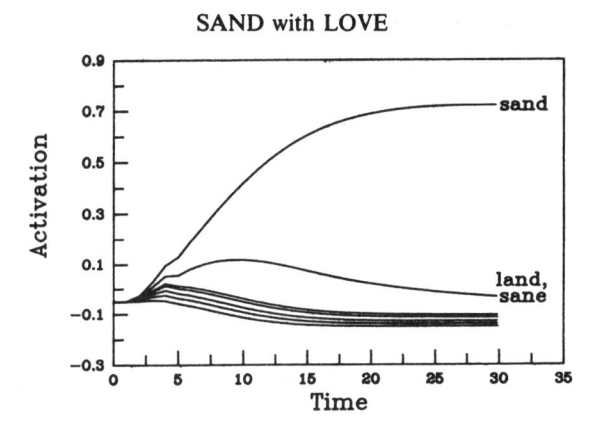

Figure 7. Activation curves for various programmable word nodes in the module to which SAND is shown, when input to the other module is LOVE.

necessary consequence of the CID mechanism. Considerably more empirical and theoretical work is required before we will be in a position to say that the CID version of the interactive activation model provides an adequate account of all aspects of these perceptual interactions. But it appears that the model is on the right track.

## THE INFORMATION PROCESSING CAPACITY OF CID

There is one apparent problem with the CID scheme. It appears that it requires a prohibitive number of nodes and connections. To point to the most serious aspect of the problem, the CID model of word perception as I have described it requires one CA node for each potential connection between a letter node and a word node. This number grows as the *product* of the number of letter level node times the number of word-level nodes. A system sufficient to process 50,000 different words of up to 7 letters in length would require 7*26 = 182 letter level nodes, 50,000 word level nodes, and nearly 1 million CA nodes. This seems like an awful lot of nodes. It would appear that the benefits of parallel processing are being purchased at a prohibitive cost.

However, the situation can be improved dramatically by switching over to a *distributed* representation, in which each word (the argument applies to letters, too, or anything else, of course) is represented by a pattern of activation over a set of nodes, rather than by the activation of a single node.

At first glance the switch to distributed representation may appear to be a major change of stance. However, there is less to the choice between local and distributed representations than meets the eye. All the matter really comes down to is whether we associate conceptual units like words with individual nodes, or with overlapping constellations of nodes. In the CID model I have used local representation for clarity and comparability with the original model, but now that the basic idea of CID is on the table, I will argue that a switch to distributed representation would be of great benefit. Distributed representation has many virtues, several or which are described in Hinton (1984) and in McClelland and Rumelhart (in press). For our present purpose, the great advantage of distributed representation is that it allows us to get by with much smaller programmable modules and far fewer CA nodes. First, I'll briefly explain how distributed representation saves us nodes even with a hard-wired connectionist mechanism. Then I'll show how it pays off in spades in CID.

I'll begin by sketching a simple connectionist mechanism for associating patterns of activity at one level with paired patterns of activity at another level. For word recognition, the levels might be letter or word, but I'll just call them levels A and B for generality. Our goal is to be able to activate the correct B pattern whenever the corresponding A pattern is shown. We assume that each association involves an A pattern and a B pattern each containing some number M of active nodes on each of the two levels. To allow A patterns to retrieve the appropriate corresponding B patterns, we simply imagine that there is an excitatory connection from each node active in the A pattern to each of the nodes active in the corresponding B pattern. Note that a local representation model in which, say, each pattern at the A level is represented by as single active node at the B level, is just a variant of this model, in which the pattern at the B level consists of just a single node.

Essentially, distributed representation can save on nodes because we can use partially overlapping codes for different items: we are no longer required to have at least one node for each alternative pattern. There are, however, limitations on the number of associations that can be stored in an A-B associator: The more patterns we store, the more likely it is that the B pattern retrieved by any A pattern will be contaminated by spurious activations at the B level.

Willshaw (1981) has analyzed the extent of contamination, under a particular set of simplifying assumptions. First, he assumed that the connection between a particular A node and a particular B node has only two states: it is on if the A and B nodes are associated in any of the patterns stored, and off otherwise. This assumption means that when the nodes for the A member of the pair are activated, each of the appropriate B nodes will receive one unit of excitation from each active A unit. Other B nodes not part of the appropriate B pattern may receive excitation from some or all of the A units, via connections that are on because they belong to other stored associations. Second, Willshaw assumed that B nodes remain inactive unless they receive excitation from all of the active A units. (This is implemented by assuming that the unit has a threshold equal to the excitation produced by a complete A pattern). With this assumption, when a known A pattern is presented, the model is able to turn on all the correct B nodes. A spurious B node will only be turned on if *all* active members of the A pattern happen to have excitatory connections to it. Third, Willshaw assumed that the A and B patterns were random selections of some fixed number of A nodes and some fixed number of B nodes.

Given Willshaw's assumptions, it is possible to calculate the average number of B nodes that will be spuriously activated for any combination of Na and Nb, the number of units in each of the two pools, Ma and Mb, the number units active in the A and B members of the pattern, and R, the number of associations known. We then need only adopt an error tolerance criterion to determine how large Na and Nb must be to accommodate R associations of pairs of patterns of size Ma by Mb. Adopting a criterion of an average of one spurious B node activation per retrieval, Willshaw derived the following relationship:

$$\sqrt{Na*Nb} = 1.2\sqrt{Ma*Mb}\sqrt{R}.$$

These relations only hold if we use distributed representations. The relation holds exactly if the number of units active in the A and B patterns is equal to the $\log_2$ of the number of units in the corresponding pool. For larger fractions of active units, the equation actually underestimates the number of patterns that can be stored for given values of Na and Nb.

The result just presented depends on the fact that, when the associations involve patterns with a reasonable number of active units in both members of the pair, the chances that the connections will result in a spurious B node receiving as much activation as an appropriate B node are remote, as long as the overall proportion of connections that are turned on is reasonable. The proportion we can get away with varies a bit with the exact values of Ma and Mb, but if half or fewer of the connections are turned on, we will generally be quite safe. Roughly speaking, then, the number of patterns we can store is just the number which keeps us from turning on more than half of the connections in the matrix.

We are now ready to examine the implications of distributed representation for CID. Based on the previous equation, we can calculate the number of programmable connections, (and therefore, the number of CA nodes we would need) to program a module to perform like the hard-wired module considered in Wilshaw's analysis. The number of connections between Na A nodes and Nb B nodes is just Na*Nb. Since we need one programmable connection for each connection, we can simply square the previous equation to get an expression for the number of programmable connections (Npc):

$$Npc = Na*Nb = 1.45*Ma*Mb*R.$$

While this represents a slight improvement over the nondistributed case, the number of programmable connections and CA nodes still appears to grow linearly with R, the number of patterns known, and with the size of the patterns.

But this result ignores the fact that in CID, we do not turn on the connections relevant to all the known patterns at one time. Let us just consider what would happen if we were able to avoid activating any unwanted connections, and could turn on only the connections relevant to the particular pattern or patterns we wished to process at one time. In this case, we could get by with far fewer nodes in each of the pools, and therefore far fewer programmable connections. The more patterns we wished to process at one time, of course, the more connections we will have to turn on, and the more risk we will run of spurious activation. Thus, the number of A and B nodes required in each programmable module, and thus the number of programmable connections, is related to the number (S) of associations we wish to be able to process *simultaneously,* rather than by the total number of associations known. The number of nodes we need also depends on the number of A and B nodes active in each member of a pair, as before. Again, allowing an average of one spurious B node activation per retrieval, the following approximate relation holds (the constant will be somewhat larger for values of S less that 3):

$$Npc = Na*Nb = 1.45*Ma*Mb*S.$$

This equation is the same as the one we had for the number of connections needed in a hard-wired associator, except that we now have S, the number of patterns to be processed simultaneously, instead of R, the number known. Replacing R with S makes a huge difference, since it can plausibly be argued that we know something like 50,000 words. We could process up to 5 of these at once (each in a different programmable module), the equation says, with *five orders of magnitude* fewer CA nodes and programmable connections, and two and a half orders of magnitude fewer nodes in each programmable module than we would need in a hard-wired module of the same capacity. Of course, the size of the central module still depends on R, but we pay that price only once, and we incur it with models which do not allow parallel processing, as well as with CID. Relative to the resource requirements of the central module, then, the extra cost of simultaneous processing via connection information distribution is modest, if we are willing to switch to distributed representation.

I have actually overstated my case a little. For one thing, the number of programmable connections required depends on the number of B level patterns fully activated in the central module by the superposition of all of the A patterns presented for processing at one time. This may be more than the number of patterns actually presented for processing (as when SAND-LANE was presented, their superposition contained all the letters of these two words and LAND and SANE besides), and will depend on the ratio of Ma/Na—for adequate performance, then, Na may have to be bigger than the equation given above suggests. Space prevents a full analysis of these matters here. Suffice it to say for the present that these complications do not eliminate the basic result that with distributed representations we can greatly reduce the resource requirements of CID. Though the degree of the savings previously indicated may be slightly exaggerated, it remains true that distributed representation still brings the resource requirements of CID into reasonable bounds.

## DISCUSSION

Thus far, we have explored a particular model embodying the idea of distribution of connection information, and we have seen how this idea provides a way of allowing a single, central representation of knowledge to be made available to each of a number of programmable processing modules, thereby turning the programmable modules into programmable connectionistic information processing structures. We have seen how this scheme provides a natural account of the errors human subjects make in processing two-word displays. Abstracting from this specific application a little, we have explored some aspects of the resource requirements of such a system, and we have found that they are not as exorbitant as one might have feared.

In this section of the paper, I step back even further from particular detailed models, and consider the idea of distributing connection information more generally. First, I discuss the essential properties of the CID mechanism. Then, I discuss possible extensions of the approach to other domains such as sentence processing. Third, I suggest reasons why some sequentiality in programming parallel processing structures might occasionally be a good thing. The paper concludes with a brief examination of the relation between interactive activation processes and connectionist implementations, in light of the CID mechanism.

### Essential Characteristics of CID

There is a sense in which CID is not as powerful a mechanism as I had hoped to discover. Although it permits parallel processing to some degree, the performance of the model degrades when multiple items are processed simultaneously. One might well ask questions then: Might another mechanism not do better? Do we need such a complex mechanism to accomplish what

CID has done? Might we not get at least as good behavior with something simpler?

Obviously, there are difficult questions to give definitive answers to. However, I think it worth considering CID somewhat abstractly for a moment, to see what its essential properties are. This will, I think, provide a little insight into these questions.

One essential feature of CID is the superposition of the patterns of activation in each programmable module. Superposition permits simultaneous access to and retrieval from the central module, but what is retrieved is not the response to either or even both individual patterns, but the response to their superposition. This method of simultaneous information retrieval automatically runs the risk of crosstalk because by its very nature it looses track of which letters appeared in each of the programmable modules. Actually, though, the amount of crosstalk depends on the complexity and similarity structure of the patterns we wish to process. In fact, if *all* the patterns stored in the central module of a CID mechanism are maximally dissimilar (that is, orthogonal), there will be no crosstalk. It is only when the known patterns overlap with each other that crosstalk becomes a problem.

The point of this discussion is simply to suggest that the *kind* of limitation we see in CID as a parallel processor is intrinsic to superposition of inputs. This may be intrinsic to parallel retrieval itself—I have not been able to conceive of an alternative (connectionist) scheme for *simultaneous* retrieval of information about two patterns. But the exact *extent* of the limitation depends on the details of the patterns and their similarity structure. Since the similarity structure of patterns assigned to inputs can be affected by the way the inputs are coded, it is possible to manipulate the extent of the crosstalk problem quite easily.

But couldn't we do just as well without CID? Isn't there another, less complex mechanism that could do the same job it does just as well? One difficulty answering this is the amorphous definition of complexity, and the difficulty of specifying in detail what counts as a similar mechanism and what as a different one. However, it is instructive to consider briefly one simpler alternative to CID, in which we distribute *activation* information from the central module instead of *connection* information. Figure 8 shows such a mechanism. It is like CID except that the central word nodes project directly back to the corresponding local word nodes. The idea is that patterns of activation arising in the local letter nodes will be superimposed as inputs to the central module, and the composite output generated by the composite input will be distributed back to the local word nodes. In such a mechanism, if two words are presented at once, the pattern of activation that would appear on the output nodes of each local module would be the same for both of the programmable modules. Indeed, it would be the same as the pattern of activation over the central word nodes. This pattern contains

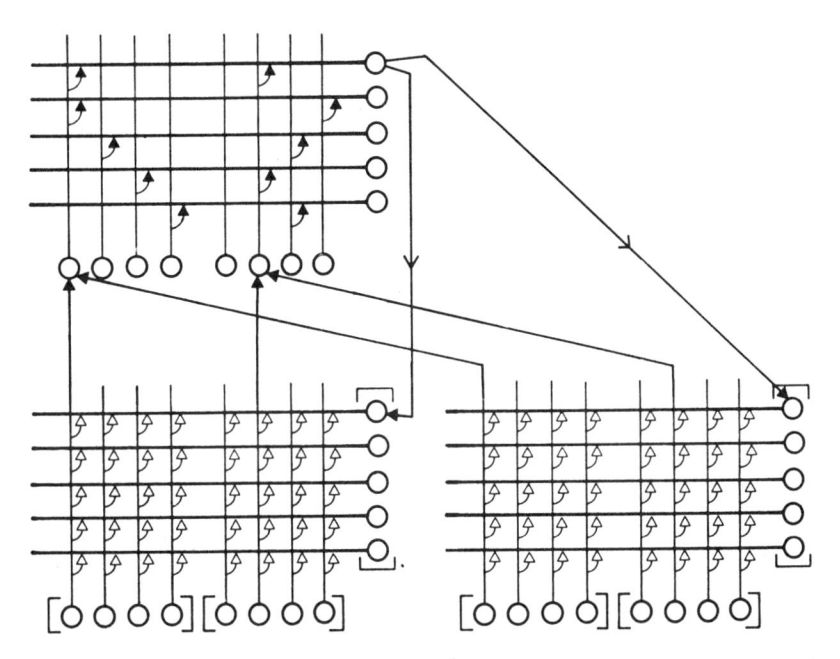

Figure 8. A mechanism that distributes activation information, rather than connection information. It is like CID in that it has local modules and a central module, but what it distributes is the pattern of activation over the central word nodes. Inputs to two central letter nodes and outputs of one central word node are shown.

equal representation of both inputs, as well as any known patterns which can be formed from the superposition of these inputs. With such a mechanism, if SAND LANE were presented, we would have no way of knowing from the activations of the word nodes in the first module whether the input to the module consisted of SAND, LANE, LAND, or SANE. Such a mechanism is slave to the composite output of the central module, and is not much good for processing more than one pattern at a time.

Just like this simpler mechanism, the information CID distributes is based on the composite of the inputs to the two programmable modules. But since CID distributes *connection* information, instead of *activation* information, the pattern at the letter level in each module still influences what the output pattern will be at the word level. As I stressed early on, *connection* information is *contingency* information. It says, if node x is active, let it activate node y. Distributing connection information allows the central module to tell the programmable modules what to do with their inputs, and this allows their output to reflect these inputs, as well as what they are told.

In summary, the essential features of CID are *superposition of inputs* to the connection activation process and *distribution of conditional information*. Some crosstalk is an inevitable byproduct of superposition, but the amount of it will depend on the amount of hardware used and the number, complexity and similarity structure of the patterns to be processed.

## Extension of CID to Other Problems and Other Levels

I claimed in the introduction that CID would be generally useful in extending the computational power of connectionist mechanisms. I have illustrated how it can be applied to the word level in a word perception model, and I generalized the idea in the discussion of the abstract pattern associator introduced in considering the resource requirements of CID. Now it is time to consider the relevance of the approach to computationally more challenging levels, such as syntactic and semantic analysis of sentences.

Sentences, as objects for processing, have one essential characteristic which individual words (at least in English) do not have. In sentences, the same structures can occur at many different levels of the representation of the same sentence. To many, this recursive characteristic of sentences seems to require a recursive processing mechanism, of the sort typically implemented in AI language processing programs.

But recursive structure does not necessarily require recursive—that is sequential—processing. The beauty of recursive processing is that the same knowledge—say, of the constituent structure of a noun phrase—can be made available at multiple levels, because the same subroutine can be called at any level, even inside itself. Connection information distribution provides a way of doing the same thing, in parallel.

CID has already given us a mechanism which makes the same knowledge (connection information) available simultaneously for processing different patterns on the same level without resorting to sequential processing. If access to the same knowledge was possible, not only from different slots on the same level, but from different levels, then recursively structured objects could be processed, in parallel, on several levels at the same time. Of course, crosstalk would still be a problem—it would tend to confound the bindings of things at different levels—but the programmable modules on the same level in our word-perception model are able to keep straight what goes with what in each of the two patterns they are processing at the same level, given sufficient resources and patterns that are sufficiently distinct. The same would be true for programmable modules accessing the same central knowledge system from different levels.

The idea of allowing the simultaneous programming of parallel processing structures at different levels thus preserves the essential positive aspect of recursive processing—access to the same information at different structural levels—without requiring us to resort to seriality.

But there is still a problem, for I seem to be assuming that we have available some sort of stack of levels, all of which can access the same central knowledge. The difficulty is that the number of levels of depth we will need cannot be specified in advance. Although we would probably be able to get by in almost all practical cases with some adequate fixed number of levels, this assumption does violence to the essential open-endedness of sentences. Any attempt to extend the idea to even more global structures, such as text structures or plans, would be doomed.

But CID provides us, at least in general terms, with a way to get by without any fixed number of processing levels. For levels are defined in terms of connections, and if we can program connections, we can in principle set up the entire hierarchical structure of the processing system on the fly, as well as any specific interactions between units imbedded in that structure.

Of course, we are several steps away from a concrete realization of this idea, and there are many complications that have to be addressed. But, I believe that programming connections will play an important part in the development of interactive activation mechanisms of sentence processing and other higher-level cognitive tasks, and I hope that CID represents a step in this challenging and important direction.

## A Little Sequential Programming May Not be a Bad Thing

I have taken the position that parallel processing is important, because it permits the exploitation of mutual constraint. But, this argument does not really apply to the simultaneous *programming* of parallel processing structures. In some cases, it is sufficient to program the processing structures sequentially, so that processing can then occur in parallel. Serial programming could be arranged by projecting from one programmable module at a time to the central module and projecting the output of the connection activation nodes back to the same place. Hinton (1981b) illustrated how this kind of thing can be done, using programmable connections. If connection activations were "sticky," a number of programmable modules could be programmed sequentially, but the resulting activations could continue to interact within (and, through higher-level structures, between) the modules for some time.

There are two advantages to programming parallel processing structures sequentially. One is that we would cut down on crosstalk in the programming process. The less we project onto the central representational structures at one time, the fewer spurious connections will be activated, and the fewer nodes and programmable connections we will need for accurate processing.

The second reason is that crosstalk is especially devastating for learning, since learning takes place in the central knowledge structures. Learning in connectionist models generally involves increasing connection strengths based on simultaneous activation (Ackley, Hinton, & Sejnowski, 1985; Rumelhart & Zipser, 1985). If several patterns are superimposed in the input to the central module, the learning mechanism would be unable to separate the simultaneous activations which actually came from the same pattern from those which came from different patterns. Thus, serial programming may be particularly important during acquisition of an information processing skill. Indeed, we would probably not expect the central representations to be robust enough to tolerate much crosstalk before they have been well learned in any case. As we learn, we may be forced to proceed sequentially for adequate performance, but this may help us learn better, so that we can eventually process in parallel.

Sequential programming of parallel processing structures allows us most of the benefits of parallel processing without the costs associated with superposition of inputs to the central knowledge system. I would not, however, adopt the view that programming is always sequential. Or rather, I would not suggest that it occurs a single unit at a time at every level. Just how much input can be handled at a time probably changes with practice—but this is a matter to be examined in further research.

**Connectionism and Interactive Activation**

I began this paper by suggesting that the interactive activation model of word perception had some important limitations as a literal description of a connectionist processing mechanism. If we thought of the model as a description of the mechanism, we would take the present paper as suggesting the replacement of the original model with another kind of model, in which nodes are dynamically assigned to roles and dynamically connected to other nodes, instead of being hard-wired as they were in the original model.

However, as I have already suggested, there is an alternative way of thinking about the interactive activation model and its relation to connectionism. In this alternative approach, we would not view the interactive activation model as a description of a mechanism at all. Rather, we would see it as a functional description of the behavior of a processing system whose actual implementation—connectionistic or otherwise—is not specified.

I believe that it is important to be able to shift between these two perspectives. As important as it is to be clear about implementation, there are two reasons why it is occasionally useful to adopt a more abstract or functional point of view. The first is that it allows us to study interactive activation models of a wide range of phenomena at a psychological or functional level without necessarily worrying about the plausibility of assuming that they provide an adequate description of the actual implementation. On this view, for example, Rumelhart and I would not necessarily be seen as assuming that there "really are" multiple hard-wired copies of each letter node, one for each position within a word, in the original interactive activation model. The existence of CID allows us to be reasonably confident in the belief that a mechanism with the information processing characteristics of a model which postulated multiple copies of letter nodes could be implemented in connectionist hardware. This would free us to consider the adequacy of the particular dynamic assumptions of the interactive activation model, or more interestingly, its claim that some aspects of apparently rule guided behavior can emerge from the interactions of units standing only for particular exemplars of the rules. Similarly, I think we should be prepared to treat CID in the same way, and examine the adequacy and utility of the functional information processing characteristics it provides. The question of implementation remains an important one, and we would certainly not want a model for which no plausible implementation could be conceived, but it is not the only question which we must consider in trying to understand cognitive processes.

The second reason for taking an abstract view of activation models is to keep in view the fact that we have only begun to scratch the surface of distributed, parallel information processing mechanisms. The original interactive activation model was a step that captured some of the flavor that such a mechanism should have, and I see CID as another. But, we still have a long way to go before we can claim to have done justice to the exquisite information processing mechanism so faintly reflected in the models we have constructed so far.

## REFERENCES

Ackley, D., Hinton, G., & Sejnowski, T. (1985). Boltzmann machines: Constraint satisfaction networks that learn. *Cognitive Science, 9,* 147-169.

Anderson, J. A. (1983). Cognitive and psychological computation with neural models. *IEEE Transactions on Systems, Man, and Cybernetics, SMC-13,* 799-815.

Crick, F., & Asanuma, C. (in press). Certain aspects of the anatomy and physiology of the cerebral cortex. In J. L. McClelland & D. E. Rumelhart (Eds.), *Parallel distributed processing: Explorations in the microstructure of cognition, Vol. II: Applications.* Cambridge, MA: Bradford.

Dell, G. S. (1985). Positive feedback in hierarchical connectionist models: Applications to language production. *Cognitive Science, 9,* 3-23.

Elman, J. L., & McClelland, J. L. (1984). Speech perception as a cognitive process: The interactive activation model. In N. Lass (Ed.), *Speech and language: Vol. X.* Orlando, FL: Academic.

Feldman, J. A. (1981). A connectionist model of visual memory. In G. E. Hinton & J. A. Anderson (Eds.), *Parallel models of associative memory.* Hillsdale, NJ: Erlbaum.

Feldman, J. A., & Ballard, D. H. (1982). Connectionist models and their properties. *Cognitive Science, 6,* 205-254.

Grossberg, S. (1976). Adaptive pattern classification and universal recoding, I: Parallel development and coding of neural feature detectors. *Biological Cybernetics, 23,* 121-134.

Hinton, G. E. (1981a). A parallel computation that assigns canonical object-based frames of reference. *Proceedings of the Seventh International Joint Conference in Artificial Intelligence,* Vol 2. Vancouver, BC, Canada.

Hinton, G. E. (1981b). Implementing semantic networks in parallel hardware. In G. E. Hinton & J. A. Anderson (Eds.), *Parallel models of associative memory.* Hillsdale, NJ: Erlbaum.

Hinton, G. E. (1984). *Distributed representations*. (Tech. Rep. No. CMU-CS-84-157). Pittsburgh, PA: Department of Computer Science, Carnegie-Mellon University.
Luce, R. D. (1963). Detection and recognition. In R. D. Luce, R. R. Bush, & E. Galanter (Eds.), *Handbook of mathematical psychology: Vol. I* New York: Wiley.
McClelland, J. L., & Rumelhart, D. E. (1981). An interactive activation model of context effects in letter perception: Part I. An account of basic findings. *Psychological Review, 88*, 375-407.
McClelland, J. L., & Rumelhart, D. E. (in press). Distributed memory and the representation of general and specific information. *Journal of Experimental Psychology: General.*
Mozer, M. C. (1983). Letter migration in word perception. *Journal of Experimental Psychology: Human Perception and Performance, 9*, 531-546.
Poggio, T., & Torre, V. (1978). A new approach to synaptic interactions. In R. Heim & G. Palm (Eds.), *Approaches to complex systems*. Berlin: Springer-Verlag.
Rumelhart, D. E., & McClelland, J. L. (1981). Interactive processing through spreading activation. In A. M. Lesgold & C. A. Perfetti (Eds.), *Interactive processes in reading*. Hillsdale, NJ: Erlbaum.
Rumelhart, D. E., & McClelland, J. L. (1982). An interactive activation model of context effects in letter perception: Part 2. The contextual enhancement effect and some tests and extensions of the model. *Psychological Review, 89*, 60-94.
Rumelhart, D. E., & Zipser, D. (1985). Competitive learning. *Cognitive Science, 9*, 75-112.
Sejnowski, T. (1981). Skeleton filters in the brain. In G. E. Hinton & J. A. Anderson (Eds.), *Parallel models of associative memory*. Hillsdale, NJ: Erlbaum.
Shepherd, G. M. (1979). *The synaptic organization of the brain* (2nd ed.). New York: Oxford University Press.
Treisman, A., & Gelade, G. (1980). A feature integration theory of attention. *Cognitive Psychology, 12*, 97-136.
von der Malsburg, C. (1973). Self-Organizing of orientation sensitive cells in the striate cortex. *Kybernetik, 14*, 85-100.
Waltz, D. L., & Pollack, J. B. (1985). Massively parallel parsing: A strongly interactive model of natural language interpretation. *Cognitive Science, 9*, 51-74.
Willshaw, D. (1981). Holography, associative memory, and inductive generalization. In G. E. Hinton & J. A. Anderson (Eds.), *Parallel models of associative memory*. Hillsdale, NJ: Erlbaum.

CHAPTER 4

# Positive Feedback in Hierarchical Connectionist Models: Applications to Language Production[1]

GARY S. DELL

*University of Rochester*

> Recent connectionist models of the perception and production of words make use of positive feedback from later to earlier levels of processing. This paper focuses on production and identifies several specific effects of phoneme-to-morpheme feedback. In addition, I argue that there is support for the use of this kind of feedback in production from experimental and naturalistic studies of slips of the tongue.

One feature of highly parallel network models of word and letter perception is the existence of positive feedback from the word to the letter level (Adams, 1979; McClelland & Rumelhart, 1981; Rumelhart & McClelland, 1982). So, not only do activated letter units send activation to all words that contain them, but the reverse happens as well. Words send their activation back to the letters that comprise them. This mutual backscratching between words and letters is an elegant mechanism for allowing lexical knowledge to augment stimulus information at the letter level. More generally, positive feedback from "later" to "earlier" levels of processing is a simple way of letting knowledge influence the identification of perceptual features and objects in models that employ connectionist principles. This paper considers the utility of this kind of positive feedback in another information processing domain, the production of ordered behavior. In particular, I will focus on the production of spoken words and consider, first, what positive feedback can contribute to production models in general, and second, what empirical justification there is for adding it to models of human language production.

[1] The author wishes to thank Susan Garnsey, James L. McClelland, and Michael Tanenhaus for helpful comments. This research was supported in part by National Science Foundation grant BNS-8406886.

Before turning to language production, I would like to clarify what I mean by positive feedback. Connectionist models dealing with the perception of words and their parts (and many other information processing models for that matter) contain a network with two or more levels of processing. For example the interactive activation model of McClelland and Rumelhart (1981; Rumelhart & McClelland, 1982) includes a feature, a letter, and a word level, each composed of a set of units or nodes as shown in Figure 1.

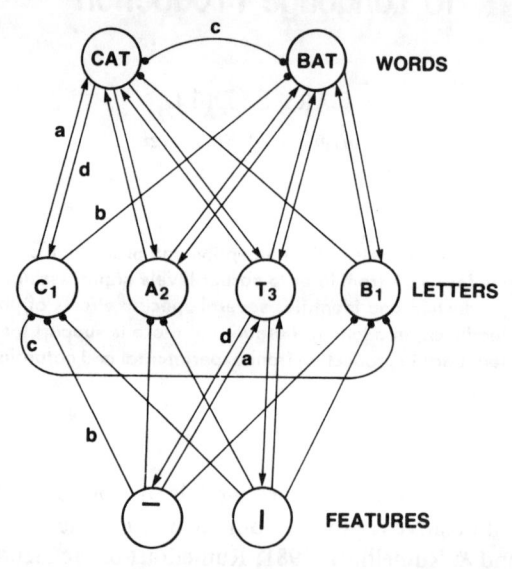

Figure 1. Word, letter, and feature nodes and their connections according to the interactive activation model. Excitatory connections are indicated by arrows and inhibitory ones by dots. The subscripts on the letter nodes indicate the letter's word position. The small letters labeling some connections are explained in the text. One should note that only the features of the letter T are shown.

As one can see in the figure each node stands for some particular word or part of a word. At any given time, each node has a potential or activation level which is a real number reflecting the extent to which that node is participating in current processing. The connections between nodes can be of four types, labeled $a$, $b$, $c$, and $d$ in Figure 1. Each connection is either excitatory ($a$ and $d$) or inhibitory ($b$ and $c$), and either bottom-up ($a$ and $b$), lateral ($c$), or top-down ($d$). An excitatory connection from node $x$ to node $y$ transmits an excitatory signal to $y$ in proportion to the activation level of $x$. An inhibitory connection does the same except that it transmits an inhibitory signal to $y$. The excitatory and inhibitory signals directed to $y$ combine and may raise or lower $y$'s activation level depending on $y$'s current level and the net input. The function of the connections is to transmit a pattern of activation in the feature nodes to the word level in such a way as to identify the most likely word given the set of activated features.

Connections of type *a,* the excitatory bottom-up ones, are the true workhorses of the network. They get the main job done by activating nodes on a higher level that are consistent with activated lower level nodes. Thus the node for *initial T* should have type-*a* connections to word nodes for *TAN* and *TENT,* but not to *BAN* and *BENT.* Because any conceivable connectionist word recognition model would require something like these connections, I will call them the primary connections.

The other kind of bottom-up connections in the interactive activation model are labeled as type *b* and are inhibitory in nature. These serve to communicate information about the presence of a feature (or letter) to those letters (or words) that do not contain it. Thus these connections do much the same thing that the primary connections do, only in an inhibitory fashion.

The two remaining connection types act to control the spread of activation rather than to directly send it from lower to higher levels. The lateral inhibitory connections (type *c*) occur between all pairs of units at the same level that are competitors. Two nodes are competitors if they collect evidence for incompatible hypotheses. For example, in a given word stimulus there can be only one initial letter. So a letter node for *initial T* would be competitor with one for *initial B.* By having a set of competitors each inhibit the others one tends to create a "winner-take-all" situation, in which one competitor, usually the one that initially had the highest activation, will end up with most of all the activation (Feldman & Ballard, 1982). The general effect is to sharpen and simplify the activation pattern at a given level.

The excitatory top-down connections (type *d*) provide positive feedback from higher to lower levels. Like the lateral inhibitory connections, they clean up the activation pattern, but unlike them, they do it with some degree of sophistication. Whereas the lateral inhibitory connections merely throw most of the activation to the "winner," the positive feedback connections, acting in concert with the primary connections, mold the activation pattern of a lower level until it meshes with information available at higher levels. Consider the following example: Assume that at a certain time the nodes for $G_1$, $C_1$, $A_2$, and $T_3$ are equally activated (the subscripts on the letters indicate the letters' word positions). Later, after activation has spread to the word level and fed back to the letter level, the pattern will change so that $C_1$ is more activated than $G_1$ simply because *CAT* is a word and *GAT* is not. Thus, through positive feedback, the activation pattern at the letter level is changed into one that makes sense when viewed from the higher word level. In general, a word node and the nodes for that word's letters and features will form a mutually reinforcing group of nodes, what Feldman and Ballard (1982) term a *stable coalition,* and over time activation will drift toward the nodes in a single such group.

Why is it good to have mutually reinforcing nodes at different levels? There are both functional and empirical reasons. It is a good idea to have letter perception constrained by lexical knowledge simply because input is often noisy. You cannot be sure that the key features of each letter will be detected and so you do not want letter identity decisions made independently of lexical knowledge. A second reason is that a model that employs positive feedback between levels will be able to explain many psychological phenomena. McClelland and Rumelhart (1981) and Rumelhart and McClelland (1982) have shown that the interactive activation model does an excellent job of simulating the perceptual advantage for letters in words over letters in unrelated contexts (the word superiority effect) and a variety of other lexical context phenomena. It is the positive feedback from the word to the letter level that makes this possible.

There is little doubt that positive feedback is valuable in hierarchical perception models, particularly if they are to model psychological processes. What about production models? Is there a role for positive feedback here, too? Highly parallel network models of language production (Dell & Reich, 1977, 1980; MacKay, 1982; Stemberger, 1982, in press) and typing (Rumelhart & Norman, 1982) have been proposed. Of these only Stemberger's and my own model have allotted a role to the kind of positive feedback being discussed here. One thing I hope to do here is clarify the function that positive feedback has in these models.

So, exactly what is positive feedback in production? Consider Figure 1, with its network composed of words, letters, and features. If we change words into morphemes, letters into phonemes, and features into phonemic or phonetic features, we have a rudimentary network for phonological encoding processes in production. There are good reasons to include other intermediate levels corresponding to syllables and syllabic constituents, but these are not important for the discussion here.

To produce a morpheme in this kind of model one would activate its node and allow activation to spread. After a period of time some decision mechanism would select the most highly activated features or phonemes, with their order being determined by either activation levels or a kind of position encoding of the nodes similar to that in the interactive activation model.

Next let us consider whether there is a role for positive feedback in a model like this. Unlike the perception model, in which the $a$ connections are the primary connections, the $d$ connections become the primary ones in a production model and the $a$ connections now provide the positive feedback, since they go from later to earlier levels. However, despite the differences in the functions of the connections in the perception and production models, the spreading activation process is similar. The presence of positive feedback in conjunction with the top-down primary connections enables the nodes for a single morpheme and its phonemes and features to act as a mutually reinforcing coalition. Earlier we saw that the tendency to make such coalitions is helpful in perception models because it filters out noise. Is there a corresponding role for these coalitions in production? I claim that there is. By allowing feedback from lower levels (e.g., phonemes) to higher levels (e.g., morphemes) one can edit out potential production errors. Specifically, I will show that this kind of feedback enables a system to avoid encoding strings of sounds that are inconsistent with higher level information. Before I elaborate on this claim, I want to present a small "demonstrator" simulation model that will point out some features of phoneme-to-morpheme feedback in a phonological production model.

## A MODEL OF PHONOLOGICAL ENCODING

This model, a no-frills version of a larger model (Dell, 1980, 1984, 1985), was designed to translate an ordered string of morphemes into an ordered string of phonemes. To keep things simple there was just a morpheme and a phoneme level, and each morpheme was a single CVC syllable. Each morpheme node had excitatory connections to the nodes representing its phonemes and each phoneme node had excitatory (feedback) connections to all the morphemes that contained that phoneme.

Each phoneme node was marked as to whether it represented an onset (initial consonant), vowel, or coda (final consonant). So, although the /k/'s in *cat* and *came* would be represented by a common node (the onset /k/), the /k/ in *tack* would be a different node (the coda /k/). The separation of onsets and codas is a form of position encoding that insures that each phoneme knows its position in the morpheme. I have discussed elsewhere how this kind of scheme can be expanded to handle the general problem of order for any morpheme or word (Dell, 1985). The model's network contained 12 phoneme nodes (4 onsets, 4 vowels, and 4 codas) and 16 morpheme nodes. Each phoneme was present in exactly 4 morphemes.

The processing in the network occurs in four stages: input, spreading activation, decision, and post-decision clean up. These stages are cycled through for each intended morpheme. In the input stage the intended morpheme is activated, that is, its activation level is incremented by an arbitrary amount, for example, 100 units. Also, upcoming morphemes in the same phrase are primed; their activation levels are increased by a smaller amount. (This was always 50 units.) For our purposes it will be sufficient to assume that this priming occurs only one morpheme in advance of the intended one.

Following input, activation is assumed to spread by the following rules. During a given period of time called a time step, each node sends some fraction of its current activation level to all nodes directly connected to it.

When the activation that is sent out reaches its destination, it adds to that node's current activation level. Because all the connections for this model are excitatory (i.e., the fraction of activation that is sent is positive), activation levels will grow without bound. To combat this tendency it is assumed that activation levels passively decay. During each time step all nodes lose some fraction of their activation. However, I should note that using decay to restrict spreading activation is, in general, not a good idea because it makes it difficult to tune parameters so that activation levels stay in useful ranges.

The spreading activation rule proposed here is among the simplest possible of such rules. There are no thresholds, saturation points, or other non-linearities. Furthermore, I will assume that the fraction of activation sent during each time step, designated by $p$, is constant for all connections regardless of their type, and the fraction of activation lost during each time step, designated by $q$, is constant for all nodes. Thus we can state a very simple rule for the spread of activation for one time step:

$$V_1 = [V_0 + V_0(pM)][(1-q)I]$$

where $V_0$ is a vector containing the activation levels of all $n$ nodes at a certain time and $V_1$ contains the same after one time step has passed. $M$ is an $n \times n$ matrix of 1's and 0's, a 1 indicating a connection from node $i$ to node $j$, and $I$ is the $n \times n$ identity matrix. After activation has spread for a certain number of time steps (parameter $r$, which reflects the speaking rate), the decision stage is entered. Here the most highly activated onset, vowel, and coda are determined and scheduled for articulation in that order. (For this model I will assume that the sounds are selected simultaneously.) Thus the model can produce 4 onsets × 4 vowels × 4 codas = 64 different strings, only 16 of which are actual morphemes from the model's point of view. In this way the model captures the generative nature of phonology. Many more strings are phonologically legal than are actual lexical items, and the model is perfectly capable of encoding these legal, but nonmorphemic, strings.

In the final stage, the post-decision cleanup, two things happen. First, the activation levels of the three selected phonemes are set to zero. This is necessary to prevent a large number of perseveratory errors. Second, the next morpheme in the set of planned morphemes becomes the intended morpheme, and the entire process cycles back to the input stage.

If a model like this encodes several morphemes in a row it may have difficulty, particularly if the intended utterance repeats sounds and is encoded at a fast rate (Dell, 1980). This difficulty reveals itself as encoding errors—slips of the tongue, in a sense. It is easy to see how errors might happen by imagining how a phrase such as *blue bug's blood* would be encoded. When the model is encoding the morpheme *bug,* the morpheme nodes for *blue* and *blood* will be somewhat activated —*blue* because it has not yet decayed and *blood* because it is primed. Thus the *bl* onset node has two sources of activation compared with the *b* onset, and depending on the model's parameters and the speaking rate, *bl* may have a higher activation level than *b,* leading to the selection of *blug* rather than *bug.* Earlier work with similar models (Dell, 1980; Dell & Reich, 1980; MacKay, 1982; Stemberger, 1982) and work in progress have shown that the models can explain the variety of errors that occur, their frequency of occurence, and the effects of changing the speech rate on error probability. In this paper, I will focus on error effects that arise directly from excitatory feedback from lower (and later) to higher (and earlier) levels of processing—particularly phoneme-to-morpheme feedback. The next section will identify these effects using the model outlined above to demonstrate them.

## EFFECTS OF PHONEME-MORPHEME FEEDBACK

If we use the demonstrator model to encode the phrase *deal back* with $p = .3$, $q = .4$, and with the speech rate at 4 time steps per morpheme ($r = 4$), it will do so correctly. This assumes that there is no other input to the network and every node starts at zero activation. Next, let us assume that for some reason there is initially some residual activation, 40 units worth, in the node for the onset /b/ and every other node starts at zero. Under these circumstances the model will err by encoding *beal back,* a phoneme anticipation slip, which is the most common kind of phonemic slip of the tongue. Other errors, such as exchanges—for example, *beal dack*—tend to occur if the speech rate is faster and the replaced sound /d/ from *deal* has a chance to bump out the /b/ in *back* (Dell & Reich, 1980). So far, none of this is very interesting. The model can, naturally enough, be made to slip by prodding it in the right way. It turns out that it takes at least 30 units of activation on /b/ at the beginning of the encoding process to create the slip of *deal* to *beal* when the next morpheme is *back.* I will call this value the anticipatory threshold for the phrase *deal back* with the above mentioned parameter values.

Next consider the phrase *dean bad.* Its anticipatory threshold using the same parameters is only 25. In the model the slip *dean bad → bean bad* (or *bean dad*) is much easier to generate—that is, it requires less stringent pre-conditions—than the slip *deal back → beal back* (or *beal dack*). The reason is that, from the model's perspective, *bean* and *dad* are legitimate morphemes and *beal* and *dack* are not. Errors that create morphemes are more likely in the model than those creating phonologically legal nonsense strings. This *lexical bias* effect comes directly from the feedback loops that develop as activation spreads between phonemes and morphemes. A pattern of activation in which the most highly activated phoneme nodes come from a single morpheme is continually reinforced. This is true whether the pattern corresponds to the correct morpheme or to some other one. If the pattern does not correspond to a morpheme it is likely to change until it does.

In this way the feedback system acts as what Baars, Motley, and MacKay, (1975) have called a lexical editor. It edits out potential nonmorpheme slips, but it does nothing to prevent slips that make erroneous, but genuine, morphemes. Like an editor, the feedback system only works well if it has enough time. When the speech rate is fast the model's errors do not show any lexical bias. This can be seen in Figure 2 in which the anticipatory threshold for a morphemic outcome (e.g., $dean \rightarrow bean$) is contrasted with that for a nonmorphemic outcome (e.g., $deal \rightarrow beal$) for various speech rates ($r=1$ to $r=4$ time steps per morpheme). At the fast rates ($r \leq 2$) the thresholds are low and do not differ as a function of lexical status of the outcome. Thus, at those rates errors would be common and they would exhibit no lexical bias. As speech slows ($r > 2$) the thresholds rise and the lexical bias effect emerges. Although only 3 time steps are necessary to create lexical bias (1 for morphemic input to reach the phonemes, 1 for phonemic activation to feed back to the morphemes, and 1 for the effects of feedback to be transmitted back to the phonemes), additional time steps increase the size of the effect. In general, the greater the opportunity for activation to reverberate between morphemes and phonemes the greater the likelihood of editing out nonmorphemes.

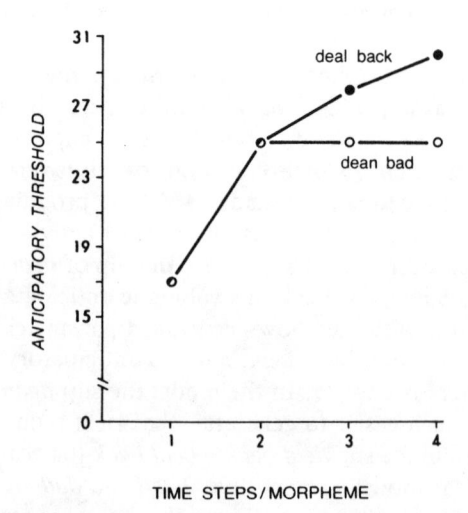

Figure 2. Lexical bias in the model's errors as a function of speaking rate. The anticipatory threshold is lower for error outcomes that create morphemes (dean bad → bean dad) than for those that create nonsense (deal back → beal dack) and this effect increases with the number of time steps per morpheme.

Another way to conceptuailze the model's lexical bias is as a production counterpart to the interactive activation perception model's word superiority effect. The latter model finds it easy to perceive letters that occur in words just as the production model finds it easy to say strings of phonemes that make words, or technically, morphemes. Both effects arise from excitatory connections in both directions between adjacent levels of processing. The influence of word-letter feedback on letter perception, however, extends beyond simple word superiority and encompasses several effects that reflect the simultaneous influence of many lexical items. For example, in the interactive activation model letters are perceived in word-like nonwords (e.g., *SIND*) nearly as well as in words. This effect is due to feedback from the many words that contain most letters of the stimulus (e.g., *SEND, MIND, SING,* etc.). Stimuli that are not word-like (e.g., *OHSG*) would not receive this benefit.

Analogous effects occur in the production model. Slips are biased toward creating morpheme-like strings as well as true morpheme strings. I have shown elsewhere that positive feedback from phonemes to morphemes leads to certain frequency asymmetries in slips (Dell, 1980), that is, a given phoneme or combination of phonemes that is present in many morphemes will tend to substitute for those that are present in fewer morphemes. For example, if the onset /t/ occurs in more morphemes that the onset /f/, then initial /f/ will slip to /t/ more than vice-versa. The same would be true for a phoneme combination. The sequence /æn/, which is present in many morphemes, would tend to replace /æb/ which is not common. A commonly recurring set of sounds will, via feedback, activate the many morphemes that contain them and these morphemes will, in turn, lead to an even higher activation level for those sounds. In this way the model is biased toward creating common phoneme combinations, and more generally, toward creating combinations that reflect the entire morphemic inventory.

So far I have shown that phoneme-morpheme feedback has what seems to be a beneficial quality. Activation patterns at the phoneme level are modified through feedback to form patterns that are likely to be correct, that is, likely to correspond to morphemes. However, not all effects of phoneme-morpheme feedback are editorial in nature. Consider the phrase *deal back,* which, as I showed earlier, has an anticipatory threshold of 30 under the previously specified conditions. If the phrase is changed to *deal beak* the threshold for the slip *deal→beal* is reduced to 28. Clearly this difference has nothing to do with the nature of the error string because it is the same in both cases (*beal*). Rather, the difference reflects the fact that *deal* and *beak* have the same vowel (/i/) while *deal* and *back* have different vowels. As *deal* is being encoded with *beak* primed, the node for the vowel /i/ will act as a pathway for activation to spread between *deal* and *beak*. The effect will be to equalize the activation levels of the two morphemes which will, in turn, lead to a greater chance of their phonemes jumping from one morpheme to the other. This effect will be called the *repeated phoneme* effect.

As with the lexical bias effect the repeated phoneme effect does not occur at fast speech rates. Figure 3 shows the anticipatory thresholds for *deal back* and *deal beak* as a function of speech rate. The effect, shown as a difference in anticipatory thresholds for the two phrases, only shows up when $r > 2$.

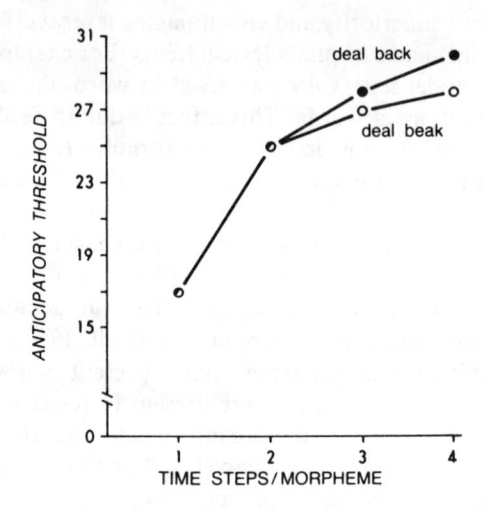

Figure 3. Repeated phoneme effect in the model's errors as a function of speaking rate. The anticipatory threshold is lower for pairs with a repeated vowel (*deal beak*) than for those with different vowels (*deal back*) and this effect increases with the number of time steps per morpheme.

Up to now I have identified two general effects of phoneme-morpheme feedback in connectionist production models, a tendency for errors to create morpheme-like strings and a tendency for phrases with similar morphemes (ones with common phonemes) to lead to slips. With these effects in mind I would now like to consider the question of the usefulness of positive feedback in production models. Does the feedback serve any valuable functions in production and, if so, do these benefits outweigh the costs incurred by adding feedback connections to a model? Following that I will examine the psychological evidence. Does the human language production system—undoubtedly the most efficient yet devised—employ positive feedback from more peripheral, "later" levels of processing to more central, "earlier" levels?

## FUNCTIONS OF POSITIVE FEEDBACK IN PRODUCTION

Positive feedback seems to be beneficial for at least two reasons. First, it allows for prearticulatory editing of the activation pattern at each processing level so that the pattern reflects constraints from higher levels. Second, some linguistic constraints such as those between particular lexical items and syntactic structures can be nicely handled by mutual excitatory connections.

The first of these functions, prearticulatory editing, has already been discussed with respect to the editing out of potential phonological slips if they do not resemble morphemes. Of course this function is only worthwhile to the extent that there is potential for these and other errors. I would like to claim that errors are always a possibility at all levels of processing in connectionist production models. Production, like comprehension, is inherently a noisy enterprise. One reason is that the basic retrieval mechanism in connectionist models is parallel spread of activation from many units to very many other units. Thus by its basic nature, this kind of model activates many more units than just the ones that are sought, a kind of "many are called, few are chosen" retrieval system. These extra units can act as a source of error because it takes time to sort out the right units from the others. If the speaking rate is rapid or variable, it is likely that this sorting out will often not be finished when a decision is required. The result would be errors if there were no editorial mechanisms in the model.

Another source of noise in production is variability in the input to the language production system. This input would be some kind of semantic-pragmatic representation of the utterance-to-be-spoken, perhaps a speech act plus a set of propositions, for example, ASSERT (COLD (THIS ROOM)). Bock (1982) calls this the *interfacing representation*. There could easily be variability in the coalition of units forming this representation simply because there are many different reasons for saying a given sentence. For example, one may say "This room is cold" to get someone to close the window, to get sympathy, or to explain why the plants died. Each reason would be associated with different computations and, perhaps, somewhat different representations of the final speech act. However, because the intended sentence is supposed to be the same regardless of its function, the possible variations in the interfacing representation will act as noise from the perspective of the next level down, the syntactic level. Thus, it would be worthwhile to have feedback from the syntactic decisions back up to the interfacing representation to prevent this noise from leading to error.

The second function that positive feedback could have in production is to link word selection with syntactic structure selection. Certain syntactic structures (e.g., those with direct objects) require the selection of certain lexical items (e.g., transitive verbs). Bock (1982) has outlined an elegant model of syntactic processing in which word and syntactic structure selection occur in parallel guided by the interfacing representation. So structures are not necessarily selected before words, nor are words selected before structures. What happens is that many candidate structures and words become activated to the extent that they are consistent with the higher level representation. The words then feed activation to the structures that they are consistent with and the structures activate words that they are consistent with. For example, transitive verbs would activate syntactic structures with direct objects and vice versa. Thus, in connectionist terms, word units and syntactic structure units are joined by mutual excitatory connections. The feedback loops that would be created would quickly select out the best structure and words to go with it. This function of feedback differs from the editing functions, only in that the two mutually interacting components (word selection and syntactic structure selection processes) are not ordered with respect to each other. In the case of editing through feedback, one level of processing, the earlier one, provides the criteria that govern the editing of the later level.

So far we have seen that positive feedback as described here would be of considerable benefit to connectionist language production models. But there are costs associated with positive feedback. One that has already been identified is that, because of positive feedback, similar items will interfere with one another. We saw that phoneme–morpheme feedback makes it somewhat more difficult to encode a string of morphemes that share phonemes, which was called the repeated phoneme effect.

Other similarity effects would occur for other types of feedback. For example, if feature–phoneme feedback is permitted, phonemes that share features will have a greater chance of substituting for each other than dissimilar phonemes. If phoneme feedback goes all the way to the nodes associated with lexical selection, similar sounding words will interact in errors, either by misordering (e.g., *cob on the corn*), or substitution (e.g., *Lizst's second Hungarian restaurant*).

Although each of these feedback effects seems to contribute to errors, I suspect the contribution is small. In the model presented earlier the repeated phoneme effect (Figure 3) is small compared with the lexical bias effect (Figure 2). At least in this case the editorial effect outweighs the similarity interference effect. However, this is a limited example and I do not want to conclusively state that positive feedback prevents more errors than it causes.

Another cost of positive feedback is the increased number of connections. Connectionist modelers must, above all, avoid computational processes that require a tremendous number of nodes and connections when scaled up to real-world size. The use of positive feedback does add connections, but the number added does not accelerate as the complexity of the model increases, at least for the kind of feedback described here. In the worst case there would be one bottom-up excitatory feedback connection (type *a* in Figure 1) for every top-down primary connection (type *d*). Thus, there is no combinatorial explosion of feedback connections as the model is made larger.

Perhaps the best way to determine if positive feedback is of value in language production models is to look at the psychological evidence. If the human language production system employs the kind of feedback described here, there are probably good reasons. The next section will identify some speech error effects that can be attributed to positive feedback. In particular I will examine alternate (nonfeedback) explanations of these effects and consider whether a strong case can be made for feedback models.

## PSYCHOLOGICAL EVIDENCE

In the previous discussion I identified two general effects of phoneme–morpheme feedback in the model: lexical bias, the tendency for errors to create morphemes, and the repeated phoneme effect, the tendency for slips to occur between morphemes with common sounds. Do these effects occur in people? With regard to lexical bias, it has been suspected since the time of Freud that slips tend to be meaningful. However, lexical bias was first demonstrated experimentally by Baars, Motley, and MacKay (1975) who used visual interference to create initial consonant misorderings in subjects' speech (exchanges and anticipations such as *deal back → beal back* or *beal dack*). They compared the probability of errors that created lexical items (*dean bad → bean dad*) with those that created nonmorphemic strings (*deal back → beal dack*) and found that errors were more than twice as likely when the outcomes were lexical than when they were not. This and other studies have established that lexical bias is a reliable effect (Berg, 1983; Dell & Reich, 1981). One of the corollary effects of the tendency for errors to create morphemes in the model was a tendency for errors to create strings that look like morphemes, that is, strings with frequently occurring phoneme combinations. This effect has, as well, been shown to be true of people's slips (Motley & Baars, 1975).

The repeated phoneme effect also has substantial empirical support. MacKay (1970) and Nooteboom (1969) have shown that phoneme exchanges are often characterized by a repeated phoneme next to the exchanging ones. For example, in the slip *left hemisphere → heft lemisphere* the phoneme /ɛ/ is next to both of the exchanging ones, /l/ and /h/. In addition, it has been demonstrated in an experiment using a similar procedure to that of Baars, Motley, and MacKay (1975) that initial consonant misorderings are more likely between two words that share a vowel (*deal beak*) than those that do not (*deal back*) (Dell, 1984).

Thus both the lexical bias and repeated phoneme effects occur in human language production. The fact that they do occur, however, does not guarantee that there is phoneme–morpheme feedback, or something like it, in production. There are other possible explanations that do not involve feedback, and these must be considered. In order to identify these alternatives it is useful, first, to distinguish between *multi-level interactive* and *single-level structural* explanations of psycholinguistic phenomena. Many empirical effects involving the processing of linguistic information are of the form: "the perception (or production) of X is influenced by Y." Numbers 1-3 below give some examples.

1. The perception of letters is influenced by lexical context (example —word superiority effect).
2. The perception of words is influenced by preceding semantically related words (example—semantic priming in lexical decision).
3. The production of a string of phonemes is influenced by whether or not that string forms a lexical item (example—lexical bias in sound speech errors).

Explanations for these kinds of effects can take two general forms. First, one can say that X-type units are processed at a different level than the one where the Y influences come from, but these influences are nonetheless made available during the processing of X. This is what I call a multi-level interactive explanation. The second kind of explanation incorporates the Y influences directly into the structure of the X-level, hence the term single-level structural explanation.

Consider the effects of semantic priming. The time to decide whether *nurse* is a word (lexical decision) or the time to read it aloud (naming) is faster if it is preceded by a semantically related word such as *doctor* (e.g., Meyer & Schvaneveldt, 1971). I will restrict the discussion to priming in naming rather than lexical decision because naming seems to be the simpler task (Stanovich & West, 1983). There is considerable debate as to why priming in naming occurs, whether it is due to an influence of the semantic level of processing on lexical access (a multi-level interactive explanation) or "intra-lexical" connections between highly related words (a single-level structural explanation). If it turns out that this priming is not semantic, but rather just a structural feature of the mental lexicon, then one can propose a simple modular account of lexical access (Forster, 1976, 1979; Seidenberg, Tanenhaus, Leiman, & Bienkowski, 1982; Tanenhaus, Leiman, & Seidenberg, 1979). In fact the issue of the "modularity of language" in general hinges upon distinguishing the two explanations of priming (Fodor, 1983).

Returning to the speech error effects under consideration (lexical bias and repeated phoneme effects) it should be clear that the account of these effects using phoneme-morpheme feedback is an interactive explanation. The alternative explanations, which I will now consider, are all single-level structural explanations. These accounts claim that either the phonemic level or the lexical level is structured so as to produce the effects. The next three sections describe some of these accounts and contrast them with the feedback account.

## A Single-level Account of the Repeated Phoneme Effect

Wickelgren (1969) proposed that repeated item effects arise from associative contextual influences between adjacent items. In his theory *deal beak* might slip to *beal deak* because the units for both /d/ and /b/ are labeled as being associated with a following /i/ sound. Hence they would be more likely to switch places than if they had come from the phrase *deal back,* in which case they would have been labeled for different following vowels.

This explanation of the repeated phoneme effect can be empirically distinguished from the feedback one. If the effect is due to associations between adjacent phonemes as Wickelgren (1969) proposed, then repeated phonemes should only be instrumental in causing slips in those phonemes adjacent to the repeated ones. So, in the phrase *rolling ball* the repeated /l/ should only induce slips in the preceding adjacent vowel sounds, (e.g., *ralling boll*), not in the nonadjacent initial consonants, (e.g., *bolling rall*). If the effect is the result of feedback from the phonemes to some higher level then the repeated /l/ could induce *bolling rall.* The scope of influence of the repeated sound should extend to all phonemes within the relevant higher order unit, and if that unit is the morpheme or syllable, the repeated /l/ can affect /r/ and /b/.

The data clearly support the interactive feedback account of the repeated phoneme effect. Repeated sounds can induce slips in nonadjacent sounds from the same syllable (or morpheme) to about the same extent as they can in adjacent sounds (Dell, 1984). Although it is not clear whether the relevant higher order unit is the syllable or the morpheme, the effect is definitely not limited to adjacent sounds. This result effectively eliminates structural explanations of the effect that appeal to a phonemic adjacency mechanism.

## A Single-level Account of Lexical Bias

Like the repeated phoneme effect, the lexical bias effect might be explainable through properties of the phonemic level alone. For example, it could be that frequently co-occurring phonemes have excitatory lateral connections among them—a phonemic analogue to the intra-lexical connections mentioned earlier. These connections would bias for activation patterns that resemble actual morphemes. Adams (1979) has, in fact, proposed a similar explanation for the effect of pseudowords on letter perception based on these kinds of connections among letters. Although I cannot conclusively reject this as an account of lexical bias I can provide some data that can be more easily explained through phoneme-morpheme feedback than through lateral connections between phonemes. These data concern the time-dependence of lexical bias, and the sensitivity of the effect to expectations and semantic factors.

It is generally true that feedback-caused effects need time to become established. Earlier, we saw that the lexical bias effect in the feedback model increased as speech slowed (Figure 2). If it can be shown that lexical bias actually is dependent on speech rate in this way it would support the feedback explanation. Although such a finding would not be inconsistent with a single-level structural explanation of the effect based on lateral phonemic connections, the feedback explanation provides a more natural account.

The data showing the time dependence of lexical bias come from an experimental study of sound errors (Dell, 1985). I used a modification of the procedure used by Baars, Motley, & MacKay (1975) to generate initial consonant exchanges and anticipations at three different speech rates. Subjects saw pairs of words, one pair at a time, at a 1-second rate. Certain critical pairs were designed so that anticipation or exchange of their initial phonemes created either words (e.g., *dean bad→bean dad*) or meaningless syllables (*deal back→beal dack*). Each critical pair was preceded by three or four interference pairs that biased the subject toward making a slip. For example, the critical pairs *dean bad* or *deal back* were preceded by *big dumb, bust dog,* and *bet dart*. Following the presentation of a critical pair, subjects saw a series of question marks. This directed them to say aloud the most recent pair. They had to complete the utterance of the pair before a certain deadline, which was either 500, 700, or 1000 ms for different groups of subjects. The probability of anticipations or exchanges of initial consonants was dependent on both the deadline (shorter deadlines led to more errors) and whether the error outcome was meaningful. However, as can be seen in Figure 4, the lexical bias effect was only present at the two longer deadlines. When subjects had only 500 ms to complete their utterance there was no lexical bias in their errors. Thus the lexical bias effect is dependent on time, as would be expected if the effect were due to phoneme-morpheme feedback. If the effect were due to some structural property of the phonological level, it is not likely that the effect could be nullified simply by speeding up the speech.

In addition to the time-dependence of lexical bias there is other evidence that favors a feedback explanation of the effect. Baars, Motley, and MacKay (1975) found that lexical bias is reduced in a context where most of the subjects' intended utterances consist of nonwords. This kind of flexibility seems more consistent with an explanation in which lexical bias is imposed on phonological encoding from outside, that is, from a separate level, than an explanation that is internal to a level. Finally, there is evidence that lexical bias in slips is sensitive to the kind of information that should only be available at morphemic or higher levels. If subjects have just read a phrase such as *damp rifle* they are more likely to make sound errors such as *get*

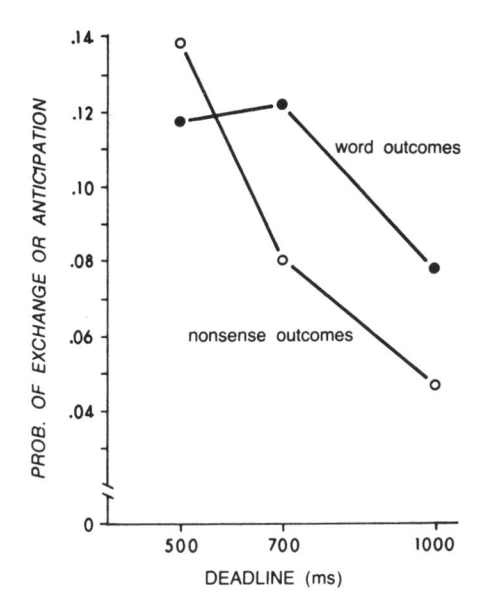

Figure 4. Empirical demonstration of the interaction of lexical bias with the speaking rate. Data are from Dell (1985) and come from 132 subjects. Each point is based on 440 error opportunities.

one → *wet gun* (Motley & Baars, 1976). This result can be interpreted as evidence for phoneme-to-semantics feedback and this is how Motley and Baars interpreted the effect. However, if we assume that the process of comprehending the phrase *damp rifle* activates morpheme nodes for *wet* and *gun* for some period of time, then the effect can be seen as just lexical bias and the feedback need only go from phonemes to morphemes, not up to the semantics. In either case, however, it is difficult to reconcile the effect with a single-level structural explanation based on lateral phonemic connections without making the connections very flexible and complex.

In conclusion, it seems that explanations of the lexical bias and the repeated phoneme effects based on properties of the phonemic level do not fare so well, at least when compared with the feedback model. Neither of the alternatives accounts for as much data in a natural fashion as the feedback model and, in the case of the contextual association view of the repeated phoneme effect, some data actually contradict it.

Thus far, the single-level alternatives to the feedback model that were considered were *phonemic,* that is, the structure necessary to explain the effects was given to the phonemic, not the lexical level. The next section considers a *lexical* explanation of the effects—one that produces much of the effects of the feedback model, but without using feedback.

## The Lexical Neighborhood Model

In the feedback model, when a particular morpheme is activated phonologically similar morphemes become activated through bottom-up feedback. This is essentially how the feedback leads to the lexical bias and repeated phoneme effects. What if the lexicon (the morpheme nodes) happened to be "arranged" so that the activation of one morpheme would directly, without feedback, lead to the activation of a set of phonologically similar neighbors? This will be called the lexical neighborhood model. This model proposes that the lexicon is structured so that phonologically similar items are "close" and that retrieving one item partially activates that item's neighbors. The only difference between this and the feedback model is that the activation of the neighbors is accomplished through feedback from the phonological level in the one case, and it is simply a structural property of the lexical level in the other case. Thus any particular feedback model could be mimicked by a lexical neighborhood model whose lexicon had the right structure.

When contrasted with feedback models, however, lexical neighborhood models have some shortcomings. First, feedback models can explain the fact that lexical bias does not occur at fast speaking rates (Figure 4). There is just not enough time for feedback to work at the fast rate. Lexical neighborhood models have no natural account of the time dependence of the effect. (One could say that the activation of the neighbors is delayed relative to the activation of intended morpheme, but this certainly does not fall out of the lexical neighborhood concept.) Second, Baars, Motley, and MacKay (1975) found that under some circumstances, lexical bias holds even when the intended utterance is composed of nonwords. How would the lexical level become activated when one intends to say a nonsense string if not through a bottom-up (i.e., feedback) route? Finally, one can fault lexical neighborhood models for not providing a distinct mechanism for the activation of the neighbors. Whatever mechanism does activate the neighbors, it will have to work exactly as does phoneme–morpheme feedback if it is to account for the data. If it must act like feedback why not just admit that it is feedback?

In conclusion, there is good evidence for feedback during production that goes from a phonemic level to a morphemic or lexical level and that this feedback affects the selection and ordering of phonemes. Alternative explanations of the evidence appear to be weaker than the feedback explanations and they are certainly not as completely specified. All of the evidence discussed so far can be explained by phoneme-to-morpheme feedback as outlined in the phonological encoding model presented earlier.

## Evidence for Other Kinds of Feedback in Production

Other speech error effects have been attributed to positive feedback, and I would like to briefly consider them. One of the most robust facts about phonological slips is the sound similarity effect—similar sounds (e.g., /t/ and /k/) slip with each other much more readily than dissimilar sounds (e.g., /z/ and /k/). Because similar sounds can be said to share features, theorists have suggested that the effect may be due to feature-phoneme feedback (e.g., Dell & Reich, 1980; Meyer & Gordon, in press; Stemberger, 1982). Common features would provide pathways for activation spreading between competing similar phonemes in much the same way that occurs in the repeated phoneme effect.

The case for feature-phoneme feedback, however, is not nearly as strong as that for phoneme-morpheme feedback. As one would expect, the sound similarity effect can be given single-level structural explanations, as well as interactive ones. For example, Shattuck-Hufnagel and Klatt (1979) argue that phonemic features are not units of production, but instead are just names of categories of phonemes. Similar phonemes are just those that are in the same category and similar phonemes slip because phoneme selection processes use these categories. In order to rule out this and other structural explanations for the sound similarity effect we need data on the effect's time dependence and its boundary conditions—the same kind of evidence that was used to argue against structural explanations for the lexical bias and repeated phoneme effects. Unfortunately, this evidence is lacking, so the case cannot be strongly made for feature-to-phoneme feedback during production.

The situation is similar with regard to phoneme-to-lexical selection feedback. Dell and Reich (1980, 1981) and Stemberger (1982, 1983) have argued for feedback from the phonemic level to lexical selection processes based on the fact that similar sounding words often substitute for one another (e.g., saying *propose* for *propel*). Again, there is a possible structural explanation that is difficult to rule out. Fay and Cutler (1977) and Fromkin (1971) have argued that lexical neighborhood models of the kind discussed earlier can easily explain phonologically related word errors—the substituted word was simply in the intended word's neighborhood. Although there are some data favoring the feedback explanation (Dell & Reich, 1981; Harley, 1984; Stemberger, 1983) there is, at present, no resolution (see Garrett, 1976).

## CONCLUSIONS

Despite the difficulty in establishing that human language production makes use of positive feedback, I think the case can be made that such feedback exists, and moreover, that its purpose is to establish coalitions made up of units from separate levels. Establishing these coalitions can prevent errors that would have gone uncorrected if production were purely top-down in nature.

# REFERENCES

Adams, M. J. (1979). Models of word recognition. *Cognitive Psychology, 11,* 133-176.
Baars, B. J., Motley, M. T., & MacKay, D. G. (1975). Output editing for lexical status from artificially elicited slips of the tongue. *Journal of Verbal Learning and Verbal Behavior, 14,* 382-391.
Berg, T. (1983). *Monitoring via feedback in language production: Evidence from cut-offs.* Unpublished manuscript.
Bock, J. K. (1982). Towards a cognitive psychology of syntax: Information processing contributions to sentence formulation. *Psychological Review, 89,* 1-47.
Dell, G. S. (1980). *Phonological and lexical encoding in speech production: An analysis of naturally occurring and experimentally elicited slips of the tongue.* Unpublished doctoral dissertation, University of Toronto.
Dell, G. S. (1984). The representation of serial order in speech: Evidence from the repeated phoneme effect in speech errors. *Journal of Experimental Psychology: Learning, Memory & Cognition, 10,* 222-233.
Dell, G. S. (1985). *A spreading activation theory of retrieval in sentence production.* (Cognitive Science Tech. Rep. No. 21). University of Rochester.
Dell, G. S., & Reich, P. A. (1977). A model of slips of the tongue. *The Third LACUS Forum,* pp. 448-455.
Dell, G. S., & Reich, P. A. (1980). Toward a unified theory of slips of tongue. In V. A. Fromkin (Ed.), *Errors in linguistic performance: Slips of the tongue, ear, pen, and hand.* New York: Academic.
Dell, G. S., & Reich, P. A. (1981). Stages in sentence production: An analysis of speech error data. *Journal of Verbal Learning and Verbal Behavior, 20,* 611-629.
Fay, D., & Cutler, A. (1977). Malapropisms and the structure of the mental lexicon. *Linguistic Inquiry, 3,* 505-520.
Feldman, J. A., & Ballard, D. H. (1982). Connectionist models and their properties. *Cognitive Science, 6,* 205-254.
Fodor, J. A. (1983). *The modularity of mind: An essay on faculty psychology.* Cambridge, MA: Bradford.
Forster, K. I. (1976). Accessing the mental lexicon. In R. J. Wales & E. Walker (Eds.), *New approaches to language mechanisms.* Amsterdam: North Holland.
Forster, K. I. (1979). Levels of processing and the structure of the language processor. In W. E. Cooper & E. Walker (Eds.), *Sentence processing: Psycholinguistic studies,* Hillsdale, NJ: Erlbaum.
Fromkin, V. A. (1971). The non-anomalous nature of anomalous utterances. *Language, 41,* 27-52.
Garrett, M. F. (1976). Syntactic processes in sentence production. In R. J. Wales & E. Walker (Eds.), *New approaches to language mechanisms.* Amsterdam: North Holland.
Harley, T. A. (1984). A critique of top-down independent levels models of speech production: Evidence from non-plan-internal speech errors. *Cognitive Science, 8,* 191-219.
MacKay, D. G. (1970). Spoonerisms: The structure of errors in the serial order of speech. *Neuropsychologia, 8,* 323-350.
MacKay, D. G. (1982). The problems of flexibility, fluency, and speed-accuracy trade-off in skilled behavior. *Psychological Review, 89,* 483-506.
McClelland, J. L., & Rumelhart, D. E. (1981). An interactive activation model of context effects in letter perception: Part 1. An account of basic findings. *Psychological Review, 88,* 375-407.
Meyer, D. E., & Gordon, P. C. (in press). Speech production: Motor programming of phonetic features. *Journal of Verbal Learning and Verbal Behavior.*

Meyer, D. E., & Schvaneveldt, R. W. (1971). Facilitation in recognizing pairs of words: Evidence for a dependence between retrieval operations. *Journal of Experimental Psychology, 90,* 227-234.

Motley, M. T., & Baars, B. J. (1975). Encoding sensitivities to phonological markedness and transition probability: Evidence from spoonerisms. *Human Communication Research, 2,* 351-361.

Motley, M. T., & Baars, B. J. (1976). Semantic bias effects on the outcomes of verbal slips. *Cognition, 4,* 177-188.

Nooteboom, S. G. (1969). The tongue slips into patterns. In A. G. Sciarone et al. (Eds.) *Nomen: Leyden studies in linguistics and phonetics.* The Hague: Mouton.

Rumelhart, D. E., & McClelland, J. L. (1982). An interactive activation model of context effects in letter perception: Part 2. The contextual enhancement effect and some tests and extensions of the model. *Psychological Review, 89,* 60-94.

Rumelhart, D. E., & Norman, D. A. (1982). Simulating a skilled typist: A study of skilled cognitive-motor performance. *Cognitive Science, 6,* 1-36.

Seidenberg, M. S., Tanenhaus, M. K., Leiman, J. M., & Bienkowski, M. A. (1982). Automatic access of the meanings of ambiguous words in context: Some limitations of knowledge-based processing. *Cognitive Psychology, 14,* 489-537.

Shattuck-Hufnagel, S., & Klatt, D. H. (1979). The limited use of distinctive features and markedness in speech production: Evidence from speech error data. *Journal of Verbal Learning and Verbal Behavior, 18,* 41-55.

Stanovich, K. E., & West, R. F. On priming by a sentence context. *Journal of Experimental Psychology: General, 112,* 1-36.

Stemberger, J. P. (1982). *The lexicon in a model of language production.* Unpublished doctoral dissertation, University of California, San Diego.

Stemberger, J. P. (1983). Inflectional malapropisms: Form-based errors in English morphology. *Linguistics, 21,* 573-602.

Stemberger, J. P. (in press). An interactive activation model of language production. In A. −Ellis (Ed.), *Progress in the psychology of language.*

Tanenhaus, M. K., Leiman, J. M., & Seidenberg, M. S. (1979). Evidence for multiple stages in the processing of ambiguous words in syntactic contexts. *Journal of Verbal Learning and Verbal Behavior, 18,* 427-441.

Wickelgren, W. A. (1969). Context-sensitive coding, associative memory, and serial order in (speech) behavior. *Psychological Review, 76,* 1-15.

CHAPTER 5

# A Developmental Neural Model of Visual Word Perception

RICHARD M. GOLDEN

*Brown University*
*Psychology Department, Box 1853*
*Providence, RI 02912*

---

A neurally plausible model of how the process of visually perceiving a letter in the context of a word is learned, and how such processing occurs in adults is proposed. The model consists of a collection of abstract letter feature detector neurons and their interconnections. The model also includes a learning rule that specifies how these interconnections evolve with experience. The interconnections between neurons can be interpreted as representing the spatially redundant, sequentially redundant, and transgraphemic information in letter string displays. Anderson, Silverstein, Ritz, and Jones's (1977) "Brain-State-in-a-Box" (BSB) neural mechanism is then used to implement the proposed model. The resulting system makes explicit qualitative predictions using both letter recognition accuracy and reaction time as dependent measures. In particular, the model offers an integrated explanation of some experiments involving manipulations of orthographic regularity, masking, case alternations, and experience with words. The similarities and differences between the model and models proposed by Adams (1979) and McClelland and Rumelhart (1981) are also discussed.

## INTRODUCTION

Recently, a number of rather novel information processing models of perception and cognition have been proposed. These models are based upon the assumption that design constraints derived from neurophysiological considerations may provide useful insights about certain psychological phenomena. Such models are sometimes known as *connectionist* models (Feld-

---

This research was funded in part by Grant BNS-82-14728 from the National Science Foundation, Memory and Cognitive Processes Section to J.A. Anderson. I am happy to acknowledge the technical advice, software, and encouragement provided by the Brown University neural modelling group. In particular, I thank Alan Kawamoto, Mike Rossen, Howard Winston, and especially Jim Anderson for their helpful suggestions throughout this project. I wish to also thank Jim Anderson, Garrison Cottrell, Peter Eimas, Jay McClelland, Gregory Murphy, and Michael Tanenhaus for their comments and criticisms of an earlier version of this article. Finally, I am grateful to the Center for Cognitive Science at Brown University for the use of their computing facilities.

man & Ballard, 1982). One particular connectionist model that considers the problem of perceiving letters within the context of words has been suggested by McClelland and Rumelhart (1981).

In the McClelland and Rumelhart model, a set of letter and word *nodes* are connected together in a pre-specified manner. When a stimulus representing a visual pattern of features is presented to the system, the relevant letter nodes are partially activated. These letter nodes then partially activate word nodes in the system that in turn reinforce the activation of the appropriate letter nodes. Because the model uses *bottom-up* and *top-down* processing simultaneously to arrive at a consistent interpretation of the incoming information, the model is often referred to as the Interactive Activation (IA) model.

The IA model has been successfully applied to a number of problems in the "letter-within-word" perception literature (McClelland & Rumelhart, 1981; Rumelhart & McClelland, 1982). A unique aspect of the IA model is that all processing within the system is based on very simple local computations similar in spirit to the types of computations that might be performed by neurons. These simple local computations then give rise to interesting global phenomena at the network level. Although the IA model successfully explains a number of experimental findings, McClelland and Rumelhart (1981) do not specifically address the developmental issues. In particular, what are the origins of the letter and word nodes in the IA model? How did the connections between the nodes within the system originate? Although some partial solutions to these questions have been recently suggested (McClelland, 1985; Rumelhart & Zipser, 1985), the problem of how experience with letters and words might specifically influence the development of the ability to detect orthographic information has not been considered.

In this paper, an alternative model of visual letter in word perception, based upon dynamic principles similar to the IA model but different representational assumptions, is suggested. More importantly, however, this alternative model provides a formal framework for considering how the effects of experience influence the visual perception of letters in the context of words. This new model is called the Letter-in-Word (LW) neural network model. The LW model consists of a collection of position-specific letter feature nodes and a learning rule that describes how correlations can develop between pairs of letter feature nodes. Position-specific letters are represented as activation patterns over the position-specific letter feature nodes. Words are represented as conjunctions of position-specific letter activation patterns. A correlation among two letter feature nodes associated with a given letter position in a word is called a *within-letter* feature correlation. A correlation among two letter feature nodes located in different letter positions in a word is called a *between-letter* feature correlation. An activation pattern over the letter feature nodes is "learned" by the system through the simultaneous perturbation of both between-letter and within-letter feature correlations. An *error-correction* property associated with the learning algorithm is also assumed. Thus, atypical letter feature correlations are learned faster than typical letter feature correlations (Figure 1).

**Sources of Orthographic Information in Visual Stimuli**
Experimental evidence obtained through studies of how people visually perceive letters in the context of words has been steadily accumulating over the past decade. Such studies have often focussed upon the various types of visual information in a word stimulus that might be used in perceptual processing. In particular, three distinct types of *orthographic* information in the visual stimulus have been described by a number of researchers.

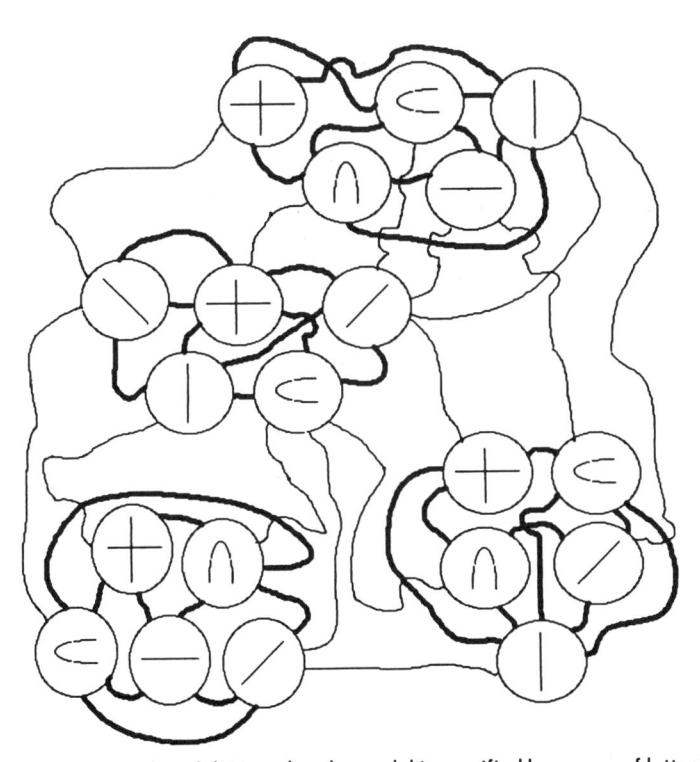

**Figure 1.** The LW neural model. Note that the model is specified by groups of letter feature detector neurons and their interconnections. The neurons in each group are dedicated to detecting letter feature information only at a specific letter position in a word. Interconnections within a group (dark lines) are referred to as *within-letter* feature correlations. Interconnections between groups (light lines) are referred to as *between-letter* feature correlations. For clarity, only a small subset of the neurons and their interconnections are shown.

*Spatial Redundancy Information.* Spatial redundancy information is one particular type of orthographic information. The spatial redundancy of a letter is simply the frequency of occurrence of that letter at a given letter position within a word. The spatial redundancy of a word is calculated by simply summing the spatial redundancy scores of all letters in a word. Mason (1975) (also see Massaro, Venezky, & Taylor, 1979; McClelland & Johnston, 1977) have found that letters presented in the context of letter strings with low spatial redundancy ratings are perceived less efficiently than letters presented in the context of letter strings with high spatial redundancy ratings. Moreover, because spatial redundancy information is an extremely good predictor of performance in such letter perception tasks, this type of information may be quite important for the visual letter-within-word perception process.

*Sequential Redundancy Information.* Sequential redundancy information is defined as the likelihood of the occurrence of a letter at a specific letter position within a word, given information about the occurrence and location of other letters in the word. The sequential redundancy hypothesis is appealing because letters within orthographically regular strings of unrelated letters (pseudowords) are perceived more efficiently than letters within orthographically irregular strings of unrelated letters (nonwords) (see for example, Baron & Thurston, 1973; Reicher, 1969). Moreover these effects have been observed to be independent of the effects of spatial redundancy information (Massaro et al., 1979). On the other hand, both McClelland and Johnston (1977) and Johnston (1978) in a series of careful studies were unable to obtain any direct evidence supporting the presence of sequential redundancy information in letter-within-word perception. These studies therefore suggest that sequential redundancy information plays only a minor role in the visual word perception process.

*Transgraphemic Information.* The third type of information in a visual letter string stimulus is sometimes referred to as transgraphemic information. A transgraphemic feature is defined as a constellation of simple contour features that specifies a useful property of visual patterns representing letter strings. Typically, experiments involving mixed-case stimuli tend to support the hypothesis that transgraphemic information is an important aspect of visual letter-within-word perception. In particular, letters within same-case words (e.g., THAT) are perceived more efficiently than letters within mixed-case words (e.g., ThAt) (Adams, 1979; McClelland, 1976; Taylor, Miller, & Juola, 1977). Note that in this case both the spatial and sequential redundancy information in the stimulus have been held constant, but the actual pattern of visual information has been disrupted.

## Development of Orthographic Information in the LW Model

*The LW Model Learning Rule.* Learning in the LW model is based upon the following two-step procedure. First, when a training stimulus is presented to the model, the model does a partial categorization of the training stimulus using the current set of letter feature correlations. An error-correction learning rule is then used to perturb the set of stored correlations using feature correlations in the training stimulus. The magnitude of the perturbation is assumed to be proportional to the difference between the model's interpretation of the training stimulus and the actual value of the training stimulus.

*The LW Model Training Procedure.* Evidence from the educational literature (for a review see Durr & Pikulski, 1981) indicates that, before children are taught to visually perceive letters in the context of words, the ability to visually discriminate individual letters is at least partially attained. Similarly, learning in the LW model begins by first training the system with a set of single letters, and then training the system with a set of word stimuli. Because letters possess only within-letter feature correlations, the model acquires knowledge of only within-letter feature correlations during the initial stages of learning. Then later, during the word learning stage, a set of between-letter feature correlations is also acquired by the system since words possess both within-letter and between-letter feature correlations.

*Developmental Consequences of the Learning Rule and Training Procedure.* If such a learning procedure is followed, then the magnitude of the within-letter feature correlations will always be greater than the magnitude of the between-letter feature correlations for the following reason. During the letter learning stage, the magnitude of error the model makes in categorizing letter strings decreases as the time period of letter learning is increased. The smaller error rate in conjunction with the error-correction learning rule then causes the feature correlations to be updated more slowly. Since letters are taught to the model before words are presented, the system acquires a set of large within-letter feature correlations during letter learning and a set of relatively small between-letter feature correlations during word learning. Also note that the ratio of the magnitude of the within-letter feature correlations to the between-letter feature correlations can be indirectly adjusted by varying the ratio of the letter learning time period to the word learning time period.

*Correspondence Between Feature Correlations and Orthographic Information.* In the LW model, the between-letter feature correlations correspond to the transgraphemic information in visually displayed letter strings. Moreover, certain subsets of the between-letter feature correlations may be considered to correspond to the sequential redundancy information in the matrix. Finally, since the within-letter feature correlations are sensitive to specific orderings of letters and are ambivalent to the transgraphemic information in a word stimulus, the within-letter feature correlations are defined as the spatial redundancy information in the matrix. Thus, the learning rule and the training procedure implicitly produce a system that is highly dependent upon spatial redundancy information, moderately dependent upon transgraphemic feature information, and weakly dependent upon sequential redundancy information. This "hierarchy" of dependencies is an important psychological modelling assumption of the LW model since it predicts the presence of effects of spatial redundancy and transgraphemic information, and the absence of effects of sequential redundancy information.

**Neural Modelling Assumptions**
A particular characteristic of processing in visual cortex is the presence of retinotopic mappings (Cowey, 1981). That is, evidence that the original spatial distribution of information in the visual world is preserved to some extent in the representation of that information by neural activation patterns. Such observations tend to support the hypothesis that a visual word stimulus may be represented as a pattern of neural activity over a set of position-specific letter feature neurons: the ith neuron in the neuronal set responding to a particular letter feature at a particular letter position in a word stimulus. Note that such a representation assumes prior processing that solves problems such as word position and word size invariance. Rumelhart and McClelland (1982) have noted that such problems also arise with their IA model and have suggested a "smearing" transformation as a possible solution. Such a solution may also be applicable to the LW neural model.

The information processing and learning assumptions in the LW model also have a neurophysiological basis. The position-specific letter feature neurons process information by computing weighted sums of the firing rates of the other neurons in the system. Moreover, the weights a given neuron assigns to incoming information evolve according to a Hebbian-like learning rule. Such a rule states that when two neurons are simultaneously active, a change occurs in the nervous system such that the two neurons become more correlated in their discharges. The validity of these modelling assumptions are considered in greater detail by Anderson et al. (1977), Kohonen (1984), and Anderson and Silverstein (1978).

**Integrating the Model with Other Subsystems**
How might this model be integrated with other subsystems that are necessary for word recognition? Kawamoto (1985) has proposed a system consisting of four groups of neurons. The first group of neurons in Kawamoto's model process the semantic properties of a word stimulus, the second group of neurons process the syntactic properties, the third group process the

phonetic properties, and the final group process the graphemic properties of a word stimulus. In a manner similar to the model proposed here, knowledge in Kawamoto's system is represented through the use of correlational synapses between and within these four subsystems. The LW model may be considered to be the graphemic subsystem of Kawamoto's model.

## COMPUTER SIMULATION METHODOLOGY

### Specific Modelling Assumptions

*Representational Assumptions.* To implement the LW model, Anderson et al.'s (1977) Brain-State-in-a-Box (BSB) model was used. In the BSB model a pattern of neural activation is represented by a state vector: the ith element of the state vector corresponding to the difference between the firing rate of the ith neuron in the system and the ith neuron's spontaneous firing rate (Anderson et al., 1977). Using the BSB state vector formalism, a word vector in the LW model may be represented as the concatenation of four letter subvectors. Each letter subvector, in turn, can be represented as a concatenation of letter feature subvectors. More specifically, since the dimensionality of a word vector was equal to 112, a unique 28-dimensional letter subvector was assigned to each of the upper case and lower case forms of the nine most frequent letters of the English alphabet. Each of these letter subvectors was constructed in a principled manner from an extension of an abstract letter feature set originally proposed by Gibson (1969, p. 88). Although the psychological validity of this letter feature set is uncertain, the Gibson feature set was useful for the purposes of simulation. Figure 2 illustrates the state vector encoding procedure. Appendices 1 and 2 provide additional details of the letter and word stimulus sets used in the computer simulations.

*Information Processing Using the BSB Model.* In addition to representing incoming information as a state vector, the BSB model represents knowledge with a matrix of real-valued synaptic weights. Information processing in the BSB model amounts to transforming the incoming state vector in such a manner so as to increase the familiarity between that vector and the matrix of synaptic weights. Neurophysiologically, this transformation may be interpreted as increasing the amount of neural activity in the system using positive feedback until all of the neurons have obtained their maximum or minimum firing rates (Anderson et al., 1977).

More formally, the explicit transformation is given by the following equation:

$$\mathbf{X}(K+1) = S(\mathbf{A}\mathbf{X}(K) + \mathbf{X}(K)) \qquad (1)$$

where $\mathbf{X}(K)$ is the neural activation pattern at discrete time interval K, $\mathbf{A}$ is a matrix of synaptic weights, and the S function sets all vector elements whose

## VECTOR CODING

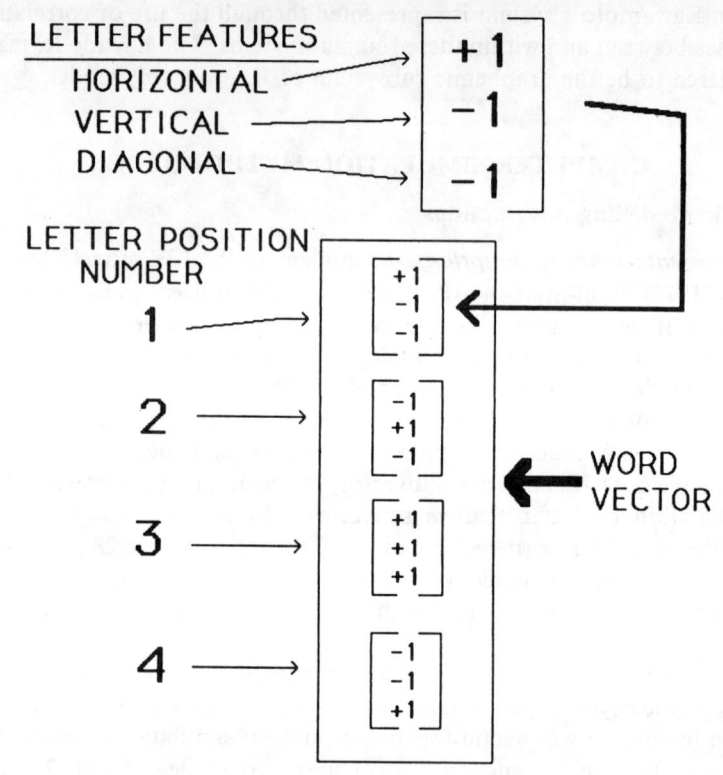

**Figure 2.** An example illustrating the basic concepts used to represent words as state vectors. A word vector is a concatenation of letter subvectors. A letter subvector is a concatenation of letter feature subvectors. In this example, the word vector is 12-dimensional, a letter subvector is 3-dimensional, and the letter feature subvector is a real number. The example illustrates in detail how a letter, located in the first letter position of a word and possessing a horizontal line segment but no vertical or diagonal line segments is, represented in the word vector.

values are above some maximum firing rate deviation equal to that maximum firing rate deviation, and all vector elements whose values are below some minimum firing rate deviation equal to that minimum firing rate deviation.

Equation 1 describes how the state vector evolves at any given time interval in the BSB model. For example, the input state vector $\mathbf{X}(0)$ may be inserted into Equation 1 to obtain the transformed state vector $\mathbf{X}(1)$. The vector $\mathbf{X}(1)$ is then inserted into Equation 1 to obtain $\mathbf{X}(2)$. These iterations continue until $\mathbf{X}(K) = \mathbf{X}(K+1)$. At this point, usually all of the elements of the system state vector have obtained their maximum or minimum values. When all elements of the state vector are saturated, the vector is referred to as a hypercube corner. If the input stimulus vector travels to the appropriate hypercube corner, then the stimulus is defined to have been properly categorized. The number of iterations required to arrive at a hypercube corner is taken as the system's reaction time. Finally, note that because the eigenvalues of the matrix tend to be large and positive, the trajectory of the state vector is always directed "outward" towards the corners of the hypercube. (More detailed descriptions of the BSB model may be found in Anderson [1983], Anderson and Mozer [1981], and Anderson et al. [1977].)

The familiarity of a state vector with the set of matrix weights is now precisely defined by the following function. Following Smolensky (1983), this function will be referred to as the *harmony* function. The harmony function for the BSB model is:

$$E(\mathbf{X}) = \sum_i \sum_j a_{ij} x_i x_j = \mathbf{X}^T \mathbf{A} \mathbf{X} \tag{2}$$

where $a_{ij}$ is the ijth element of the **A** matrix in Equation 1, and $x_i$ is the ith element of the system state vector **X**.

Assuming the vector **X** is a hypercube corner, note that each term of the harmony function measures how closely a given synaptic weight in the matrix *matches* two specific elements of the corner vector. Thus, larger values of the harmony function suggest a given vector is more familiar to the matrix, while smaller values of the harmony function suggest a given state vector is less familiar. Golden (1986) has formally demonstrated that the BSB model transforms the input vector so as to make the vector more harmonious with the synaptic weights of the matrix where *harmonious* is given a precise meaning using Equation 2.

*Learning in the LW Model.* The learning rule used in the LW model is very simple. A "scaled-down" version of a training vector is cycled K times through the BSB model algorithm (Equation 1) to obtain a vector **Z**. The BSB model's response (the **Z** vector) is then compared with the original training vector through a subtraction, and the resulting vector is used to update the matrix weights using a Hebbian (see Anderson et al., 1977) learning rule (Figure 3). The ith on-diagonal matrix element represents the correlation of the firing rate of the ith neuron in the system with itself. Thus, to prevent the on-diagonal elements of the matrix from growing without bound, the on-diagonal elements were not updated (see Anderson et al., 1977).

More formally a hypercube corner training vector, $\mathbf{C}_i$, is randomly selected from the set of training stimuli. The vector $\mathbf{C}_i$ is then normalized, perturbed with random noise, and passed through Equation (1) exactly K times to obtain a new vector $\mathbf{X}_i(K)$. Let $\mathbf{Z}_i = \mathbf{X}_i(K)$. The matrix of synaptic weights is then updated using the following learning algorithm:

$$\mathbf{A}_{new} = \mathbf{A}_{old} + \gamma[(\mathbf{C}_i - \mathbf{Z}_i)(\mathbf{C}_i - \mathbf{Z}_i)^T] \tag{3}$$

## THE LEARNING RULE

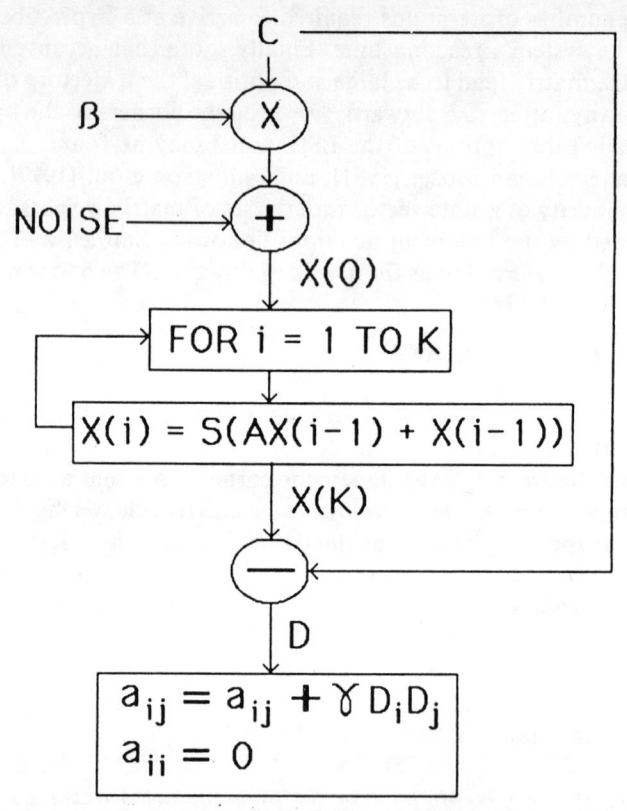

**Figure 3.** The learning rule used to update the letter feature correlations. The training vector, **C**, is multipled by a constant $\beta$ $(0<\beta<1)$ and noise is added to the vector. The resulting vector, **X**(0), is cycled K times through the BSB model to obtain the vector **X**(K). The difference between **X**(K) and **C** is then computed and used to update the matrix weights with a Hebbian learning rule.

where $\gamma$ is the learning constant and **[B]** indicates that the main diagonal elements of the matrix **B** should all be set equal to zero. The initial value of the **A** matrix is a matrix of zeroes. Related learning rules that have been successfully used in the BSB model are discussed in detail by Anderson et al. (1977) and Anderson and Mozer (1981).

Consider now the following cost function that directly measures the harmony of the set of training vectors with the set of matrix weights:

$$J(\mathbf{A}) = \Sigma \mathbf{C}_i^T \mathbf{A} \mathbf{C}_i \tag{4}$$

where $\mathbf{C}_i$ is the ith hypercube corner vector in the set of training stimuli, **A** is the BSB model matrix, and the summation is taken over the set of training vectors.

Empirical studies of $J(A)$ as the learning rule modifies the weights of the **A** matrix indicate that the learning algorithm is increasing the value of the cost function (Equation 4) for the simulation parameter values reported here. Or in other words, the matrix weights are modified at each learning trial so as to increase the harmony of the matrix with the set of training vectors. This basic idea of constructing an *energy landscape* in some principled manner, and then using that landscape during the recall process has been suggested by Ackley, Hinton, and Sejnowski (1985). Appendix 3 describes some explicit conditions that indicate when the learning rule (Equation 3) must maximize $J(A)$.

## Computer Simulation Details of the Training Procedure

Different *computer subjects* were trained using a set of 196 four-letter words coded as state vectors. Half of the vector set consisted of the upper case forms of the 98 word stimuli (see Appendix 2), the other half of the vector set consisted of the lower case forms of the 98 word stimuli. The individuality of each subject was determined by a unique random number seed that specified the learning sequence order and the random perturbations of the training stimuli.

During the training period, each computer subject learned state vectors whose elements had been corrupted by independent and identically distributed Gaussian noise (see Equations 1 and 3). The noise was added only to the non-zero elements of the position-specific letter training vectors. This noise represented the natural variation of stimuli in the environment plus the internal noise of the nervous system. The standard deviation of the noise/neuron was 0.02 units which was roughly 10% of the signal amplitude/neuron (0.19 units). The maximum deviation of a neuron's firing rate from its spontaneous firing rate was 1.0 units. The minimum deviation was −1.0 units. The value of the training constraint $\gamma$ was always 0.005.

The simulation subjects were initially trained for 1500 learning presentations using state vectors representing randomly selected position-specific letters. Thus, because only vectors representing four-letter words are considered here, three fourths of the elements of each position-specific letter training vector were all set equal to zero. The subjects were then tested using the reaction time and/or letter recognition accuracy testing procedures that will shortly be described. For the next 200 learning presentations, subjects learned stimuli selected at random from the set of upper case and lower case word stimulus vectors and were tested again. The subjects were then trained for an additional 800 learning presentations with the word stimuli, and tested a final time.

## The Reaction Time Testing Procedure

In the testing phase, each of 1176 test stimuli was presented in sequence to the BSB model. Each stimulus vector was cycled through Equation 1 until each stimulus was categorized. Only reaction times for stimulus vectors that had been correctly categorized were recorded. A correctly categorized stimulus was defined as a hypercube corner whose four-letter subvectors were all correctly categorized. This particular dependent variable was selected because of its similarity to reaction time measures used in same/different matching tasks for the "same" trials, and letter string search tasks when the target letter is not present.

## The Letter Recognition Accuracy Testing Procedure

The reaction time testing procedure examines the behavior of the system under normal operating conditions. To compare the pattern of letter recognition errors produced by the model with the pattern of errors produced by people, a special testing procedure was devised to force the system to make letter recognition errors. The format of the testing procedure was modelled after Reicher's (1969) visual forced-choice paradigm. In his procedure, a letter string is presented for a brief time period. The letter string is then immediately replaced with a masking stimulus and two letters. One letter appears above one of the letter positions where the orginal stiumulus had been presented, while the other letter appears below that letter position. The subject is then asked to decide which of two letters was presented. The intensity of the display, the stimulus and mask onset times, and various other factors are usually adjusted such that the subject responds correctly about 60% to 90% of the time.

In the specific procedure implemented here, independent and identically distributed Gaussian noise was added to each element of the initial state vector. The signal amplitude per neuron was also reduced. After the first few iterations of the BSB algorithm (Equation 1), a *mask* vector composed of four Xs was added to the system state vector. The system was not trained upon position-specific X vectors during the letter training stage. The addition of the mask and noise were specifically intended to make the system produce letter recognition errors. The magnitude of the signal vector, the magnitude of the mask vector, the magnitude of the noise vector, and the number of iterations before the mask was applied were selected by the experimenter to avoid floor and ceiling effects in pilot studies. The dependent variable in this procedure was the average percentage of letters correctly categorized at a specific letter position averaging over all four-letter positions.

Finally, in all of the following computer simulation experiments, the treatment effects were always significant (typically, $p < .0001$). Thus, only cell means are reported.

# EXPERIMENT 1: ORTHOGRAPHY AND MIXED-CASE EFFECTS (REACTION TIME DATA)

Nine computer subjects were trained using the learning procedure and tested using the reaction time testing procedure. The test stimuli consisted of a set of 1176 same-case and mixed-case stimuli generated from the upper case and lower case forms of the set of 98 words, 98 pseudowords, and 98 nonwords. To generate a mixed-case stimulus, a same-case stimulus was altered by assigning the second and fourth letters of the stimulus to their alternative case values. As described in Appendix 2, an unrelated letter string was classified as a pseudoword or a nonword based upon its spatial redundancy rating. The spatial redundancy ratings were assigned on the basis of an analysis of the word stimuli learned by the model. Table 1 shows the spatial redundancy rating distribution for words, pseudowords, and nonwords.

**Simulation Results and Discussion**
Juola, Schadler, Chabot, and McCaughey (1978) found that second graders, fourth graders, and college students searched for letters within words faster than nonwords, while no effects of orthographic redundancy were observed in kindergarteners. Moreover, trends in the data indicate that, while for college students, letters within words were perceived faster than letters within pseudowords, the reaction times in the word and pseudoword conditions for the second and fourth graders were essentially identical.

The results of the computer simulations also demonstrate these developmental phenomena (Figure 4). First, as the number of learning presentations increases, the overall reaction time of the system for word, pseudoword, and nonword stimuli decreases. Second, reaction times decrease very rapidly during the first 200 learning trials, and then decrease at a much slower rate as the system obtains additional experience with word stimuli. And third, reaction times for words were 1.551 iterations faster than nonwords at 2500 learning presentations, while at 1700 learning presentations the reaction time difference between words and nonwords was only 1.302 iterations. At 2500 learning presentations, pseudowords were categorized in 0.449 iterations more than words and 1.102 iterations fewer than nonwords. At 1700 learning presentations, pseudowords were categorized in 0.411 iterations

TABLE 1
Spatial Redundancy Ratings of Stimulus Set

|  | M | SD |
|---|---|---|
| Words | 62.0 | 14.0 |
| Pseudowords | 65.7 | 6.9 |
| Nonwords | 33.4 | 5.7 |

# EXPERIMENT 1 - REACTION TIME DATA

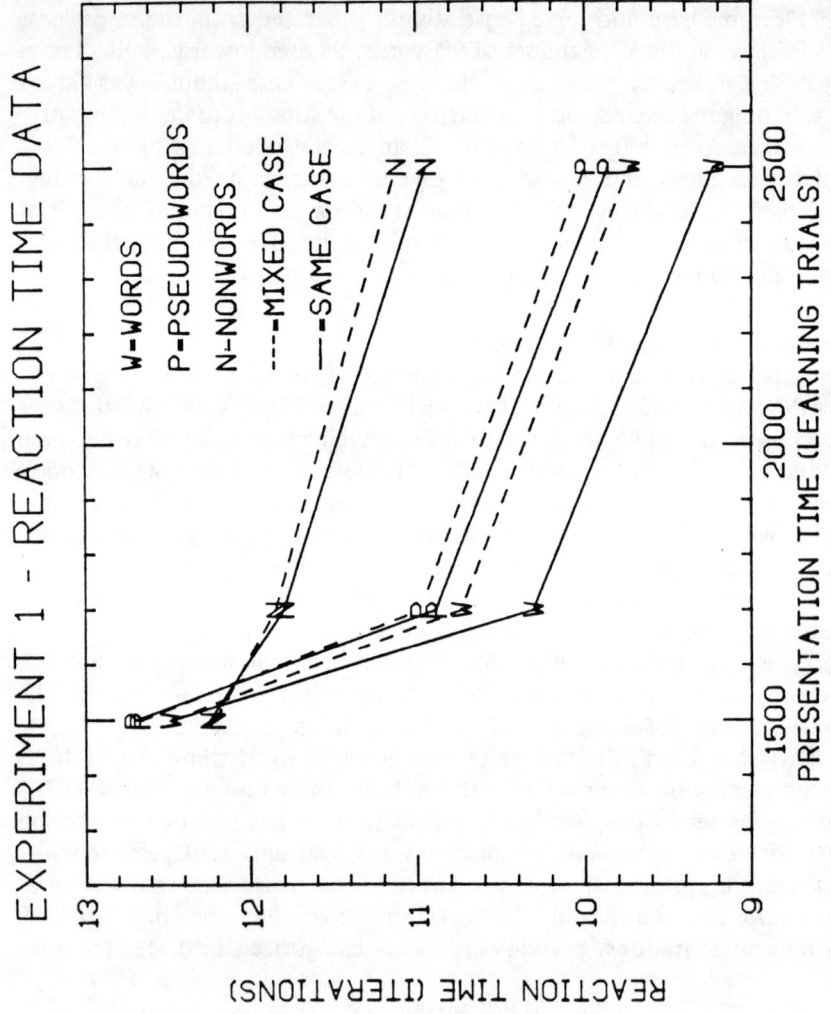

**Figure 4.** Reaction time plotted as a function of case type, orthography, and learning trials for Experiment 1. Solid lines indicate same-case stimuli. Dashed lines indicate mixed-case stimuli.

more than words and 0.891 iterations fewer than nonwords. Thus, as experience with words increased, the word-pseudoword, pseudoword-nonword, and word-nonword reaction time differences also increased in magnitude.

Adult reaction time studies of the effects of orthographic redundancy and alternating case are also compatible with the computer simulation results. For adult subjects, letters in words are perceived faster than letters in nonwords (Mason, 1975; Taylor et al., 1977), and letters in pseudowords are perceived slower than words, yet faster than nonwords (Taylor et al., 1977). In addition, for mixed-case stimuli, words are recognized faster than pseudowords which are recognized faster than nonwords, and the reaction time advantage for same-case over mixed-case stimuli decreases as stimuli become less orthographically regular (Taylor et al., 1977; Pollatsek, Well, & Schindler, 1975).

The reaction times for the same-case and mixed-case stimuli of the computer simulations after 2500 learning presentations are shown in Figure 4. For mixed-case stimuli, word are categorized 0.268 iterations faster than pseudowords which are categorized 1.122 iterations faster than nonwords. In addition, the reaction time difference between mixed-case and same-case stimuli was 0.503 iterations for words, 0.141 iterations for pseudowords, and 0.181 iterations for nonwords. Thus, the simulation results qualitatively agree with the experimental data in all respects with the exception that the alternating case effect for pseudowords was smaller in magnitude than the alternating case for nonwords.

## EXPERIMENT 2: ORTHOGRAPHY AND MIXED-CASE EFFECTS (ACCURACY DATA)

In Experiment 2, the nine computer subjects used in Experiment 1 were tested with the same set of 1176 stimuli using the letter recognition accuracy testing procedure. The masking stimulus was added to the state vector at the beginning of the fourth iteration of the BSB algorithm. The signal amplitude per neuron was reduced to 0.95 units, and the noise-per-neuron standard deviation was 0.05 units. The magnitude of the masking stimulus was equal to unity.

### Simulation Results and Discussion

The development of the ability to distinguish between pseudowords and nonwords develops very rapidly in children, usually during the first 2 years of grade school (Lefton & Spragins, 1974; Lefton, Spragins, & Byrnes, 1973; Rosinski & Wheeler, 1972). First grade children recognize letters in pseudowords as accurately as they recognize letters in nonwords, while third graders, fifth graders, and adults recognize letters in pseudowords more accurately than letters in nonwords (Lefton & Spragins, 1974). Inspection of the types of categorization errors made by the model, as the system's experience with word vectors was increased, reveals a developmental error pattern similar to that obtained in children (Figure 5). At 1500 learning presentations, letters in pseudowords and nonwords were perceived with equal accuracy because the model, at that time period, had not acquired any knowledge about words. At 1700 learning presentations, the magnitude of the pseudoword-nonword effect was 9.94%, while at 2500 learning presentations, the magnitude of the pseudoword-nonword effect was 10.8%.

The model also accounts for the basic patterns of letter recognition errors observed in adult subjects. Adults recognize letters in words more accurately than letters in nonwords (see e.g., Adams, 1979; Johnston, 1978; McClelland, 1976; Reicher, 1969), and perceive letters in pseudowords more accurately than letters in nonwords (see e.g., Adams, 1979; McClelland, 1976). Moreover, for mixed-case stimuli, letters in words are perceived more accurately than letters in pseudowords which are, in turn, perceived more accurately than letters in nonwords (Adams, 1979; McClelland, 1976). In addition, for words and pseudowords, same-case stimuli are more accurately perceived than mixed-case stimuli (Adams, 1979; McClelland, 1976), while the same-case/mixed-case advantage is not apparent for nonwords (McClelland, 1976).

Consider now the performance of the model after 2500 learning presentations. For both same-case and mixed-case stimuli, letters in words were categorized more accurately than letters in pseudowords, which were categorized more accurately than letters in nonwords. Furthermore, the relative magnitudes of the same-case over mixed-case effect for words, pseudowords, and nonwords were 6.08%, 3.85%, and 2.65% respectively. Thus, the letter recognition error patterns characteristic of adult subjects were quite similar to the data obtained from computer simulations of the model with the exception of the same-case over mixed-case advantage observed for the nonword condition.

## GENERAL DISCUSSION OF EXPERIMENTS 1 AND 2

In this and following discussions, the effects of the matrix upon the various types of stimulus vectors presented to the model are qualitatively considered. The basis of this qualitative analysis is the observation that the BSB model transforms a testing stimulus such that the resulting state vector is more harmonious with the set of matrix weights. Stimuli that are more harmonious with the set of matrix weights tend to be categorized more quickly and more accurately. The harmony existing between a vector and a matrix is given a rigorous definition in Equation 2.

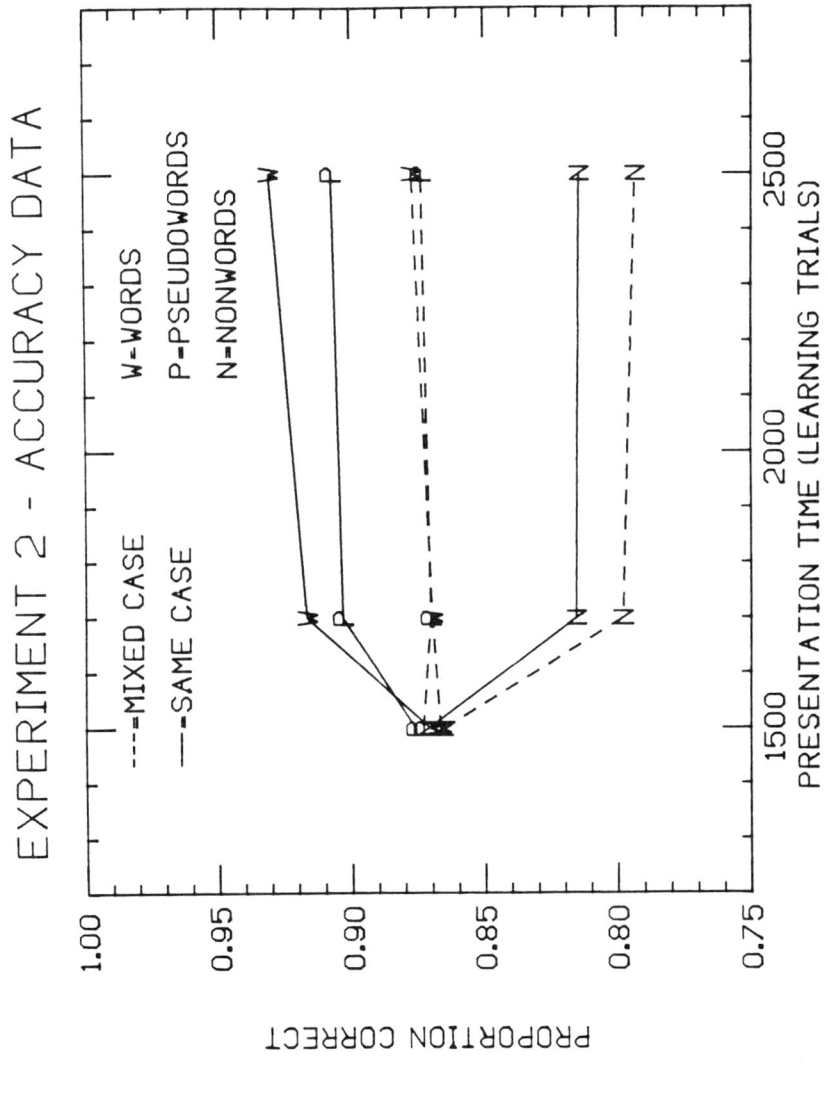

**Figure 5.** The proportion of correctly categorized letters plotted as a function of case type, orthography, and learning trials for Experiment 2. Solid lines indicate same-case stimuli. Dashed lines indicate mixed-case stimuli.

### An Explanation of Orthographic and Case Effects in the LW Model

By definition, a nonword is a state vector that has not been "learned" by the system. Such a state vector can nevertheless be categorized by the BSB model since the within-letter feature correlations are used to categorize each letter of the nonword independently.

The harmony of a nonword stimulus tends to be less than a word stimulus for two reasons. First, the within-letter feature correlations will be less harmonious with nonwords possessing low spatial redundancy ratings. And second, the between-letter feature correlations will be less harmonious with nonwords possessing low transgraphemic feature ratings. The advantage for letters in words over letters in pseudowords is explained in a similar manner. Also note that, within the framework of this model, the superiority of letter recognition for same-case relative to mixed-case stimuli is analogous to the work-pseudoword advantage.

### An Explanation of Developmental Effects in the LW Model

The development of the ability of the model to more efficiently extract orthographic information from a word stimulus as the system gains experience with the set of word training vectors is considered formally in Appendix 3. More informally, the learning rule modifies the matrix weights at each learning trial such that the harmony of the matrix with the set of training stimuli tends to increase. The rapid acquisition of the ability to detect orthographic information occurs because letter feature correlations acquired from only a few words are used to categorize many other words possessing those letter feature correlations.

## EXPERIMENT 3: EFFECTS OF VISUAL LETTER FRAGMENT MASKING

The model proposed here differs from several other models in the letter in word perception literature (e.g., Adams, 1979; McClelland & Rumelhart, 1981) in that the letter feature, letter, and word levels of the system are highly interactive. This distinction suggests that the proposed model might be able to offer an explanation regarding why the word superiority effect (Baron & Thurstone, 1973; Estes, 1975; Reicher, 1969; Wheeler, 1970) is so dependent upon the nature of the masking stimulus. Johnston and McClelland (1973) found that when a letter fragment masking stimulus was used, subjects perceived letters in words more accurately than letters alone. However, when a clear mask stimulus was used, the word superiority effect was considerably diminished. It is also important to note that, in both the clear and letter fragment mask conditions, the overall performance of subjects was kept fairly constant. This observation suggests that in the letter fragment mask condition, a high quality stimulus is presented for a short period of time and is then disrupted by a masking stimulus. On the other hand, in the clear mask condition, a degraded stimulus is presented for a much longer time period (see McClelland & Rumelhart, 1981, for a more detailed discussion).

To explore this phenomenon, a set of five computer subjects were trained with letter and word stimuli according to the training procedure previously described. The subjects were tested with the letter recognition accuracy testing procedure after 2500 learning presentations. The test stimuli consisted of the upper case word stimuli used in Experiment 1. Each word stimulus was then used to generate four position-specific letter stimuli: the letter in the kth position of the word stimulus generating a position-specific letter vector with non-zero elements only in the region of the kth subvector. For example, the word stimulus THAT generated letter stimuli T000, 0H00, 00A0, and 000T where a 0 indicates that region of the state vector should be filled with zeros.

A series of pilot studies suggested the following signal to noise ratios and mask onset times. For the letter fragment mask condition, the standard deviation of the added noise/neuron was 0.05 units, the signal amplitude/neuron was 0.133 units, and the mask was added to the system state vector at the beginning of the sixth iteration of the BSB model testing equation (Equation 1). The magnitude of the letter fragment mask was 5.29 units. For the clear mask (i.e., no mask) condition the standard deviation of the added noise/neuron was 0.095 units, and the signal amplitude/neuron was 0.0855 units. These parameters were selected such that the overall performance degradation in the clear mask condition was approximately equal to the overall performance degradation in the letter fragment mask condition. Note that, for the clear mask condition, the variance of the additive noise is larger than the signal power.

## Simulation Results and Discussion

Table 2 shows the results of the computer simulations. The overall performance of the system was about 65% in both the letter fragment and clear mask conditions. For the letter fragment mask condition, letters in words were categorized 23.2% more accurately than letters presented alone. For the clear mask condition, the magnitude of the word superiority effect was

TABLE 2
Letter Recognition Accuracy as a Function of Stimulus Type and Masking Condition (Experiment 3)

|  | Words | Letters |
| --- | --- | --- |
| Letter fragment mask | 0.69 | 0.56 |
| Clear mask | 0.70 | 0.60 |

only 16.7%. Qualitatively, these simulation results are in agreement with the experimental literature (Johnston & McClelland, 1973).

In the LW model, the word superiority effect occurs because both the between-letter and within-letter feature correlations help a letter in a word become categorized, while only within-letter feature correlations can aid the model in categorizing single letters. The qualitative differences between the clear mask and letter fragment mask conditions in this experiment are due to the fact that the performance of the model is very robust with respect to test stimuli perturbed with random noise, yet the model is very sensitive to the effects of a letter fragment mask. Moreover, to obtain equal levels of overall performance, the stimulus must be so degraded in the clear mask condition that the system's sensitivity to the orthographic information in the stimulus becomes diminished. This loss of sensitivity results in a decrease in the magnitude of the word superiority effect.

The reason why the model proposed here is so robust with respect to random noise, but so sensitive to letter fragment masks can be explained as follows. Let a test stimulus vector be perturbed with independent and identically distributed Gaussian noise, and then compute the harmony of the random vector. If the on-diagonal elements of the matrix defining the harmony function are all equal to zero, then the expected value of the random vector's harmony turns out to be equal to the harmony of the original deterministic test vector. On the other hand, if a letter fragment mask vector is added to the test stimulus vector, the harmony of the perturbed test vector will be quite different from that of the original test stimulus vector. Or in other words, the trajectory of a test stimulus vector perturbed with random noise is not severely affected as it travels toward a hypercube corner since the noise acts equally upon each component of the vector on the average. A letter fragment mask vector, however, is constructed from precisely those vector components that are most influential in determining the trajectory of the state vector through the hypercube.

## EXPERIMENT 4: A DETAILED EXAMINATION OF MIXED-CASE EFFECTS

Word superiority effect experiments involving mixed-case and same-case stimuli are important because they directly address the issue of whether transgraphemic features should be included in models of visual word perception. If spatial and sequential redundancy were the only types of information used in word perception, then the recognition of letters in same-case words (e.g., THAT) should be as efficient as the recognition of letters in mixed-case words (e.g., ThAt). A number of experimenters have found a same-case over mixed-case advantage for words and pseudowords and a noticeable absence of this effect for nonwords using both reaction time (Pollatsek et al., 1975; Taylor et al., 1977) and letter recognition accuracy studies (McClelland, 1976). A notable exception to this consensus was an experiment involving a wide range of font styles and case types by Adams (1979). Adams found that all stimuli (words, pseudowords, and nonwords) were perceived equally well in the same-case relative to the mixed-case condition. She interpreted her findings to suggest that case alternations affect only the processing of individual letters and not the conjunctions of letters.

Adams's results, however, have an alternative interpretation. The suggestion is made here that the use of stimuli comprised of such a wide variety of font styles effectively destroys much of the transgraphemic information. The word–pseudoword and pseudoword–nonword effects in her experiment were presumably due to effects of spatial redundancy. If this hypothesis is correct, then increasing the number of case alternations should tend to decrease the efficiency of letter in word perception. Exactly this hypothesis was tested in a same–different visual comparison task by Taylor et al. (1977). Taylor et al. found that increasing the number of case alternations for words and pseudowords increased reaction time.

To examine the effects of mixed-case stimuli in more detail than Experiments 1 and 2, an additional set of mixed-case stimuli was constructed possessing only a single case alternation (e.g., ThAT). The five computer subjects from Experiment 3 were then tested after 2500 learning presentations with the reaction time testing procedure. The goal of these additional simulations was to explore the relationship between the effects of orthography and the effects of increasing the number of case alternations in words, pseudowords, and nonwords in greater detail.

**Simulation Results and Discussion**

The results of the computer simulations are displayed in Figure 6. Like Taylor et al.'s (1977) study, a case alternation by orthography interaction was obtained. In addition, the simulation shows a sudden increase in reaction time with the first case alternation for words in the same manner as the experimental data (Taylor et al., 1977). The model diverges from the experimental data, however, in predicting for nonwords an increase in reaction time as the number of case alternations are increased. This problem with the model (a problem also observed in Experiments 1 and 2) might be remedied by experimenting with varying the ratio of letter learning trials to word learning trials. Alternatively, an effect of case type upon nonwords might be present (but not easy to detect) in human subjects.

## EXPERIMENT 5: HIGHER ORDER INFORMATION IN THE LW MODEL

It is generally believed that the perceptual system should make use of "higher order" information during the processing of a visual stimulus. On the other hand, McClelland and Johnston (1977) were unable to find any evidence for

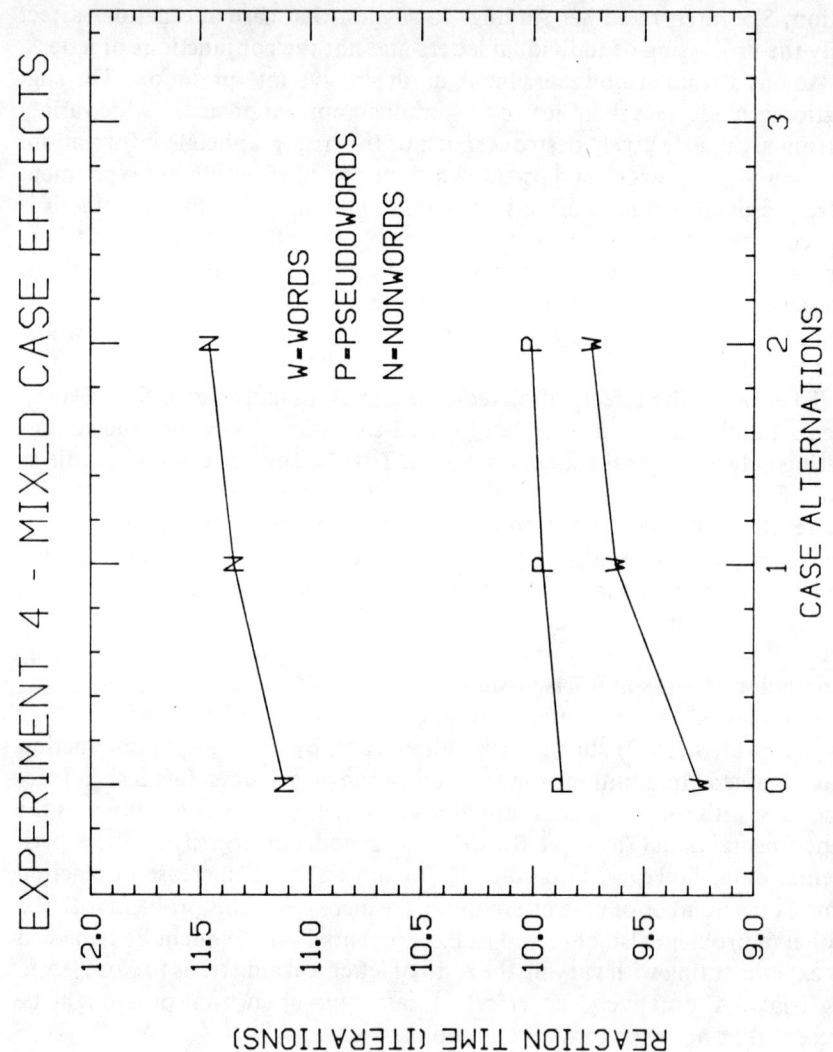

Figure 6. Reaction time plotted as a function of orthography and case alternations for Experiment 4.

position-specific bigram detectors. In this section, the LW model is demonstrated to be very insensitive to the effects of sequential redundancy information. Following these demonstrations, a more sophisticated proposal for thinking about higher order information in word perception is suggested based upon the eigenstructure of the BSB model matrix.

## Absence of Sequential Redundancy Effects

*Selective Removal of Between-Letter and Within-Letter Correlations.* The model proposed here is predominantly a position-specific letter detector model whose processing is guided by transgraphemic features. The model does not depend upon sequential or bigram information for processing. To illustrate this contention, all of the between-letter feature correlations in the matrix associated with one of the computer subjects in Experiment 3 were set equal to zero. After "surgery" the computer subject was still able to correctly categorize each of the 98 upper case word stimuli. The average reaction time was about nine iterations. The between-letter feature correlations were then replaced and the within-letter feature correlations were removed. Under these conditions, the computer subject was unable to categorize any of the 98 word stimuli in less than 300 iterations.

*Absence of Effects of Position-Specific Bigram Stimuli.* In another experiment, two sets of unrelated letter strings were constructed based upon statistical regularities computed from the set of words actually learned by the system. The first group of unrelated letter strings (the *hi-bigram group*) possessed a mean position-specific letter frequency count of 42.9 and a mean position-specific bigram frequency count of 5.89. The second group of unrelated letter strings (the *lo-bigram group*) had a mean position-specific letter frequency count of 42.0 and a position-specific bigram frequency count of 1.29. For purposes of comparison, the set of word stimuli had a mean letter frequency count of 62.0 and a mean bigram count of 11.4. There were 56 stimuli (both upper case and lower case forms) in both bigram groups. The reaction time testing procedure was used to evaluate the performance of the five subjects tested in Experiment 3 using a within-subjects design as usual. For the hi-bigram group, the mean reaction time was 10.93 iterations. For the lo-bigram group, the mean reaction time was 10.81 iterations. Thus, the lo-bigram group was actually slightly *faster* than the hi-bigram group in processing the stimuli. Note, for purposes of comparison, that the reaction time difference between the pseudoword and nonword conditions in Experiment 1 was about 1.0 iterations, while the corresponding difference in this experiment was 0.12 iterations in the reverse direction.

*An Explanation of the Absence of Sequential Redundancy Effects.* The absence of sequential redundancy effects in the model are due to three factors. First, the learning algorithm is a highly nonlinear error correction procedure. Frequently used matrix weights are updated less often than rarely used matrix weights. This tends to remove slight statistical frequency differences in the set of training stimuli. Second, the recall algorithm is a nonlinear categorization procedure that also reduces the magnitude of slight frequency differences. And third, the within-letter feature correlations dominate the between-letter feature correlations to such an extent that manipulations affecting only the between-letter feature correlations tend to be "washed-out" by the BSB model recall algorithm (Equation 1).

**Perceptual Invariants in the BSB Model**
Despite the lack of sequential redundancy effects, however, higher order structural information is present in the BSB model. To see this, rewrite Equation 2 as follows. Assume the **A** matrix is symmetric.

$$E(\mathbf{X}) = \Sigma\Sigma \, a_{ij} x_i x_j = \mathbf{X}^T \mathbf{A} \mathbf{X}$$
$$= \Sigma \lambda_k \mathbf{X}^T \mathbf{e}_k \mathbf{e}_k^T \mathbf{X} = \Sigma \lambda_k \, (\mathbf{e}_k^T \mathbf{X})^2 \qquad (5)$$

where $\mathbf{e}_k$ is the kth eigenvector of the **A** matrix and $\lambda_k$ is the kth eigenvalue.

Equation 5 demonstrates that the harmony function may be alternatively interpreted as directly measuring how compatible a given state vector is with the **A** matrix's eigenstructure. Anderson and Mozer (1981) (also see Anderson et al., 1977) note that the eigenvectors with the largest eigenvalues of the matrix in the BSB model have many of the desirable properties associated with what Gibson (1969) and other theorists have referred to as *perceptual invariants*. Like perceptual invariants the eigenvectors are typically unique and independent. Moreover, Equation 5 shows that the BSB model is a system that enhances the components of a stimulus vector associated with the eigenvectors possessing large eigenvalues.

# EXPERIMENT 6: WORD BIAS EFFECTS IN LETTER FEATURE PERCEPTION

A critical assumption of the model proposed here, that differentiates this model from many other models in the letter-in-word perception literature, is that the orthographic information and visual information that specify a word stimulus are allowed to freely interact. Massaro (1979) has attempted to directly study this critical assumption. In his experiment, strings of four letters were presented to subjects followed by a masking stimulus. The subjects' task was to identify a particular target letter in the letter string. Moreover, the target letter was an ambiguous stimulus that could be interpreted as either one of two letters. The magnitude of the visual ambiguity was also varied. Massaro (1979) found that the other letters in the letter string could be selected to bias the interpretation of the ambiguous target letter.

To illustrate how the model can account for the effects of orthographic bias upon the perception of letter feature information, a special set of stimulus vectors was constructed. Referring to Appendix 1, note that if four elements of the letter subvectors that represent E and T are all set equal to zero, that the resulting letter subvector is ambiguous. The coding details of the resulting ambiguous letter subvector E/T are provided in Appendix 1. A set of 15 word stimuli that possessed the letter T in the final letter position, and a set of 15 word stimuli that possessed the letter E in the final letter position were then selected from the set of upper case words learned by the model. The final letter of each of these word stimuli was then replaced with the ambiguous letter subvector E/T. For purposes of control, a set of 30 nonword stimuli with E/T in the final letter position were also constructed.

All 60 stimuli were then presented to the five computer subjects that were tested in Experiment 3. The letter recognition accuracy testing paradigm was used. The input signal to noise power ratio was identical to that used in Experiment 3 for the letter fragment mask condition. The magnitude of the letter fragment mask, however, was equal to unity. Analysis of the simulation data revealed that the model always managed to properly reconstruct either a T or an E in the final position of the letter string. Thus, only the proportion of times that the simulation reconstructed an E in the final letter position of the stimulus will be reported.

For nonwords (control condition), the model interpreted E/T as E 13% of the time. For word stimuli that biased the model to reconstruct a T in the final letter position, the letter subvector E/T was interpreted as an E only 6% of the time. For word stimuli that biased the model to reconstruct an E in the final letter position, the letter E was reconstructed 24% of the time. Massaro (1979) observed similar qualitative effects in his experiment.

The behavior of the model may again be readily understood in terms of the within-letter and between-letter feature correlations. The between-letter feature correlations tend to create a bias in the ambiguous region of the letter subvector E/T in the initial stages of processing. The within-letter feature correlations and between-letter feature correlations then use this bias to reconstruct the missing region of the ambiguous letter subvector.

## RELATED MODELS OF WORD PERCEPTION

A good review of additional theories of letter-within-word perception may be found in Henderson (1982). Paap, Newsome, McDonald, and Schvaneveldt (1982) also have proposed an interesting formal model of word perception that uses an empirically derived set of correlations to represent the information in a word stimulus. To emphasize more effectively certain unique aspects of the model proposed here, some comparisons of the Letter-in-Word neural model will be made only with Adams's (1979) and McClelland and Rumelhart's (1981) theories of letter-in-word perception.

## Adams's Model of Letter-Within-Word Perception

*A Description of Adams's Model.* Adams (1979) has proposed that associations between letter detector units are formed by a Hebbian learning assumption. As words are presented to her model, pairs of letter detectors that are simultaneously active become bonded together. More specifically, Adams (1979) suggested that the association strength between a pair of letter units could be estimated from a table of bigram frequencies. When a word is presented to Adams's model, the component letter units of the word are directly activated and indirectly receive activation from one another via pair-wise letter unit associations. In contrast to this behavior, when a nonword is presented to her model, only the component letter units of the nonword are directly activated since the pair-wise letter unit associations are very weak. Adams's model is interesting because it parsimoniously explains the word-nonword effect and makes a specific claim regarding how the ability to visually perceive words develops with experience. More specifically, she suggests that the associational strength between pairs of letter units is proportional to the frequency of occurrence of that pair of letter units.

*A Comparison of Adams's Model to the LW Model at the Psychological Level.* The psychological assumptions underlying Adams's model of letter-in-word perception are very similar to the psychological assumptions of the LW model. In both models, learning is assumed to proceed through the extraction of regularities from an environment, and effects of orthography are considered a result of differential association strengths. The two models differ, however, in several important ways.

First, the functional units in Adams's model are position-independent letter units, while the functional units in the LW model are position-specific letter feature units. The position-specific code in the LW model is used to explain effects of spatial redundancy. In Adams's model, effects of spatial redundancy are assumed to arise from inter-letter association strengths. Massaro et al. (1979) in a very careful study, however, demonstrated independent effects of both spatial redundancy and sequential/transgraphemic redundancy thus casting doubt on Adams's interpretation of the origins of spatial redundancy effects. Second, her letter units are *case-independent*, while the BSB model does not identify the same letter in two different fonts as the same letter. Thus, her model cannot explain why same-case stimuli are perceived more efficiently than mixed-case stimuli as the stimuli are made more orthographically regular. Third, Adams's model assumes that the processing of orthographic information is independent of the visual letter feature information in a word. Thus, if Adams's model is assumed, then the interaction between orthographic and visual information obtained by Massaro (1979) presumably is due to the *response bias* stage of processing. Fourth, her model is based upon position-independent letter bigram information, which is notably absent in visual word perception tasks (McClelland & Johnston, 1977). The LW model essentially uses only transgraphemic and spatial redundancy information. And finally, although Adams offers an explanation regarding how letter feature correlations develop, her model does not make any statements regarding how the system's letter units initially originated.

## The Interactive Activation Model of Word Perception

*A Description of the IA Model.* The IA model consists of a set of feature detection units or *nodes,* a set of letter nodes, and a set of word nodes. Each letter node for a specific letter position within a word has a bi-directional excitatory connection with all words possessing that letter in that letter position, and a bi-directional inhibitory connection with the remaining word nodes in the system. In addition, every word node inhibits all other word nodes and every letter node inhibits all other letter nodes through an additional set of inhibitory connections. This within-level or lateral inhibition helps the system to arrive at consistent interpretations by preventing more than one word node from becoming too active at any given time. Each node is also associated with a specific number or *activation level* that varies continuously between a minimum and maximum level according to a sigmoidal (S-shaped) function. This sigmoidal function will sometimes be referred to as the IA model's saturating nonlinearity. At each time interval, each node simultaneously computes a weighted sum of the activation levels of the other nodes in the system and the connection strengths from those nodes. The weighted sum is called the *net input* to the ith node. The net input, current activity level, resting level, and decay rate of the ith node are then used to update the activity level of the ith node.

When a word is presented to the model, the letter feature nodes become initially activated. The letter feature nodes then activate the letter nodes which can activate word nodes. Partially activated word nodes then excite their component letter nodes at the letter level of the system while inhibiting the remaining letter nodes. These interactions between the letter and word nodes continue until eventually the system's state settles upon a consistent interpretation of the input at the word and/or letter levels of the system.

*Implementing the IA Model Using the BSB Model Formalism.* The qualitative dynamic behaviors of the IA model and the BSB model are very closely related. To examine these similarities more closely, arrange the letter and word nodes of the IA model in any order within a high dimensional state vector. Also ignore the IA modelling assumption that the activation levels of the active nodes in the system tend to return to their resting levels. The connection strengths between all the letter and word nodes in the IA model may then be arranged using a BSB model synaptic weight matrix. For example, let the word THAT be the first element of the IA model's state vector and let the symbols T1, H2, A3, and T4 be the second, third, fourth, and fifth components of the state vector. (The notation T4 indicates the letter T in the fourth letter position.) Now consider the IA modelling assumption that states when a word node is activated, the component letters of that word node are also activated. This modelling assumption would be implemented by assigning an equal positive weight to the elements of the matrix that represent the correlations between THAT and T1, THAT and H2, THAT and A3, and THAT and T4. In a similar manner, between-level inhibitory and within-level inhibitory connections may be represented in the matrix using negative valued weights. Thus, by selecting the weights of the corresponding BSB model matrix appropriately, the pre-wired hierarchical structure of the IA model may be realized.

The equations governing the dynamic behavior of the IA model, when represented using matrix algebra, are therefore quite similar to the testing dynamic equations of the BSB model. Both systems cycle their respective state vectors through a connectivity matrix, and obtain useful categorization behaviors using a saturating nonlinearity. Only two significant qualitative differences exist with respect to the mechanisms involved in both models. The activation levels of the nodes in the IA model tend to return to their resting levels, and the saturating nonlinearity used in the IA model is considerably more sophisticated than the BSB model's nonlinearity.

*Comparison of the IA Model and the LW Model at the Psychological Level.* Although the mechanisms underlying the dynamic behavior of the BSB and IA models are very similar, the models differ in their "psychological interpretation" of how letter and word stimuli are represented over a given set of nodes (or neurons). In the LW model, the entire activation pattern represents the incoming visual stimulus. In the IA model, the incoming visual stimulus only activates the letter nodes within the system state vector. When a word is presented under normal operating conditions in the LW model, a completely specified state vector is presented to the system that is then merely amplified. In the IA model, the initial state vector will only have the letter nodes activated. Consequently, when a word is presented under normal operating conditions in the IA model, only a partially specified state vector is presented to the system and the model must "reconstruct" the word node regions of the state vector. The incoming visual stimulus in the LW model, on the other hand, partially activates *all* of the nodes in the system.

It should be noted at this point that the hierarchical vector coding scheme employed in the IA model is powerful enough to permit the reconstruction of completely missing letters in words. On the other hand, the use of only pair-wise letter feature correlations to represent word stimuli in the LW model coupled with the weak between-letter feature correlation contribution, tends to make the LW model very poor at this task.

Two specific points must be made regarding this computational limitation of the LW model. First, this problem arises because of the non-hierarchical coding scheme in the LW model and is not an intrinsic problem with the BSB neural network mechanism. Furthermore, it is presently unclear to what extent people, on the basis of visual information alone, can accurately reconstruct an entire letter stimulus. In fact, the conspicuous absence of sequential redundancy effects in the experimental literature (Johnston, 1978; McClelland & Johnston, 1977) suggests that this computational limitation of the LW model in visual letter-in-word perception may also be shared by people.

The second major difference between the LW and the IA models is that feedback from the orthographic information level to the letter feature level is not permitted in the IA model, but is essential to the LW model. Massaro (1979) has attempted to test experimentally this critical assumption. In his experiment, Massaro was unable to find any direct evidence for an interaction between orthography and perceptual discrimination accuracy. Massaro interpreted his results as evidence for discounting models like the LW model that allow feedback from the orthographic information levels to the letter feature levels. There are some problems, however, with this interpretation. The LW model, for example, assumes a priori the existence of an abstract position-specific letter feature coding. If perceptual discrimination accuracy effects occur at a level of processing prior to the construction of this code, then the lack of an interaction between perceptual discrimination accuracy and orthography is not damaging to the validity of the LW model. Moreover, Massaro did find a highly consistent interaction between orthographic and letter feature information when the dependent variable was the subject's interpretation of the stimulus. Massaro interpreted this latter effect as being associated with the response bias stage of processing. A more parsimonious explanation, however, might assume that this latter effect arose from interactions between the letter feature and orthographic levels in the LW model as demonstrated in Experiment 6.

The third difference between the LW and the IA model is that the IA model assumes (like Adams's model) that the individual letter units are case independent. Models that assume a case independent coding cannot explain why a font type by orthography interaction has been observed (McClelland, 1976; Pollatsek et al., 1975; Taylor et al., 1977). Instead, such models are predicated upon the assumption that the case type by orthography interaction in these studies were obtained because the words, pseudowords, and nonwords stimuli were not tested at comparable levels of performance (Adams, 1979).

It is worthwhile at this point to mention a testable prediction of the LW model that is related to the assumption of case independence. In the LW model, alternating case effects are obtained from exactly the same type of mechanism as orthographic effects. Thus, the development of the effects of alternating case should parallel the development of the ability to detect orthographic information (see Experiments 1 and 2). As experience with words increases, the LW model predicts that the advantage of letters within same-case stimuli, relative to mixed-case stimuli, should increase at a fast rate in the initial stages of development and more slowly in the later stages. A theory suggesting that the effects of alternating case are located at the "level of single letter discriminability" (Adams, 1979, p. 154; also see McClelland and Rumelhart, 1981), would not necessarily be discredited if this developmental effect were not observed.

The final difference between the LW and the IA models, however, is that the LW model is a learning model. The IA model is initialized by providing the system with a set of letter and word nodes and a set of connection weights. The LW model is initialized by providing the system with a set of independent position-specific letter feature nodes, a learning rule, and an environment. Parameter values in the LW model are therefore a function of how the model is educated. For example, in the IA model, the magnitude of the word superiority effect could be indirectly increased by adjusting the *word-to-letter level* inhibition parameter in the IA model. Unfortunately, however, such a manipulation is impossible in human subjects. In the LW model, however, the magnitude of the word superiority effect can be manipulated indirectly by varying the ratio of the number of letter learning trials to the number of word learning trials. This latter manipulation is possible, at least in principle, from a psychological perspective.

## SUMMARY

A developmental neural model of visual letter-within-word perception has been proposed. The model was derived from some commonly accepted principles of information processing within the nervous system, yet the evaluation of the model was based primarily upon the system's ability to explain a variety of psychological phenomena.

In this particular model, a word is represented as a pattern of neural activity over a set of position-specific feature neurons. This pattern of neural activity is then amplified, using positive feedback, until all neurons within the neural network have saturated. Spatial redundancy information plays a critical role in the amplification process. The contribution of transgraphemic information is smaller in magnitude, but still exerts a detectable influence. The model learns a particular neural activation pattern representing a word by modifying a set of pair-wise correlations that indicate how the set of position-specific letter feature neurons are allowed to interact.

The model was then evaluated with respect to the experimental literature. The LW model provides a rationale for the importance of spatial redundancy and transgraphemic feature effects in visual word recognition. The LW model also explains why sequential redundancy effects have not been particularly easy to find, and how the ability to detect orthographic information develops with experience. In addition, the model makes explicit accuracy and reaction time predictions about a variety of experimental phenomena.

Both the LW and the IA models offer a unique perspective upon the nature of the processes involved in the visual perception of letters in the context of words. The detailed application of the IA model to a wide range of experimental phenomena seems to suggest that the basic dynamic principles of the IA model may be fundamentally correct. The LW model, using similar dynamic principles, suggests how experience with word stimuli can affect performance in a variety of visual word perception tasks.

## APPENDIX 1

### VECTOR ENCODINGS OF LETTER STIMULI

The following table indicates the assignment of specific letter subvectors to letters. The assignment of a vector coding to a letter was based upon an extension of Gibson's (1969, p. 88) abstract letter feature set. The absence or presence of a letter feature is indicated by a two-dimensional subvector that may take on either the values $(-1, -1)$ or $(1,1)$, respectively. An additional pair of letter features were introduced at the end of each letter subvector to indicate the relative size of that letter. For example, the first four elements of the letter subvector representing A are given by $(+1, +1, -1, -1)$ indicating the presence of a horizontal but the absence of a vertical line segment. These four elements are concisely represented in hexadecimal notation with the number C by treating negative vector elements as zeros and positive vector elements as ones. The letter subvector encodings using hexadecimal notation are provided below. The symbol * indicates a set of four vector elements that are all equal to zero (see Experiment 6).

| E | F003F3F | e | C0CF030 |
|---|---------|---|---------|
| T | F00333F | t | F03F0FF |
| A | CF033CF | a | 00C3030 |
| O | 00C030F | o | 00C0300 |
| N | 33000CF | n | 30300C0 |
| R | 33C30CF | r | 303F0F0 |
| I | 30003CF | i | 30003F0 |
| S | 000CC0F | s | 000CC00 |
| H | F0033CF | h | 30300CF |
| X | 0F0330F | E/T | F003*3F |

## APPENDIX 2

### LETTER STRING STIMULI SELECTION

Word, pseudoword, and nonword letter strings were used as stimulus vectors in the following experiments. The word stimuli were selected based upon moderate frequency of occurrence in the English language, and were constructed using only the nine most frequent letters in the English alphabet. The word stimuli were then scrambled, and the scrambled letter strings ranked using a spatial redundancy table obtained by analyzing the original set of word stimuli (see Mason, 1975, for additional details).

**Word Stimuli** (ordered row-wise by decreasing frequency):
THAT THIS INTO THAN THEN HERE AREA SEEN RATE SOON NEAR EAST
SORT REST HEAR HAIR SENT NOTE TEST ONES SHOT NONE RISE HEAT
THIN ROSE NINE TONE RAIN ARTS SITE SETS NOSE ONTO TREE SEAT
HERO REAR ASIA HANS IRON ANNE EASE HATE RARE EARS OHIO HOST
SEES HORN ROOT SONS TONS NOON STAR TORN HITS TIRE NEAT RENT
NEST TENT TOES THEE EARN HERS SINS HIRE TIES TORE HATS NEON
SHOE ROAR TROT ROSS TEAR SEAS SORE HINT HOOT HOSE IONS THOR
TOSS TRIO SANE ANNA ANTS HEIR OATS RENO RIOT STIR TART OATH
SITS TEEN

**Pseudoword Stimuli** (ordered row-wise by decreasing spatial redundancy):
TENE TERE TETS TORS SENE TOSE TEOS TEIS SESE SONT SEST TETN
RETS TASE TENR SOST THNE TONR TEAN TESN TOSN SOET SARE TEHE
SETN TOEN NETS TISE TESR REAT ROES TOOR OENS AETS HONR SOSN
TOER THOS NORT SOTR NEES NOOS SAES ROIT NOES RETN HAST TERA
SOHE RAAE NOOT SONO HESR EONT SOSR EORS NEET TNOE SAET TROE
AEES NOET HEOR TATR REAN HAET TNOS AOST HEER SOER REON SIST
NARE EORT SOHT TRIE TRAT RAST HIST HOTO TEAH TEHN SIET SETA
TOSH TNET SNAE HTNE RAET HERA HETA TOOH TAHT TAER SROE
HASN SESA SERH

**Nonword Stimuli** (ordered row-wise by decreasing spatial redundancy):
IRES AHSN OHNR TSRA OISN OTSS EASR SRTA SNTA OTAS ESSE ITSE
IHTN INOT TROI OAHT SNEO ERET EOSH ORIT ESAE NSET HTRO HREO
HTNA HSTO TSRI RAIH ESST ERTN STRA ANAN EISR ESAT HTTA EHSR
ISST HTNI EHOR RTTO OHTO ANEN ERAN RNEO EHER ONEN ISET HSEO
ERON STRI EATH ONNO OTNR ETTN IHER ESTN INEN ARER OTSN ENNO
ATTR AISA HTSI EHRA EHTA ORTO OSSN IRNO OSTR AHRI OTEN ASEN
EIRH ISSN IRTO ENNA OTNO OSNO ARNI ORRA EHRI ENRA ERRA ENTA
OTER ITNO OSER RTOI OHOI NROI ITER ONTI ETRA ESTA OIHO ESSA
OTAH ASAI

# APPENDIX 3
## AN ANALYSIS OF THE LEARNING RULE

The learning algorithm described in the text is rewritten here for convenience:

$$\mathbf{A}_{new} = \mathbf{A}_{old} + \gamma[(\mathbf{C}_i - \mathbf{Z}_i)(\mathbf{C}_i - \mathbf{Z}_i)^T] \quad (1)$$

where $\mathbf{A}_{new}$ is the updated matrix, $\mathbf{A}_{old}$ is the original matrix, $\gamma$ is the learning constant, $\mathbf{Z}_i$ is the system state vector after iterating K times through the

BSB model, and $C_i$ is both a training vector and hypercube corner. The square brackets indicate that [**B**] is equal to the **B** matrix with the exception that the on-diagonal elements of [**B**] are all identically equal to zero.

The conditions under which (1) maximizes the following cost function,

$$J(\mathbf{A}) = \sum_j \mathbf{C}_j^T \mathbf{A} \mathbf{C}_j \tag{2}$$

where $\mathbf{C}_j$ is the jth hypercube corner that is to be taught to the system, are now considered.

## Analysis

Pre-multiply and post-multiply Equation 1 by the quantity $\mathbf{C}_j$, to obtain:

$$\mathbf{C}_j^T \mathbf{A}_{new} \mathbf{C}_j = \\ \mathbf{C}_j^T \mathbf{A}_{old} \mathbf{C}_j + \gamma \mathbf{C}_j^T [(\mathbf{C}_i - \mathbf{Z}_i)(\mathbf{C}_i - \mathbf{Z}_i)^T] \mathbf{C}_j \tag{3}$$

Now both sides of Equation 3 are summed over all members of the training set yielding:

$$\Delta = J(\mathbf{A}_{new}) - J(\mathbf{A}_{old}) = \\ \gamma \sum_j \mathbf{C}_j^T [(\mathbf{C}_i - \mathbf{Z}_i)(\mathbf{C}_i - \mathbf{Z}_i)^T] \mathbf{C}_j \tag{4}$$

If the on-diagonal elements are not removed during learning, then the summation in Equation 4 is over a set of non-negative terms. Thus, in this case, $J(\mathbf{A})$ must increase in value after each learning trial.

To study the behavior of the learning algorithm when the on-diagonal elements are set to zero, Equation 4 is rewritten as follows:

$$\Delta = \gamma (\mathbf{C}_i - \mathbf{Z}_i)^T [\mathbf{Q}](\mathbf{C}_i - \mathbf{Z}_i) \tag{5}$$

$$\mathbf{Q} = \sum_j \mathbf{C}_j \mathbf{C}_j^T. \tag{6}$$

Note that the values of the right sides of Equations 4 and 5 are equivalent despite the fact that in Equation 5 we set the on-diagonal elements of the **Q** matrix equal to zero.

Now let the magnitude of $\mathbf{Z}_i$ be equal to Z. Let $\Phi$ be the angle $\mathbf{Z}_i$ makes with $\mathbf{C}_i$. Let **c** be equal to the unit vector that is parallel to $\mathbf{C}_i$. Then we can write:

$$\mathbf{Z}_i = Z(\cos \Phi) \mathbf{c} + Z(\sin \Phi) \mathbf{r} \tag{7}$$

where **r** is a unit vector orthogonal to **c**.

Let C be the magnitude of $\mathbf{C}_i$ and let $b = Z/C$. Now put Equation 7 into Equation 5 to obtain:

$$\Delta = \gamma C^2 (1 - (b(\cos \Phi))^2 \mathbf{c}^T [\mathbf{Q}] \mathbf{c} \\ - 2\gamma C^2 (b(\sin \Phi)(1 - b(\cos \Phi)) \mathbf{r}^T [\mathbf{Q}] \mathbf{c} \\ + \gamma C^2 b^2 (\sin \Phi)^2 \mathbf{r}^T [\mathbf{Q}] \mathbf{r}$$

Let $\lambda_{min}$ be the absolute value of the smallest eigenvalue of $[Q]$. Let q be the magnitude of $[Q]c$. We then have:

$$\Delta \geq \gamma C^2(1 - b(\cos \Phi))^2 \, c^T[Q]c \\ - \gamma C^2 bq(2|\sin \Phi| - b|\sin 2\Phi|) \\ - \gamma C^2 b^2 (\sin \Phi)^2 \lambda_{min} \quad (8)$$

A new variable called $\tau$ is now formed by dividing the positive term in Equation 8 by the sum of the absolute values of the three negative terms.

$$\tau = \frac{(1 - b(\cos \Phi))^2 \, c^T[Q]c}{b^2(q|\sin 2\Phi| + \lambda_{min}(\sin \Phi)^2) + 2bq|\sin \Phi|}$$

Whenever $\tau > 1$, $J(A)$ must increase after the matrix is updated. Also note that if $\Phi = 0$, then Equation 1 reduces to a gradient descent algorithm that minimizes $J(A)$. To see this, note that the gradient of Equation 2 with respect to $A$ is simply $\Sigma \, C_i C_i^T$. This suggests that the learning algorithm may be interpreted as a suboptimal gradient descent algorithm. Also note that the magnitude of a weight change is directly proportional to the square of the distance between $Z_i$ and $C_i$. This latter observation provides some insight into the origins of the error-correction properties of the algorithm.

Now since $c^T[Q]c/q \cong 1$ and since $q >> \lambda_{min}$, an approximate expression for $\tau$ may be found that is *independent* of the properties of the stimulus set. Using both of these approximations, the following expression is obtained:

$$\tau = \frac{(1 - b(\cos \Phi))^2}{b^2|\sin 2\Phi| + 2b|\sin \Phi|} \quad (9)$$

The function in Equation 9 has been carefully studied. For $|\Phi| > 135$ degrees, $\tau$ is always greater than unity. In addition, for $b < 0.3$, $\tau$ is always greater than unity. For $|\phi| \leq 135$ degrees and $b > 0.3$, the behavior of $\tau$ is more complex since the error-connection properties of the algorithm interact strongly with the gradient descent properties. For example, with $b = 0.4$, the $Z_i$ vector may deviate 25 degrees from $C_i$ and $J(A)$ is guaranteed to increase. On the other hand, for $b = 0.7$, the allowable deviation is less than 3 degrees. For the simulations reported here, $\Phi$ ranges from about 25 degrees in the initial stages of letter learning to about 5 degrees in the final stages of word learning.

## REFERENCES

Ackley, D.H., Hinton, G.E., & Sejnowski, T.J. (1985). A learning algorithm for Boltzmann machines. *Cognitive Science, 9,* 147-169.

Adams, M.J. (1979). Models of word recognition. *Cognitive Psychology, 11,* 133-176.

Anderson, J.A. (1983). Cognitive and psychological computation with neural models. *IEEE transactions on systems, man, and cybernetics, 5,* 799-815.

Anderson, J.A., & Mozer, M.C. (1981). Categorization and selective neurons. In G. Hinton & J.A. Anderson (Eds.), *Parallel models of associative memory* (pp. 213–236). Hillsdale, NJ: Erlbaum.

Anderson, J.A., & Silverstein, J.W. (1978). Reply to Grossberg. *Psychological Review, 85,* 597–603.

Anderson, J.A., Silverstein, J.W., Ritz, S.A., & Jones, R.S. (1977). Distinctive features, categorical perception, and probability learning: Some applications of a neural model. *Psychological Review, 84,* 413–451.

Baron, J., & Thurston, I. (1973). An analysis of the word-superiority effect. *Cognitive Psychology, 4,* 207–228.

Cowey, A. (1981). Why are there so many visual areas? In F.O. Schmitt, F.G. Worden, G. Adelman, & S.G. Dennis (Eds.), *The organization of the cerebral cortex* (pp. 395–413). Cambridge, MA: MIT press.

Durr, W.K., & Pikulski, J.J. (1981). *Reading research and the Houghton Mifflin reading program.* Boston, MA: Houghton Mifflin.

Estes, W.K. (1975). The locus of inferential and perceptual processes in letter identification. *Journal of Experimental Psychology: General, 104,* 122–145.

Feldman, J.A., & Ballard, D.H. (1982). Connectionist models and their properties. *Cognitive Science, 6,* 205–254.

Gibson, E.J. (1969). *Principles of perceptual learning and development.* New York: Meredith Corporation.

Golden, R.M. (1986). The "Brain-State-in-a-Box" neural model is a gradient descent algorithm. *Journal of Mathematical Psychology, 30,* 73–80.

Henderson, L. (1982). *Orthography and word recognition in reading.* New York: Academic.

Johnston, J.C. (1978). A test of the sophisticated guessing theory of word perception. *Cognitive Psychology, 10,* 123–153.

Johnston, J.C., & McClelland, J.L. (1973). Visual factors in word perception. *Perception & Psychophysics, 14,* 365–370.

Juola, J.F., Schadler, M., Chabot, R.J., & McCaughey, M.W. (1978). The development of visual information processing skills related to reading. *Journal of Experimental Child Psychology, 25,* 459–476.

Kawamoto, A. (1985). *Dynamic processes in the (re)solution of lexical ambiguity.* Unpublished doctoral dissertation, Brown University, Providence.

Kohonen, T. (1984). *Self-organization and associative memory.* Berlin, Germany: Springer-Verlag.

Lefton, L.A., & Spragins, A.B. (1974). Orthographic structure and reading experience affect the transfer from iconic to short-term memory. *Journal of Experimental Psychology, 103,* 775–781.

Lefton, L.A., Spragins, A.B., & Byrnes, J. (1973). English orthography: Relation to reading experience. *Bulletin of the Psychonomic Society, 2,* 281–282.

Mason, M. (1975). Reading ability and letter search time: Effects of orthographic structure defined by single-letter positional frequency. *Journal of Experimental Psychology: General, 104,* 146–166.

Massaro, D.W. (1979). Letter information and orthographic context in word perception. *Journal of Experimental Psychology: Human Perception and Performance, 5,* 595–609.

Massaro, D.W., Venezky, R.L., & Taylor, G.A. (1979). Orthographic regularity, positional frequency, and visual processing of letter strings. *Journal of Experimental Psychology: General, 108,* 107–124.

McClelland, J.L. (1976). Preliminary letter identification in the perception of words and nonwords. *Journal of Experimental Psychology: Human Perception and Performance, 2,* 80–91.

McClelland, J.L. (1985). Putting knowledge in its place: A scheme for programming parallel processing structures on the fly. *Cognitive Science, 9,* 113-146.

McClelland, J.L., & Johnston, J.C. (1977). The role of familiar units in perception of words and nonwords. *Perception & Psychophysics, 22,* 249-261.

McClelland, J.L., & Rumelhart, D.E. (1981). An interactive activation model of context effects in letter perception: Part 1. An account of basic findings. *Psychological Review, 88,* 375-497.

Paap, K.R., Newsome, S.L., McDonald, J.E., & Schvaneveldt, R.W. (1982). An activation-verification model for letter and word recognition: The word superiority effect. *Psychological Review, 5,* 573-594.

Pollatsek, A., Well, A.D., & Schindler, R.M. (1975). Familiarity affects visual processing of words. *Journal of Experimental Psychology: Human Perception and Performance, 1,* 328-338.

Reicher, G.M. (1969). Perceptual recognition as a function of meaningfulness of stimulus material. *Journal of Experimental Psychology, 81,* 275-280.

Rosinski, R.R., & Wheeler, K.E. (1972). Children's use of orthographic structure in word discrimination. *Psychonomic Science, 26,* 97-98.

Rumelhart, D.E., & McClelland, J.L. (1982). An interactive activation model of context effects in letter perception: Part 2. The contextual enhancement effect and some tests and extensions of the model. *Psychological Review, 89,* 60-94.

Rumelhart, D.E., & Zipser, D. (1985). Feature discovery by competitive learning. *Cognitive Science, 9,* 75-112.

Smolensky, P. (1983). Harmony theory: A mathematical framework for stochastic parallel processing. *Proceedings of the National Conference on Artificial Intelligence,* Washington, DC.

Taylor, G.A., Miller, T.J., & Juola, J.F. (1977). Isolating visual units in the perception of words and nonwords. *Perception and Psychophysics, 21,* 377-386.

Wheeler, D.D. (1970). Processes in word recognition. *Cognitive Psychology, 1,* 59-85.

CHAPTER 6

# Computing with Connections In Visual Recognition of Origami Objects*

DANIEL SABBAH

*IBM T. J. Watson Research Center*

This paper summarizes our initial foray in tackling Artificial Intelligence problems using a connectionist approach. The particular task chosen was the visual recognition of objects in the Origami world as defined by Kanade (1978). The two major questions answered were how to construct a connectionist network to represent and recognize projected line drawings of Origami objects and what advantages such an approach would have. The structure of the resulting connectionist network can be described as a hierarchy of parameter or feature spaces with each node in each of the feature spaces representing a hypothesis about the possible existence of a specific geometric feature of an Origami object. The dynamic behavior of the network is a form of iterative refinement or relaxation whose major characteristic is to prefer more globally interesting interpretations of the input over locally pleasing ones. Examples from the implementation illustrate the system's ability to deal with forms of noise, occlusion and missing information. Other benefits are an inherently parallel approach to vision, limitation of explicit ordering of the search involved in matching model to instance and the elimination of backtracking due to the sharing of partial results as the search progresses. Extensions and problems are also discussed.

## INTRODUCTION

The strongest case we can make for connectionist models of computation is to demonstrate a system that solves a difficult but tractable problem in Artificial Intelligence. Our initial attempt, summarized in this paper, was to construct a connectionist system to recognize geometric properties of objects

* Special thanks go to Jerry Feldman for his never ending patience, advice and wisdom. Many thanks to Dana Ballard for a constant and lively sharing of information throughout the development of these ideas. Thanks to Sanjaya Addanki for helping make this paper somewhat readable. Finally, thanks to the people of Rochester and IBM for various forms of support, financial and otherwise.

in a well-defined visual domain, Origami world (Kanade, 1978). We chose the world of Origami objects for several reasons: First, because it has well-defined geometric properties, second, because it defines a richer set of objects than a typical Blocks world; witness the Origami duck. Finally, because it allows for quantitative shape recovery necessary to construct and match rotationally invariant representations of Origami objects.

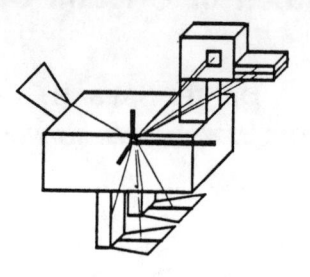

Figure 1. Origami Duck

Any vision system that performs a "recognition" function, connectionist or otherwise, must have internal models of the world to be recognized and must control the matching process between input (scene being analyzed) and these internal models. Any interesting recognition system must degrade gracefully in the presence of noisy, incomplete or partially occluded input. So, the problem comes down to posing and answering three key questions:

- How does one represent Origami objects in a connectionist system?
- How does one control the recognition process to take full advantage of the potential for massive parallelism available in the connectionist approach?
- Furthermore, how does one make this control process robust enough to deal with imperfect and occluded input in predictable and "pleasing" ways?

To answer these questions, we looked to research from two different sources; one was interdisciplinary research in animal vision and the other was, not surprisingly, computer vision. We looked to animal vision (Hubel & Wiesel, 1979; Marr, 1978) since animals are the only existing proof we have that such problems can not only be solved but can be solved exceedingly quickly (relative to current computers or programs). Our findings supplied the original motivation for the principles underlying the entire connectionist approach (Feldman & Ballard, 1981) and also helped in making several design choices within our system:

- Small units that have only one unique output per computation. The complexity of the output is limited although the fan-out is very large.

Furthermore, this output is essentially a measure of current excitation or activation and is based on input from other units.
- Networks of such units that are extensively interconnected and use patterns of connections rather than explicit transfer of symbolic information to compute.
- Units in the network that are arranged in a hierarchy that respond to increasingly more complex stimuli.
- Spaces of units that respond to regular variation in feature stimulus parameters.
- An ability to reverse or change level of activation for individual units through lateral inhibition and decay.

These principles are obviously a gross oversimplification of animal vision. The point, however, is that they act solely as guidance in hypothesizing models of computation for performing equivalent function with equal speed and accuracy.

**Hough Transforms and Iterative Refinement**

At the same time, we were examining the Hough transform (Ballard, 1981a, 1981b; Hough, 1962) as a parallel model of computing geometric parameters in computer vision. The idea is that object shape can be described in terms of geometric parameters, hence an object in an image can be represented as a point in the $n$-dimensional space defined by the $n$ describing parameters (the parameter space). For example, line objects in an image can be completely described by two parameters, $(\rho,\theta)$ where $\theta$ means slope or the angle the line makes with the origin and $\rho$ is a measure of the distance from the origin. A specific point in $(\rho,\theta)$ space corresponds to a line in image space. An approach to recognition that takes advantage of this representation consists of first transforming the input image to a parameter space and then looking for the appropriate point in that space. The Hough transform is one such method and consists of four steps:

1. Uniform decomposition of the input image into small regions; in our simple line example, regions might be formed by decomposing the input image into nonoverlapping squares, each consisting of $3 \times 3$ pixels.
2. Individually examining these small regions for evidence of the describing geometric parameters; extracting $(\rho,\theta)$ for every edge found in every square region.
3. Integrating the local evidence to form the parameter space transform of input image; essentially, every edge can vote for its specific $(\rho,\theta)$ in the line parameter space. A point in the space receiving

many votes accumulates (integrates) those votes, increasing its local density.
4. Scan the resulting transform for points of highest density; $(\rho,\theta)$ points having accumulated the largest number of votes. Note that if the input image contains a perfect line then the transform would consist of one point with as many votes as edges in the image.

Recognizing objects more complex than simple lines depends on the observation that such complex objects can be seen as being composed of simpler component features that in turn can be described by yet simpler features, and so forth. Since component features are in turn described parametrically, we can construct a hierarchy of parameter spaces paralleling exactly the composition of complex objects to be recognized. The transform of one feature space then becomes the input image to the next parameter space. For example, cubes might consist of faces which might consist of sides of faces (lines) which might consist of our old friends, edges. Such a hierarchy of parameter or feature spaces is termed a *conceptual hierarchy*.

Unfortunately, this strictly bottom-up approach works well only on perfect images. In order to deal with imperfections such as noise arising from poor transducers, missing information, and/or occlusion, we must introduce information from top-down or feedback and competing sources. In order to properly control the integration of information from all sources it is necessary to introduce and further extend another idea in computer vision, iterative refinement or relaxation (Hinton, 1977; Hummel & Zucker, 1980; Rosenfeld, 1976; Waltz, 1975).

The rest of the paper discusses how we meld these principles into a connectionist system for recognizing Origami objects, the behavioral characteristics of such a system and what we learned and continue to learn from this approach.

## REPRESENTING ORIGAMI OBJECTS

A short digression is necessary to define Origami objects and their geometry before we can fully motivate the composition of the conceptual hierarchy. Figure 2 illustrates some of the simpler Origami objects used in this paper. Figure 3 shows some of the more complex objects that can be defined.

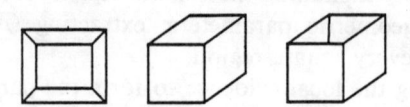

Figure 2. Simple Origami Objects

W-folded paper object (Kanade)

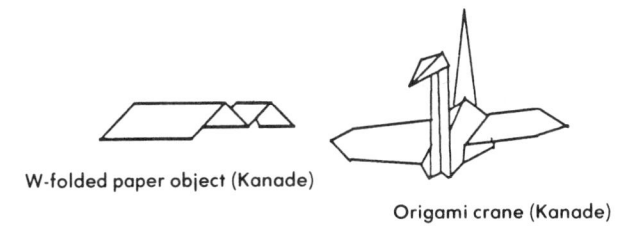

Origami crane (Kanade)

Figure 3. More Complex Origami Objects

We use the same basic model of understanding line drawings as in (Kanade, 1978). Objects in the line drawing are assumed to be composed of noninterpenetrating bounded planar surfaces in 3-d space. Planar implies a surface of constant orientation. A scene is defined to be a line drawing resulting from the orthographic projection of a set of bounded planar surfaces onto a 2-d picture plane. Regions are considered "filled in" as opposed to wire frames and hidden lines are eliminated from the input line drawing. As does Kanade, I assume trihedral objects and all the implied legal joint types. Unlike Kanade, imperfect input and general occlusion is allowed. A more precise statement of the task to be performed by the connectionist system is to efficiently represent and quickly recognize predefined Origami objects from imperfect line drawings that are projected onto the picture plane.

*Quantitative Shape Recovery.* In order to accomplish this task with some level of generality, it is necessary to describe these predefined objects in a transformationally invariant manner, hence, the need for shape recovery. Kanade (1979) reduces this problem to one of extracting skewed symmetry parameters. Skewed symmetry manifests itself in this case as a skewing of the rectangular faces of the object resulting from orthographic projection onto the image plane where the degree of skew constrains possible orientations of the face relative to the viewer. For our purposes, the number of possible orientations reduces to two, one for convex and one for concave interpretation. The result is a representation invariant to 2-d rotation and translation. A more detailed discussion of this heuristic can be found (see Kanade, 1979; Kender, 1980; Sabbah, 1982).

*Geometry of the Origami Face.* The skewed face object, a parallelogram, is described by two noncollinear vectors emanating from an origin forming a basis for a 2-d space as shown in Figure 4.

The angle formed with the horizon, (a) represents the skewed symmetry transverse axis as well as the rotation parameter. The larger angle, (b), represents the axis of symmetry. The origin of the basis is considered the centroid and is the central point of reference for this face. No discrimination is made between concave and convex interpretation at this point.

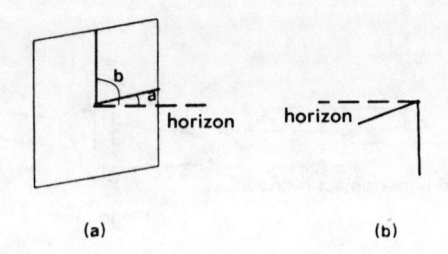

Figure 4. The Representation for a Skewed Face

***Geometry of the Origami Object.*** The representation of a given object is limited to projections onto the picture plane from a single fixed, general viewpoint. It consists of a list of vectors called object vectors that originate at a centrally chosen point called an object centroid providing an object centered frame of reference that rotates and translates with the object in 2-d. Extending the representation to include explicit view parameters would make it invariant to 3-d transformations, however, we chose to keep things relatively simple at this point.

The object vectors describe the locations for the centroids of the visible and partially visible faces composing the object. Face vectors originating at the face centroids fully describe each face according to its projected 2-d representation. Also implicit in the representation is the existence and proper location for the various joints indicating where faces meet. An example is the representation of the planar chair in Figure 5, and its corresponding representation in the subsequent figure.

Figure 5. Origami Planar Chair

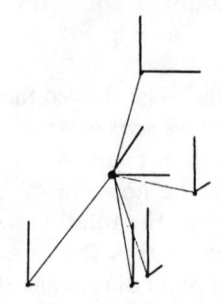

Figure 6. Corresponding Shape Representation

All vectors in the Figure 6 are defined relative to the centroids. All face centroids and all object vectors are defined relative to the object centroid. The convex and concave interpretations for each face are explicitly chosen as part of the object representation.

The representation allows for complete shape reconstruction of the projected line drawing for all visible and partially visible faces composing the object and therefore, the object as perceived. Joint features are also included in the reconstruction to avoid considering these two objects as equivalent. Skewed symmetry is by no means the only heuristic for extracting relative orientation. A more general approach to shapes based on the primitive ideas presented here can be found in (Ballard & Sabbah, 1981).

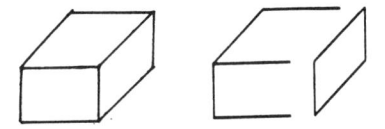

Figure 7. Equivalent Faces — = Same Object

## NETWORK DEFINITION AND CONSTRUCTION

We now describe the parameter spaces that form the conceptual hierarchy. The features described by each space are designed, not surprisingly, to progressively constrain the recognition of objects. They are all fairly straightforward and directly reflect the geometry described in the previous section. Figure 8 names the features used and how they relate to each other in the hierarchy. The table following illustrates the geometric parameters used to represent some of the more complex features. A complete description can be found in (Sabbah, 1982).

Levels and Hierachy in the Origami World

| 3d Origami Objects | | abstraction level increases |
|---|---|---|
| Complex Joints | 2d shapes | |
| Ljoints and Tjoints | | |
| Rays | Lines | |
| Edge Segments | | |

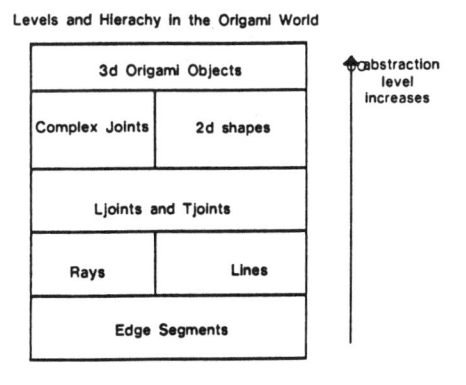

Figure 8. The Origami World Hierarchy

**Skewed Rectangle**

(m,n) — shape centered origin for parallelogram

(a,b) — skewed symmetry parameters measured relative to horizon (0,180) both CO and CX interpretations are included

(dx,dy) — parallelogram shape parameters

**Origami Polyhedra**

(Ox,Oy) — object centered origin (centroid)

$w_b$ — orientation relative to horizon

$(w_1 w_2 w_3)$ — orientation of axes in polyhedra representation

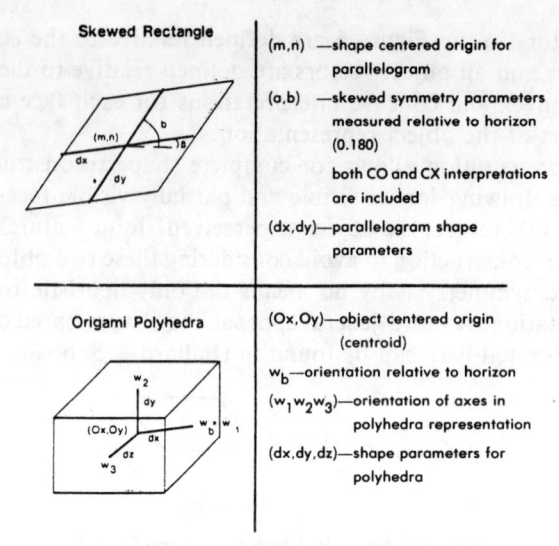

(dx,dy,dz) — shape parameters for polyhedra

Figure 9. Complex Feature Definitions

We briefly describe the constraining role played by each figure in the recognition process.

- Unit length edge segments form the most primitive feature recognized in the conceptual hierarchy. It is assumed that edges are already extracted from the static, digitized image of an Origami world scene. Extracting the first level of input is a difficult problem in itself and is not discussed here. The input image is considered similar to Marr's Primal Sketch (Marr, 1978). No assumption is made about completeness or correctness of the primal sketch input. It may be noised in certain ways or missing arbitrary amounts of information.
- Lines are considered infinite and are composed of edge segments with similar orientations. The edges may or may not be completely connected.
- Rays are semi-infinite half lines and are used primarily to establish the existence of simple joints(ljoints and tjoints). Rays describe a precise location and direction for an ljoint and its component line segments.
- Ljoint features point to the existence of a corner of a face and constrain its skewed symmetry axes.
- Tjoints determine points of possible occlusion.
- Cjoints represent the meeting of two or more partially visible faces. They partially constrain the existence of the more primitive, simple objects such as rectangular solids.
- Skewed Faces have already been discussed. Faces in the correct orientations, along with the complex joints at the proper projected

locations imply the existence of the Origami objects recognized in this experiment.

The conceptual hierarchy is obviously not limited to these features. As more complex assemblages, such as rectangular solids, are incorporated they can form primitives for new levels of the hierarchy (say to form the Origami Duck).

**Forming The Connectionist Network**

In order to build a full blown connectionist network from our current description of the conceptual hierarchy, it is only necessary to observe that each set of parameter values describing an object or visual feature of the Origami world can be associated with a small, simple unit. This unit computes a measure of confidence in the existence of that feature in the input scene (i.e., accumulates votes). For example, if we define an ljoint by the parameters: *[(x,y), w,v]* where *(x,y)* determines ljoint location in 2-d space; w determines orientation of one of the sides; v the size of the angle opening and if we further specify those parameters to range over discrete location values on a (10,10) grid with (36) possible orientations for the *(w,v)* parms, then we can describe a specific ljoint—a point in the parameter space: [(5,5), 30,60], form a conjecture or 'hypothesize' about its possible existence in the input and measure current confidence in this hypothesis by assigning a real number to represent it. Each such unit/point in parameter space is a node of the connection network.

The edges of the network are formed by applying a transform to units in one parameter space to map them into another (e.g., Hough transform). A unit in one space mapping onto a unit in another space is said to be 'connected' to that unit. Votes as current confidence levels are transmitted along these edges or connections. Examples follow.

**Connection Pattern Examples**

The first example reflects the standard bottom-up Hough transform.

*Bottom-up Connections.* The existence of an ljoint (or corner of a face) in the input implies the possible existence of a number of skewed faces as illustrated in the figure below.
To actually generate the set of connected units, we use a simple trigonometric mapping. The parameter values of the ljoint fix the *(a,b)* parameters of the skewed face. *(dx,dy)*, on the other hand, are allowed to roam free and for every instantiated combination of *(dx,dy)*, the corresponding values for *(m,n)* are computed. The process is graphically illustrated below.

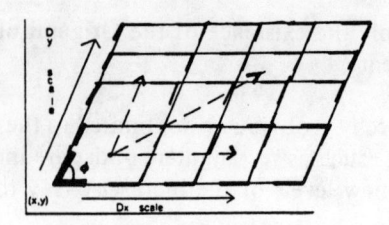

Figure 10. Composition of Ljoint into Possible Faces

Even though this mapping is one to many (i.e., somewhat underconstrained), it is still much smaller than if one was to attempt to infer possible skewed faces from, say, edge information. Hence, the benefits of a layered, hierarchical approach.

*Top-down Connections.* As we discussed in the introduction, bottom-up connections are insufficient if we assume imperfect input. The first of the two additional mappings necessary reflects a more complex object feeding back on its component features. For example, we can take a parameterized face and decompose it into its four constituent ljoints (corners).

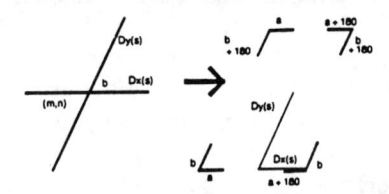

Figure 11. Decomposition of Skewed Face into Ljoints

These four ljoints are now in a position to receive feedback confidence from the face object.

*Lateral, Inhibitory Connections.* The final set of connections is among mutually antagonistic features in the network. The simultaneous existence of such features violates physical constrains of the domain and symmetric, inhibitory connections enforce these constraints by insuring that no two contradictory features accrue high confidence in the same presentation of input. For example, observing that two very similar yet distinct corners cannot occupy the same physical space at the same time leads to generation of mutually inhibitory connections between logically neighboring ljoints of slightly different parameters (i.e., + or $-10°$ or 1 unit difference in location). The effect of these connections is to sharpen feature parameters manifested as "fuzzy" or noisy corners of a face.

At higher levels of abstraction, Inhibitive connections have a slightly different task. The same "metarules" guide the generation of inhibitive connections (namely, two features cannot occupy the same space at the same time) but the algorithms do not involve simple enumeration around small parameter neighborhoods. Instead, objects that are very similar (i.e., consist of a majority of the same subfeatures) and whose instantiated parameters imply that they are violating the "same space" metarule are put in mutual inhibition. At this point, it becomes more difficult to generate the inhibitory connections with simple geometric mappings. An example is the box versus rectangular solid. The only difference in line drawing interpretation is the addition of a line *(AB)* in the box drawing as seen in Figure 12.

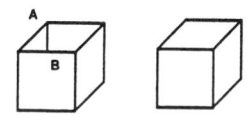

Figure 12. Box vs. Cube

Some form of spatial and temporal reasoning becomes necessary at these higher levels to dynamically generate inhibitory behavior. Such reasoning behavior is not currently implemented although primitive ideas for spatial reasoning exist. The only topic left to discuss before showing actual recognition behavior is how units/parameter values vote and accrue confidence.

## EXTRACTING GLOBALLY CONSISTENT INTERPRETATION

Pushing the Hough analogy, we can say that, with perfect input, the network should behave like an efficient lookup table-mapping the input image through a series of transforms, picking out those values/units that reflect proper geometric interpretations at all levels. Simple voting suffices for this case, however, with imperfect input, the dynamics of integrating competing and feedback information render simple voting inadequate.

We approached this problem by empirically testing a straightforward evidence integration formula and modifying or extending it only when problems arose. The first attempt consisted of summing $(+, -)$ normalized information on current confidence level from all sources:

$$al_{n+1} = al_n + (E_{n+1}) - (I_{n+1}) \qquad (4.1)$$

Where:

- $al_n$ is a number that reflects current confidence (e.g., density of votes) referred to as *activation level*. This measure of activation is bounded above by an arbitrary maximum, 10 in this case, representing that point at which enough supporting evidence is thought to exist from supporting sources (e.g., component features and feedback) to conclude that the geometric feature represented by the parametric values associated with the unit exists in the input image. Activation level is also bounded from below by 0 (one can also imagine negative or positive "priming").

Accumulation of evidence is considered partial in any one voting cycle since it takes multiple voting cycles or iterations to filter evidence up the hierarchy—hence the subscripts representing time *(n, n + 1)* and memory.

- *($E_{n+1}$)* represents the sum of all the enhancement inputs seen in this iteration normalized by expected number of positive inputs (e.g., number of component plus feedback features). Enhancement inputs are received from all connected units/parameter values on every iteration.
- *($I_{n+1}$)* represents the largest negative input from mutual inhibitory connections in this iteration. It is not normalized since predicting the number of active inhibitors at any one time is not practical and not necessary.

This formula suffered immediate problems. Any unit in the network receiving any form of partial evidence (there were many due to the underconstrained nature of the bottom-up connections) would eventually saturate over the many iterations in the relaxation process. The first modification made to the formula was to include a threshold for activation. Existence of this threshold implies that a certain percentage, say > 50%, of supporting evidence must exist before any activation is allowed to accumulate. Hence, the first instance of decay in activation:

$$Al_{n+1} = al_n + (E_{n+1}) - (I_{n+1}) - NT_{\Delta t} \qquad (4.2)$$

The threshold, *NT* is subtracted on every iteration or accumulated over time and then subtracted. It is referred to as a noise threshold since it eliminates spurious unit noise in the network. McClelland and Rumelhart (1980) deal with this problem by completely interconnecting their networks such that units representing hypothesis A have inhibitory connections to all units representing hypothesis ≠ A. Complete interconnection to suppress unit noise is not practical in a network this complex. Instead, only very similar units (> 50% similarity in subfeatures) inhibit one another.

The second problem we find is that weak or missing input will throw evidence accumulation out of synch rendering component units incapable of properly driving activation of the upper levels of the hierarchy. In order to expand the window for evidence accumulation, we added an additional normalization factor (> > 1) to be increased as the input image becomes poorer. Addition of a saturated state also proved necessary to insure proper activation of the upper levels of the hierarchy. A saturated unit communicates max activation level regardless of additional input for a fixed period of time proportional to the time required to reach max activation. The time in saturation state is calculated relative to the normalization factor.

At this point the formula becomes:

$$al_{n+1} = al_n + \left(\frac{E_{n+1}}{k_{time}}\right) - \left(\frac{I_{n+1}}{kinh_{time}}\right) - NT_{\Delta t} \quad (4.3)$$

Where the normalization factors, $k_{time}$ and $kinh_{time}$ are parameters of the relaxation and dependent on the "poorness" of the input. The formula is now further modified such that when activation level reaches max, $al_n$ is equal to 10 for a fixed period of time.

Through further experimentation with occluded and noisy input, we found that proper integration of all factors could be arbitrarily delayed due to indecision on the lower levels. On the other hand, the integration window could not be arbitrarily extended since units would never reach saturation and upper levels would never get properly activated to feed back and disambiguate.

A solution to this problem is to add the notion of a reversible decision. The idea is to make good guesses on the lower levels, drive the upper levels with partially incorrect or poor choices to be later reversed by a combination of feedback and inhibition. This reversible decision is brought about by adding another instance of decay in the integration formula with the following characteristics:

- Decay + strong feedback after saturation => continued saturation.
- Decay with no feedback after saturation => slow decay to 0.
- Decay with no feedback and inhibition => fast decay to 0.

The final form of the evidence accumulation formula now becomes:

$$al_{n+1} = al_n + \left(\frac{E_{n+1}}{k_{time}}\right) - \left(\frac{I_{n+1}}{kinh_{time}}\right) - decay_{\Delta t} \quad (4.4)$$

Where decay plays two different roles:

- The first is the noise threshold.
- The second is to exhibit the greater than normal decay seen after saturation.

For the sake of convenience, unit behavior is divided into four states: resting, active, saturated and decay. The only one not already described is resting state where the unit is not activated ($al=0$) and is not interesting to the computation. This dynamic voting results in network behavior that can be described as groups of connected units forming feedback loops that represent globally consistent interpretations of the input. Complete groupings are termed *stable coalitions* (Feldman & Ballard, 1981) and consist of all objects and their component features at all levels. Conflicting interpretations compete for supremacy in "winner-take-all" networks (Feldman & Ballard, 1981) where one interpretation at high confidence eventually suppresses all others; namely, the interpretation that belongs to the largest stable coalition (i.e., the one satisfying the largest number of constraints).

In addition, strictly local decisions are reversed quickly through decay, inhibition, and/or lack of feedback. Saturation for units in highly active stable coalitions is extended for a long period of time dependent on the number of levels of the hierarchy involved. The more global the interpretation, the more levels of the hierarchy involved, the longer the saturation behavior.

Such behavior is highly nonlinear and it is difficult to formally analyze convergence properties. Empirically, though, we can show that the network has the desired characteristics to deal with noisy and occluded input. We illustrate this with several examples from the implementation.

**Illustration of Network Behavior**

Much more detail on the implementation program can be found in (Sabbah, 1982). In short, the program simulates units communicating in the connectionist network in parallel on a standard uniprocessor. Only active units are allocated space and time so only elements corresponding to the actual transforms are instantiated. Even then, the most complicated examples resulted in over $O(10^4)$ active units which swamped the processor.

We choose four examples:

1. Perfect input with a complex object, the chair.
2. The Noisy Face which illustrates the systems ability to quickly reverse and suppress strictly local decisions.
3. The Occluded Rectangular solid where partial information from occlusion cues, feedback and mutual inhibition combine to produce proper perception behavior.
4. The Box versus Cube example. Here, two very similar objects compete with self-occlusion and the resulting partial views being a strong factor.

*The Chair.* The program receives perfect input when an accurate, complete parametric description of all visible features on the lowest level of abstraction, edges, are passed as input with maximum confidence. The input drives units on the lowest level to saturation and the entire system is forced into the proper coalition. No inhibition is necessary beyond the implicit inhibition of unit thresholds. Figure 13 below is a line drawing of the chair presented to the system as input.

Figure 13. Origami Chair Input

Response of the network will be graphically depicted as geometric representations of the features as they are "seen" by the system at several interesting times in the relaxation. The more intense the lines in the line drawing, the greater the confidence. The depictions will only include features from the higher levels of the relaxation otherwise interesting information will be lost in the resulting mess. Originally, results were captured on a color graphic computer and photographed in (Sabbah, 1982).

Figure 14. Origami Chair Behavior (1)

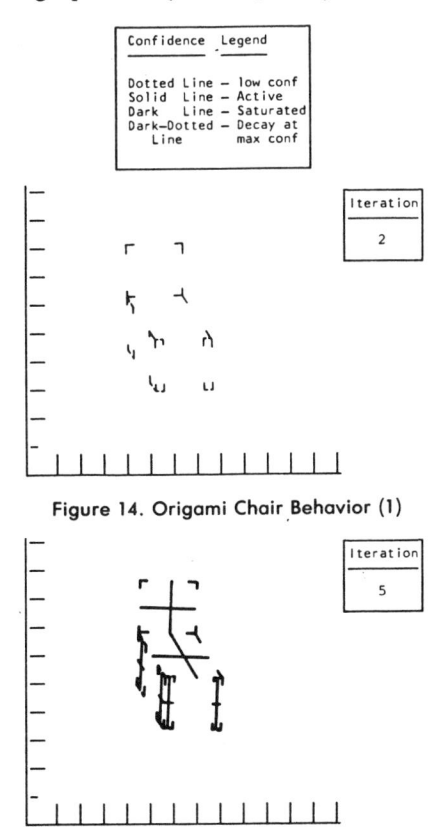

Figure 15. Origami Chair Behavior (2)

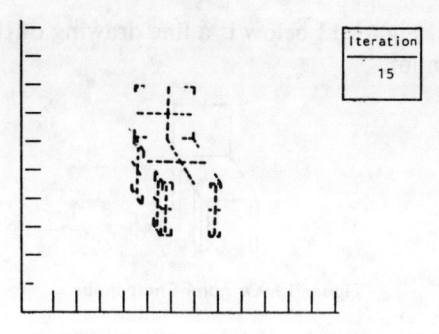

Figure 16. Origami Chair Behavior (3)

When the input is perfect, the method does not require a large amount of processing power and only features present in the input are activated. Essentially, the network performs a hardwired lookup function fairly efficiently.

*The Noisy Face.* The situation in this example is depicted in Figure 17 below. The lower left-hand corner of the face is noised and bottom up information points to spurious corners with higher confidence than the corner falling in line with the rest of the face. Such a situation can occur with fuzzy transfer functions or actual noise in the image.

Figure 17. Face with Noised Corner

The network reacts to suppress the noise but not before it is close to (and in some instances) actually at maximum activation for incorrect corner instances. It makes an incorrect decision initially and changes its mind.

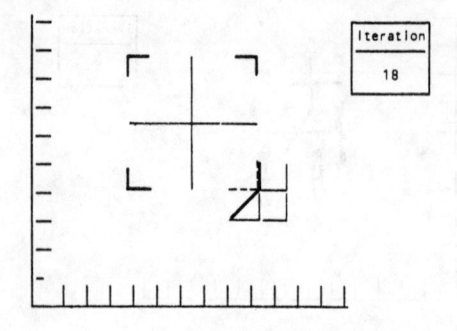

Figure 18. Noised Face Behavior (1)

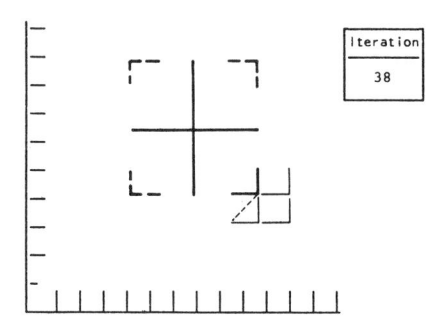

Figure 19. Noised Face Behavior (2)

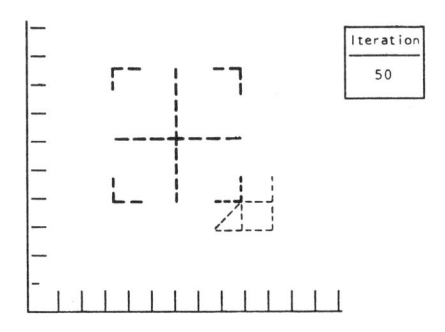

Figure 20. Noised Face Behavior (3)

An important factor is how quickly the network corrects the error in perception. The combined effects of decay, inhibition from the "correct" corner and lack of feedback (as opposed to strong feedback for the "correct" corner) force suppression in .5 times the iterations that any one factor would have caused. In this case, the combined effects (through communication) of partial results quickly prune the search tree of an incorrect hypotheses.

Also note that without feedback, the noise would still have been eliminated by mutual inhibition but the hypothesis with the strictly stronger bottom-up input would have been chosen. Feedback forced the more globally interesting face interpretation. In a more complex drawing with other noised features, recognition would have been very difficult if this were not the case since many of the feature parameters would have been slightly incorrect.

*Occlusion and Missing Information.* Missing information in the input occurs when some of the normally visible features provided as input in the perfect case are completely nonexistent or weak. This may be a result of an actual loss of information in the input image. More often, it is a result of occlusion by other objects or self-occlusion.

Reaction of the perception system depends on the abstraction level of the missing feature or features and/or the number of missing features (therefore, the percentage of the overall constraint not satisfied). Noise threshold is an important factor since it determines the percentage of the constraint that must be satisfied before the hypothesis is allowed to accumulate confidence and therefore the amount of missing information that will be tolerated. Lowering the threshold will allow more speculation by the system (i.e., activation of more hypotheses) but must be offset by extensive inhibitory networks. In the case where missing information is local and restricted to the lower levels of abstraction, the rate of convergence is slowed but results are not altered. Examples may be found in (Sabbah, 1982).

Dealing with occlusion is more difficult and requires that we apply the same assumption used by humans in determining occluded shape (Palmer, 1982), namely that shapes (features) projecting into occluded areas follow rules of good form and continuity of form. Since we assume objects to be composed of bounded planes skewed by projection, "reasonable" guesses can be made by projecting partially visible face objects behind the contours of the occluding object. To accomplish this correctly we need information from three sources:

- tjoint features to provide the occlusion cues that signal possible occlusion (Waltz, 1975).
- Bottom-up input in the form of visible corners (ljoints), lines and edges belonging to partly occluded faces to constrain partially visible forms.
- A measure of the occluding contour and its spatial location in the visual field since many of the "guesses" computed solely from tjoint and corner information project past these bounds and must be eliminated.

An example of the faces activated as reasonable guesses is illustrated in Figure 21.

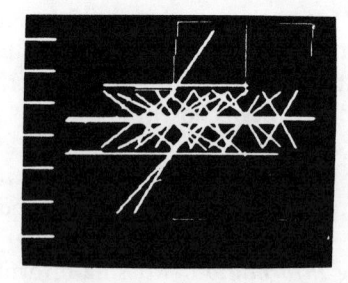

Figure 21. Occluded Faces Activated by "Reasonable Guess"

The brighter face parameters represent reasonable guesses made by hypothesizing possible faces within the occluding boundary.

In order to produce the proper behavior, activation from all three sources must be present. Absence of activation from any one source; tjoints, corners or occluding face, will eliminate that face as a reasonable hypothesis. As a result, the connections are conjunctive in nature and are called conjunctive connections (Feldman & Ballard, 1981).

Once the proper subset of faces behind each occlusion boundary has been activated, the faces in the subset are allowed to mutually inhibit one another in a winner-take-all net. The ultimate winner is determined by feedback; the only possible determining factor since multiple views are not currently allowed.

In the following example, we see the combined effects of these new connections. A random face partially occludes a rectangular solid. The line drawing input to the perception network is:

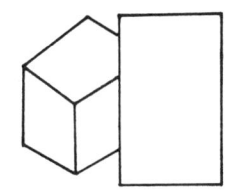

Figure 22. Face Occluding Solid

The resulting behavior is depicted in Figure 23. The face parameters with low confidence behind the occluding face are the hypothesized guesses of the perception network.

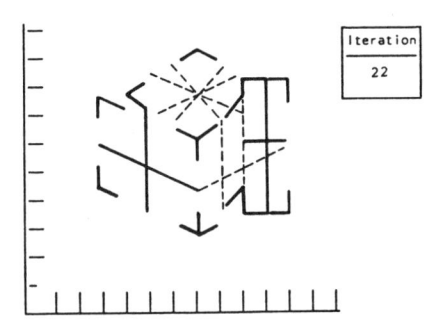

Figure 23. Occluded Solid Behavior (1)

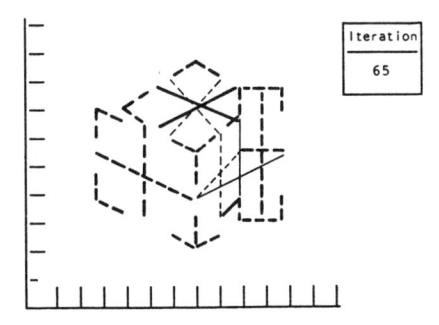

Figure 24. Occluded Solid Behavior (2)

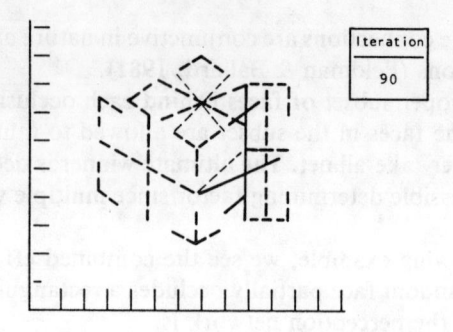

Figure 25. Occluded Solid Behavior (3)

Note that the more global interpretation of the rectangular solid eventually feeds back enough activation to the proper face interpretations so they eliminate the other elements of the winner-take-all networks and saturate.

*Self Occlusion.* In some sense, self occlusion is not a problem. The representation of an object incorporates a fixed view and we can therefore include only those visible features in the specific instance of that view. Unfortunately, some of the views produce partially visible features as in the following box object. The two faces toward the front of the box partially occlude the two back faces.

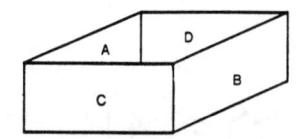

Figure 26. Partially Occluded Faces in the Box Object

Our current approach is to deal with this problem by making partial self occlusion a subset of general occlusion. All visible and partially visible faces are included in the representation at the object level. Occlusion behavior in the network activates the partially occluded faces and feedback from the object representation causes the proper instances of the face to saturate.

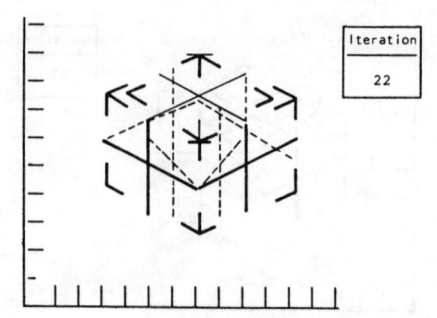

Figure 27. Self-Occluded Box Behavior (1)

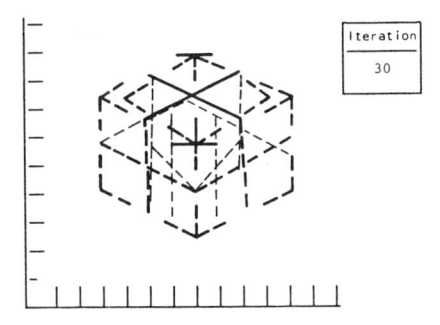

Figure 28. Self-Occluded Box Behavior (2)

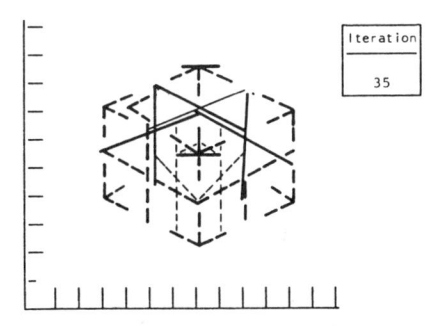

Figure 29. Self-Occluded Box Behavior (3)

Although this example works correctly, trying to predict the normalization parameters that make the network behave properly is a continual nuisance when competing alternatives are extremely close in activation level. Although they do inhibit one another strongly, they still continue to accrue activation at a slow, steady rate from bottom-up activation which may eventually force both into saturation before additional partial information from feedback becomes the crucial and proper deciding factor. Obviously, this is very undesirable since the alternatives are contradictory and one must suppress the other. In this example, most relaxation parameter settings result in the rectangular solid saturating followed closely by the contradictory box interpretation also saturating. Bottom-up evidence exists for both since the line drawing for the rectangular solid is actually a subset of the one for the box.

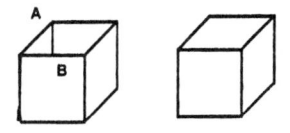

Figure 30. Box vs Rectangular Solid Line Drawings

This behavior is undesirable but we clearly cannot eliminate the ability of bottom-up input to drive high levels of confidence since it is absolutely necessary to the eventual activation of more global stable coalitions, and, therefore, eventual feedback loops for all interpretations. Furthermore, we cannot force arbitrarily long and global slowdown in integration of bottom-up input. In most cases, the decision is clear and the perceptual network need not ponder that long. The only viable solution is to have the system monitor itself and spot only those situations that require a local modification in relaxation parameters. For closely competing alternatives, monitoring units on the abstraction level can dynamically increase the size of normalization factors until such time as proper feedback is present and a true stable coalition is formed.

## DISCUSSION

What we have learned by experimenting with Origami world is that it is feasible to solve hard Artificial Intelligence problems using a connectionist model of computation embodying principles gathered from interdisciplinary study of animal visual systems and computer vision. Furthermore, such a connectionist approach has several advantages:

- Completely parallel activation of competing and cooperating hypotheses sharing partial information from varied sources (feedback, bottom-up, inhibition). As a result:
  - The need to order search is greatly reduced (in our small world, it is actually eliminated). For example, the need for search order heuristics such as the "parallel line" heuristic of Origami World (Kanade, 1979) is eliminated. Such heuristics are used solely to force the choice of *only one* next alternative in a sequential search. Choosing or odering at this level forces local optima sometimes at the cost of more globally optimal interpretations.
  - Because search order is eliminated and partial results are constantly being communicated, the need to backtrack over poor initial guesses in search is eliminated leading to benefits of, what Kornfeld has termed, Combinatorial Implosion (Kornfeld, 1979). What we mean here by implosion is that the number of units under active consideration at any one time in much less than the complete representation implies. For example, in the case of the occluded solid only $O(10^4)$ units were activated out of the $O(10^8)$ possible interpretations (8 unique parameters describing rectangular solids, 6 parameters for faces, 4 for ljoints, etc...) resulting in an implosion of $O(10^{-4})$.

- We have also extended relaxation in computer vision to deal with information gathered at multiple levels in a hierarchy.

These advantages result in a system capable, not only of dealing with perfect Origami figures, but with noised, incomplete and occluded input in a predictable and pleasing manner (graceful degradation of behavior as input degrades).

**Problems and Extensions**

Unfortunately (maybe fortunately), this system only scratches the surface of the true perception problem. Line drawings are useful but distinctly limited. Major extensions to the representation must be made to even approximate real-world objects. We are currently trying to develop such a representation based on Extended Gaussian Images. Such a representation must at least deal with complex, rigid 3-d shapes undergoing arbitrary 3-d transformations. First approximations exist (Ballard & Sabbah, 1981; Ikeuchi, 1981) but they don't yet deal with such issues as motion or nonrigid objects nor is the formation of conceptual hierarchies straightforward. To be truly "real-world" we must also include and therefore incorporate additional constraining information from other sources such as color, texture, shading, etc. Once again we are working on smoothly incorporating this information into a conceptual hierarchy.

As more of these sources of information are included, the problem of controlling proper and stable convergence behavior multiply. Such information is characteristically computed independently and yet must be available at the right time and place to properly constrain perception. Assumptions and conditions which guarantee "the right place at the right time" can only be properly understood through formal analysis of convergence. A promising approach is to apply or extend Baudet's chaotic relaxation in a network of multiprocessors (Baudet, 1978). This form of relaxation explicitly defines parameters fixing amount of asynchrony that will be tolerated before convergence breaks down and also allows for restricted non-linearity in the iteration update formula. Both exist in the present form of the relaxation.

*The Role of Visual Attention.* Parallel formulation, no matter how massive, does not eliminate the combinatorial explosion problem associated with underconstrained search. If we assume a visual system recognizing complex objects in the manner described and we further assume that the problems of representation and convergence are solved, there still remains the problem of "spreading activation" due to weakly constrained mappings between feature spaces. As abstraction level increases the complexity of the object increases, sometimes dramatically. If, for example, we try to map faces directly to Origami duck objects; the number of connections becomes huge since any one instance of a face can be mapped to an extremely large number of ducks. Therefore, activation of the large number of faces composing the duck causes activation of a huge number of units on the object level, most with little evidence for existence. This almost random spreading activation is a form of combinatorial explosion for this system. Additional levels of hierarchy and more concise representation partly alleviate it, however, a trade-off is made: The more levels of hierarchy, the larger the absolute number of units; the smaller the number of levels, the more underconstrained (i.e., larger) the connection set. In either case, we see combinatorial explosions in number of units or size of connection sets or both.

The solution to this dilemma is to incorporate a mechanism for selective attention. First, we propose to extract coarse visual shape, then we choose to attend only to specific parts of the visual field with more computational effort as time allows or interest dictates. Hence, a role for hierarchies of resolution. However, setting current resolution is a dynamic process to be internally dictated by context and feedback in the perception system. We are currently experimenting with mechanisms for implementing focus.

*Learning and Recognizing New Objects.* Two primitive types of learning are possible here. The first is simple and only requires the ability to recruit units (Feldman, 1982). Anytime the system finds a specific combination of units on an abstraction level "interesting," a unit on the next level up can be recruited to represent that interesting combination and form the apex of a stable coalition. This form of learning has the distinct disadvantage that only the instance of the object represented by the units recruiting is "learned." The ideal would be to extract, from a small number of instances, exactly that optimal set of constraints that defines all possible instances of, for example, chairs.

This second form of learning requires, at least, that transformation parameters be completely decoupled from shape representation. We could then understand the concept (i.e., abstract constraint) that defines a chair or any other object independently of specific transformations of its components. Hinton has already proposed similar modifications in (Hinton, 1981). Of course, the actual process of extracting exactly the optimal shape representation that captures all possible chairs is still a mystery. At the very least, it requires that the system be able to reason function from geometric shape and other properties (color, motion, texture...).

## REFERENCES

Ballard, D. H. (1981a). Generalizing the Hough transform to detect arbitrary shapes. *Pattern Recognition, 13,* 111-122.

Ballard, D. H. (1981b). Parameter nets: A theory of low level vision. (Tech. Rep. No. 75). Computer Science Department, University of Rochester.

Ballard, D. H., & Sabbah, D. (1981). On shapes. *Proceedings of the Seventh International Joint Conference on Artificial Intelligence,* 607-612.
Barrow, H. G., & Tenenbaum, J. M. (1976). MSYS: A system for reasoning about scenes. (Technical Note 121). SRI International, Menlo Park, CA.
Baudet, G. M. (1978). Asynchronous iterative methods for multiprocessors. *Journal of the Association for Computing Machinery, 25,* 226-244.
Davis, L. S., & Rosenfeld, A. (1980). Cooperating processes for low level vision: A survey. (Tech. Rep. No. 851). Computer Science Department, University of Maryland.
Feldman, J. A. (1982). Memory and change in connection networks. *Biological Cybernetics, 46,* 27-39.
Feldman, J. A., & Ballard, D. H. (1981). Connectionist models and their properties. *Cognitive Science, 6,* 205-254.
Hanson, A. R., & Riseman, E. M. (Eds.). (1978). *Computer vision systems.* New York: Academic.
Hanson, A. R., & Riseman, E. M. (1978a). Segmentation of natural scenes. In Hanson, A. R., & Riseman, E. M. (Eds.), *Computer vision systems.* New York: Academic.
Hanson, A. R., & Riseman, E. M. (1978b). Visions: A computer system for interpreting scenes. In Hanson, A. R., & Riseman, E. M. (Eds.), *Computer vision systems.* New York: Academic.
Hinton, G. (1977). *The role of relaxation in vision.* Unpublished doctoral dissertation. University of Edinburgh, Edinburgh.
Hinton, G. (1981). Shape representation in parallel systems. *Proceedings of the Seventh International Joint Conference on Artificial Intelligence,* 1088-1096.
Hough, P. V. C. (1962). Method and Means for Recognizing Complex Patterns. U.S. Patent No. 3069654.
Hubel, D. H., & Wiesel, T. N. (1979). Brain mechanisms of vision. *Scientific American, September 1979,* 150-162.
Hummel, R. A., & Zucker, S. W. (1980). On the foundations of relaxation labelling processes. (Tech. Rep. 80-7). Computer Vision and Graphics Lab, McGill University.
Ikeuchi, K. (1981). Recognition of 3-d objects using the extended gaussian image. *Proceedings of the Seventh International Conference on Artificial Intelligence,* 595-600.
Kanade, T. (1978). A theory of origami world. (Tech. Rep. No. CMU-CS-78-144). Department of Computer Science, Carnegie Mellon University.
Kanade, T. (1979). Recovery of three-dimensional shape of an object from a single view. (Tech. Rep. No. CMU-CS-79-153). Department of Computer Science, Carnegie Mellon University.
Kender, J. R. (1980). Shape from texture. (Tech. Rep. No. CMU-CS-81-102). Unpublished doctoral dissertation. Carnegie Mellon University.
Kornfeld, W. (1979). Combinatorially implosive algorithms. (Artificial Intelligence Laboratory Memo 561). Massachusetts Institute of Technology.
Marr, D. (1978). Representing visual information. In Hanson, A. R., & Riseman, E. M. (Eds.), *Computer vision systems.* New York: Academic.
McClelland, J. L., & D. E. Rumelhart (1980). An interactive activation model of the effect of context in perception: Part I. (Tech. Rep. No. 91). Center for Human Information Processing, University of California at San Diego.
Palmer, S. E. (1982). Symmetry, transformation, and the structure of perceptual systems. In J. Beck & F. Metelli (Eds.). *Representation and organization in perception,* Hillsdale, NJ: Erlbaum.
Rosenfeld, A., Hummel, R., & Zucker, S. W. (1976). Scene labelling by relaxation operations. In *International Electronics and Electrical Engineers Transactions on Man, Machines and Computers, 6,* 420-433.
Rumelhart, D. E., & McClelland, J. L. (1981). An interactive activation model of the effects of context in perception: Part II. (Tech. Rep. No. 95). Center For Human Information Processing, University of California at San Diego.

Sabbah, D. (1981). Design of a highly parallel visual recognition system. *Proceedings of the Seventh International Conference on Artificial Intelligence,* 722–727.

Sabbah, D. (1982). A connectionist approach to visual recognition. (Tech. Rep. No. 107). Unpublished doctoral dissertation. University of Rochester.

Waltz, D. (1975). Understanding line drawings of scenes with shadows. In P. H. Winston, (Ed.), *The psychology of computer vision.* New York: McGraw Hill.

Zucker, S. W. (1980). Labelling Lines and Links: An Experiment in Cooperative Computation. (Tech. Rep. No. 80-5). Computer Vision and Graphics Laboratory, McGill University.

CHAPTER 7

# Massively Parallel Parsing: A Strongly Interactive Model of Natural Language Interpretation*

DAVID L. WALTZ
JORDAN B. POLLACK
Coordinated Science Laboratory
University of Illinois

This is a description of research in developing a natural language processing system with modular knowledge sources but strongly interactive processing. The system offers insights into a variety of linguistic phenomena and allows easy testing of a variety of hypotheses. Language interpretation takes place on a activation network which is dynamically created from input, recent context, and long-term knowledge. Initially ambiguous and unstable, the network settles on a single interpretation, using a parallel, analog relaxation process. We also describe a parallel model for the representation of context and of the priming of concepts. Examples illustrating contextual influence on meaning interpretation and "semantic garden path" sentence processing, among other issues, are included.

## INTRODUCTION

**The Problem of Integration**

The interpretation of natural language requires the cooperative application of many systems of knowledge, both language specific knowledge about word use, word order and phrase structure, and "real-world" knowledge

---

* This work is supported by the Office of Naval Research under Contract N00014-75-C-0612. This article has benefited from discussion with various people at the University of Illinois, including Bill Brewer, Gerald DeJong, Rick Dinitz, Marcy Dorfman, David Farwell, Georgia Green, La Raw Maran, Jerry Morgan, Andrew Ortony, and, especially, The Great Lorenzo. Thanks also to Cathy Cassells for limitless editing and typing, and to Pollack's program for printing the pictures.

about stereo-typical situations, events, roles, contexts, and so on. And even though these knowledge systems are nearly decomposable, enabling the circumscription of individual knowledge areas for scrutiny, this decomposability does not easily extend into the realm of computation; that is, one cannot construct a psychologically realistic natural language processor by merely conjoining various knowledge-specific processing modules serially or hierarchically.

These particular forms of process integration, which happen to be convenient to use on modern computers, turn out to have a profound effect on the mind of the modeler, as Pylyshyn (1980, p. 124) points out:

> Now, what is typically overlooked when we [use a computational system as a cognitive model] is the extent to which the class of algorithms that can even be considered is conditioned by the assumptions we make regarding what basic operations are possible, how these may interact, how operations are sequenced, what data structures are possible, and so on. Such assumptions are an intrinsic part of our choice of descriptive formalism.

*Ambiguity.* Convenient processing assumptions lead to problems in building models for cognition. Consider ambiguity, perhaps the most ubiquitous problem in natural language processing. Humans experience an increased processing load with ambiguous language (Mackay, 1966), which suggests that humans compute multiple readings (at least in some sense). However, the "serial frame of mind" allows basically two approaches for dealing with ambiguous sentences: *backtracking* as used in Augmented Transition Networks (Woods, 1970), or *delay,* as used in Marcus' (1980) "wait and see" parser. And although lexical access appears to be an automatic process cotemporal with syntactic and semantic processing (Marslen-Wilson & Tyler, 1980) most natural language systems still work with small dictionaries and simple ad-hoc heuristics which choose word meanings before assigning structure.[1]

*Single Interpretation.* Another interesting phenomenon in language interpretation is that humans can usually entertain only one interpretation of an ambiguous sentence at a time, but can easily "flip" between interpretations. Consider the following short sentences which can be interpreted either as statements or commands:

(S1) Trust shrinks.
(S2) Respect remains.
(S3) Exercise smarts.

The fact that we can interpret the first sentence either as a general statement about dwindling confidence or as advice to place one's faith in psychiatrists,

---

[1] This, finally, is changing. See Small (1980) or Charniak (1983).

suggests that the human capacity to disambiguate language is not unlike the faculties involved in visual disambiguation—like foreground/background perception of the Necker Cube.

*Comprehension Errors.* Errors in comprehension, like Garden Path Sentences have in the past been explained by purely structural principles like early closure (Kimball, 1973) or Minimal Attachment and Right Association (Frazier, 1979), or by the breakdown of simple and limited serial mechanisms (Marcus, 1980; Milne, 1982; Shieber, 1983). However, they have more complete and natural explanations as side-effects of strongly interacting processes as we demonstrate in the section on Errors in Comprehension, p. 61. Garden path effects can occur at all levels of language processing. Consider the following sentences:

(S4) The astronomer married the star. (Charniak, 1983)
(S5) The plumber filled his pipe.
(S6) The sailor ate a submarine.

Readers usually report these as "temporarily anomalous," i.e., as a "semantic garden path sentence." A plausible explanation for the cognitive doubletake (and mild humor) caused by the first sentence is that the priming power of "astronomer" on the wrong sense of "star" is initially stronger than the logical power of case frame selectional restrictions; in the end, the selectional restrictions force the interpretation of "star" as a person.

*Nongrammatical Text.* People are able to interpret nongrammatical language, whether it is naturally occurring (due to poor grammar, foreign speakers, noise interference, interruptions, self-corrections, etc.) Most work on this topic has taken the approach of relaxing certain contraints in the parsing process—in the LSP project (Sager, 1981), a failed parse was retried without agreement constraints on syntactic features; in the PLANES project (Waltz, 1978), a semantic grammar was used which accepted very ungrammatical input as meaningful. Others have formalized the notion of constraint relaxation for handling ill-formed input (Goodman, 1984; Kwasny & Sondheimer, 1981). We believe that this ability in humans, to semi-independently judge meaningfulness and grammaticality, is yet more evidence of the modularity of knowledge but the integration of processing.

**Integrating Knowledge Sources**

All these phenomena indicate the need for a theory of language processing which posits, instead of the simple passing of semi-complete results between processing components, strong interaction between those components; so strong, in fact, that all decisions are interdependent. We believe that most theories of language processing advanced over the years have been seriously flawed because they have drawn on a limited set of computational ideas which cannot effectively deal with interdependent decisions, and because of peculiarities of English and the history of linguistics research which have led to the assumption of the autonomy of syntax in natural language understanding.

These observations are, of course, not entirely new. In the early 1970s, Schank argued that semantics, not syntax should have the central role in programmed theories of natural language processing (Schank, Goldman, Rieger, & Riesbeck, 1973; Riesbeck & Schank, 1976). Steven Small (1980) was another worker in AI who questioned the traditional serial integration of language processing. Small suggested that rather than having separate modules for syntax and semantics, each word should be its own expert, and built a system of interacting discrimination nets, reminiscent of Hewitt's work on actor formalisms (Hewitt, 1976). Cottrell and Small (1983) have published research on interacting distributed processing of word senses, case roles, and semantic markers, very similar in spirit to our effort (see section titled "Case Frames and Selectional Restrictions," pp. 64–65). And, of course, the HEARSAY II speech understanding system (Fennel & Lesser, 1977) used a parallel production system for integrating multiple knowledge sources.

## STRUCTURES OF THE MODEL

We wish to go further than these models of integrated processing; we want a machine to make its interdependent decisions smoothly—we want a system which runs like an ecological system, rather than like a Rube Goldberg device. We have chosen to work with the twin processes of *Spreading Activation* and *Lateral Inhibition* in which decisions are spread out over time, allowing various knowledge sources to be brought to bear on elements of the interpretation.

### Spreading Activation and Lateral Inhibition

The term "Spreading Activation" has been used to describe many different programs and models, but all can be basically divided into two classes. *Digital Spreading Activation* is a class of marker-passing algorithms which perform a breadth-first search for shortest paths on a relational network. *Analog Spreading Activation* takes place on a weighted network of associations, where "activation energy" is distributed over the network based on some mathematical function of the strength of connections. As an example of the digital kind, Quillian (1968) describes a technique for finding relationships between two concepts stored in a semantic network by repeatedly marking adjacent nodes with "activation tags" containing a path back to the source concept. As examples of the analog kind, Collins and Loftus (1975) model semantic priming on a network with decaying activation, and, more recently, McClelland and Rumelhart (1980) model word recognition on an analog spreading activation network.

Both forms of spreading activation suffer from the danger of "overkill." In the digital form, this means the problem of false positives, where a very large number of uninteresting paths may be found.[2] The analog form of spreading activation has the potential problem of "heat death," where the entire network becomes uniformly activated. This can be handled with some form of damping or decay, or via lateral inhibition, which spreads negative energy just as spreading activation spreads positive energy. With damping or decay, weights must be carefully chosen to avoid having too many or too few portions of the network active. Lateral inhibition seems to have fewer disadvantages, and is our method of choice.

Besides effectively dealing with the problem of overkill, lateral inhibition is useful for the coordination of distributed decisions. According to Feldman:

> Lateral inhibition at lower organizational levels is one of the most ubiquitous information-processing mechanisms in animals: it is essential that opposing action systems do not execute simultaneously. Low-level visual processing makes very heavy use of mutual lateral inhibition, and this appears to be true for other sensory systems as well (1981, p. 52).

Consider a graph with weighted nodes and links, and an iterative operation which recomputes each node's activation level (i.e., its weight) based on a function of its current value and the inner product of its links the activation levels of its neighbors. An activation (positive) link between a pair of nodes will cause them to support each other while an inhibition (negative) link will attempt to allow only one of the pair to remain active at any given time.[3] The net effect is that, over several iterations, a coalition of well-connected nodes will dominate, while the less fortunate nodes (those which are negatively connected to winners) will be suppressed.

We exploit this behavior several ways in our parser: by putting inhibitory links between nodes which represent well-formed phrases with shared constituents (which are, thus, mutually exclusive), we ensure that only one will survive. Similarly, there are inhibitory links between nodes representing different lexical categories (i.e., noun or verb) for the same word; between

---

[2] Quillian recognized that this was a problem for preposition words in his network, and used a simple heuristic of avoiding paths through dense clusters to circumvent it (1968, p. 156). Charniak (1983, p. 188) discusses a similar heuristic.

[3] The pair may, under certain circumstances, balance each other; both may have zero activation simultaneously.

concept nodes representing different senses of the same word (i.e., submarine as a boat or as a sandwich); and between nodes representing conflicting case role interpretations. There are activation links between phrases and their constituents, between words and their different meanings, between roles and their fillers, and between corresponding syntactic and semantic interpretations. The net effect is that, over several iterations, a coalition of nodes representing a consistent interpretation will dominate, while the less fortunate nodes will be suppressed.

## Integration by Parts

While we have not yet built an entire language processing system, we have done several computational "experiments," modeling various components of comprehension, as well as certain interactions between them. We will describe the different components we have modeled with activation and inhibition and how they fit together. The figures in this section have nodes represented by named and shaped shapes—the darker the shade, the higher the activation level. Activation links are shown with arrowheads, while inhibition links end with small circles. All the figures are snapshots taken of a system which runs a proportional update activation function, the same as McClelland and Rumelhart's (1980) scheme, on a range from 0.0 to 1.0 without decay. All activation links are at $+.2$ and all inhibition links at $-.45$.

*Lexical Access and Priming.* One of the earliest uses for spreading activation was to model semantic priming (Collins & Loftus, 1975; Ortony & Radin, 1983). It is very natural to think in terms of weighted connections between related concepts as being responsible for the "associations" that drive the facilitation of word meanings in context. AI models, with the exception of Small's (1980) Word Expert Parser, have basically ignored this aspect of language processing as "lower than syntax," and have usually called a subroutine to pick out the lexical category and meaning of a word upon demand.

Both approaches are a little off. Recent evidence about lexical access (Seidenberg, Tanenhaus, & Leiman, 1980; Swinney, 1979) shows that when a word is encountered, all its meanings are facilitated (phase 1) but then rapidly, the local semantic and syntactic context eliminate all but the "correct" one (phase 2). Spreading activation can easily be responsible for phase 1, but it provides no way to achieve phase 2; opportunistic selection by subroutine call can, in many cases, pick out the right lexical item, but does not accurately model the automatic process humans do. Using lateral inhibition between competing word senses, and between word senses and the local syntactic and semantic context, can effectively simulate the two stage hypothesis without two processing stages.

Figure 1 shows a four-level activation network for the sentence:

(S7) John shot some bucks.

Figure 1a shows the initial state, and 1b shows the network after about 50 cycles. The first level shows a syntactic parse tree for the sentence, the second level shows the input words, the third level shows a cluster of meanings for each word with mutual inhibition links between all the meanings, and crossed activation and inhibition links between the lexical categories and the meanings. For example, the input word "shot" is connected to four senses:

1. TIRED—an adjective, approximately meaning[4] "worn out"
2. FIRE—a past-tense verb, approximately meaning "to shoot with gun"
3. BULLET—a noun, approximately meaning "unit of ammunition"
4. WASTE—a verb, approximately meaning "squander a resource"

The fourth level, in large italics, represents an image of a context system which is discussed in the section titled "Context: Introduction" on p. 65. Its purpose is to coordinate sense selection along the sentence.

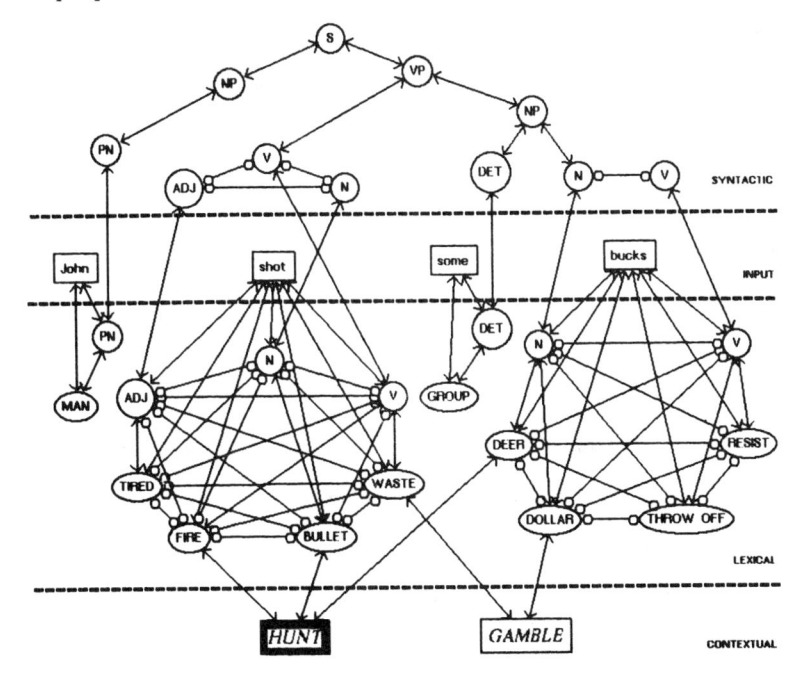

Figure 1a. Initial Network for "John shot some bucks" in the context of hunting.

[4] The nodes in this diagram, and in this paper in general are intended to have a schematic internal representation. We see the behavior of our networks as a coarse description of an even larger, more parallel representation of schemata and semantic features.

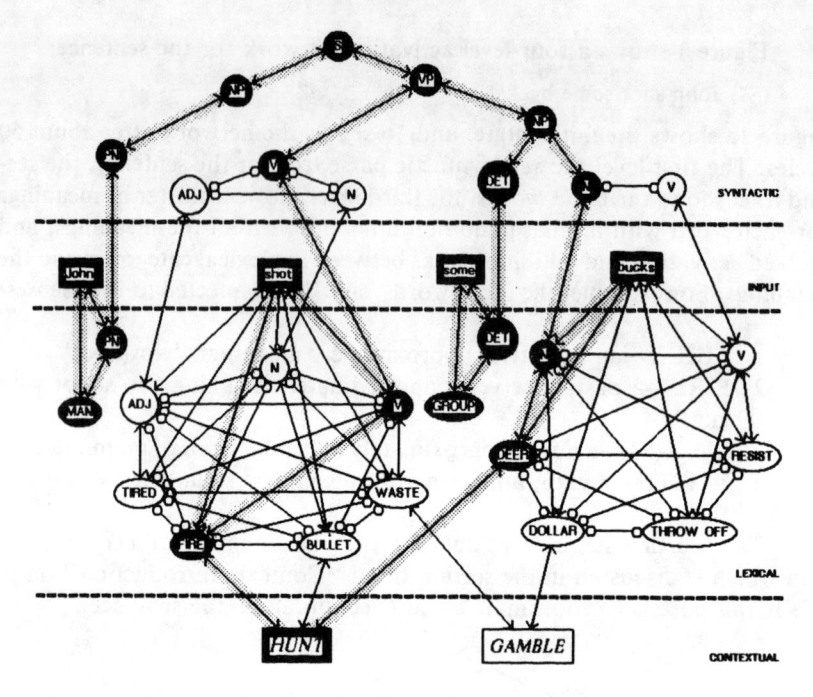

Figure 1b. Network after 50 cycles of spreading activation and lateral inhibition.

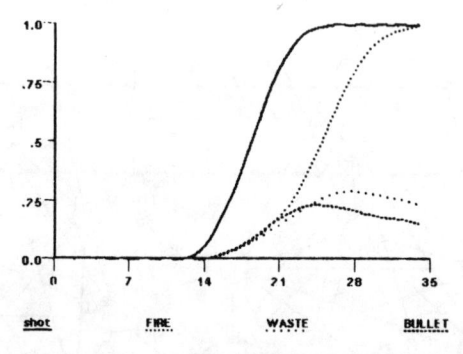

Figure 1c. Graph of the activation profile of "shot" and several of its senses. The horizontal axis represents time, in cycles (and no claims are being made at this point for a mapping to milliseconds), and the vertical axis represents activation levels.

The activation of this network (by an unshown auxiliary network which sequences the words) yields an interesting profile of the processing of "John shot some bucks" in the context of hunting. The graph in Figure 1b shows the activation profiles over time of the word "shot" and some of its meanings. Meanings are activated immediately after the word, and the syntactic demand for a verb, along with the contextual pressure of hunting, lead to the rapid demise of all but the "FIRE" sense of "shot."

*Syntax—Autonomy and Integration.* It is clear that humans can make relatively autonomous judgements about the grammaticality and meaningfulness of sentences. This is clearly demonstrated by the ability to assign structure to nonsense sentences such as:

(S8) Colorless green ideas sleep furiously.[5]

as well as to assign some meaning to ungrammatical strings. The autonomy of syntax is important because it allows a very complex element of natural language to be studied in isolation from other elements. Unfortunately, it has been misunderstood at an implementational level, with programmers having drawn the wrong conclusion: that natural language can be processed by a "syntactic faculty"[6] that assigns structure before meaning. This misunderstanding and the failures that accompanied it, along with the early successes of meaning-primitive based systems (Schank et al., 1973), have led many AI researchers to assume a "rejectionist" position with regard to syntax.[7]

Obviously, syntax is not the framework upon which an entire NLP system should be based, but neither should it be dispersed into the farthest reaches of subroutine calls. We put syntax on an equal footing with other sources of language knowledge. Notice that the misunderstanding of syntactic autonomy stems from the intellectual bottleneck of serial processing discussed earlier—when computations must be serialized, some decisions must be made before others. In a parallel and strongly interactive framework, syntax can be integrated in such a way as to allow relatively independent judgements of grammaticality, as well as to influence and be influenced by judgements of meaningfulness.

Our approach to relatively independent syntactic processor (Pollack & Waltz, 1982) is based on merging ideas from breadth-first chart parsing (Kay, 1973) with parallel relaxation by lateral inhibition. The output from a chart parser for a context-free grammar[8] is taken to be a network with activation links between mother and daughter nodes, and inhibition links between "in-laws"—nodes which both dominate a common daughter. When the network is iterated, it seeks an equilibrium state with an active coalition of phrase-markers representing a well-formed parse.

Figure 2 shows two stable states of a network representing the parses of the sentence

(S9) John ate up the street.

---

[5] From Chomsky, (1957, p. 15).
[6] See Fodor (1982) for a history of faculty theories of mental abilities.
[7] Though a comeback of sorts seems to be underway—see Berwick, (1983).
[8] Although common wisdom says that CFG's are inadequate to represent the syntax of natural languages, recent work on Generalized Phrase Structure Grammars (Gazdar & Pullum, 1982) holds the promise of overcoming the inadequacies.

The network settles on one of the states based on the initial node and link weights, and on any external activation or inhibition applied to the nodes, i.e., from the lexical or semantic level. The important thing to note is that there is no "homunculus" searching for a consistent tree, just a local competition for superiority. The weighting scheme where the words are activated sequentially from left to right prefers to settle on the shortest tree it can find, Figure 2a.

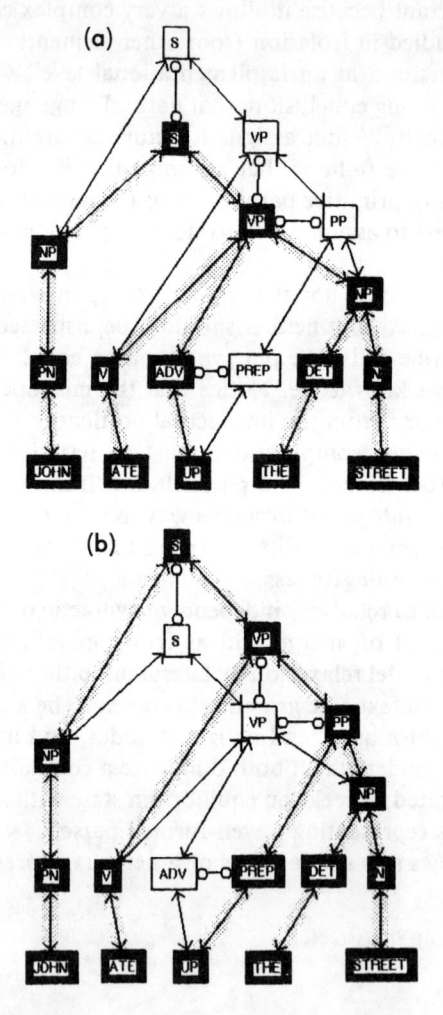

Figure 2. Stable coalitions of the ambiguous syntax of "John ate up the street." The reading of "up" in 2a (John ate all of the asphalt) is preferred but semantic information can easily overcome this preference.

There has been a great deal of research into exactly how the "human syntactic processor" works. Data on which decisions humans make in the structural interpretation of sentences in null contexts have generated many principles and strategies for parsing (Frazier, 1979; Ford, Bresnan, & Kaplan, 1982; Kimball, 1973), and this data has been used as validation of a whole host of grammatical formalisms and parsing mechanisms, including the Sausage Machine (Frazier & Fodor, 1978) ATN's (Wanner, 1980), and different varieties of deterministic parsers (Church, 1980; Marcus, 1980; Milne, 1982; Shieber, 1983). There are a number of problems with building a syntactic parser to explicitly encode these strategies as the starting point for a language understanding system. We list three major ones here. First, humans can understand sequences of words with ill-formed syntactic structure. Second, when beginning with a syntactic parser, there is no *good* way to smoothly integrate semantic and lexical strategies, so the different strategies are usually subjugated to the status of subroutines called by some control program of dubious psychological and linguistic credentials (Winograd, 1972). Third, since garden path sentences have been taken as validation of the structural principles, it would be good if they always caused humans to garden path. Not so, Crain and Steedman (1981) have shown how easy it is to prevent backtracking by adding context to garden path sentences. For example, if

(S10) Cotton is grown in several of the Southern States.

precedes

(S11) The cotton clothing is made of is grown in Mississippi.

the latter is no longer a garden path sentence.

Since structural preferences are apparently so much weaker than other contextual forces, is it not better to view them as side-effects of the organization of the syntactic processor, rather than as global guiding principles? In a parallel parser such as the one described above, preferences between competing phrases are a side-effect of lateral inhibition; the global guiding principle is the "universal will to disambiguate."

It is interesting to note that although our model is motivated by considerations of strong interaction, the syntactic modeling is closely related to the HOPE program of Gigley (1982) which was motivated by the neurolinguistic concerns of modeling aphasiac language degradation. Currently, however, we have no goals of lobotomizing our program to see how it degrades (but see Marcus, 1982 for one such experiment).

*Errors in Comprehension.* Because our system operates in time, we are able to model effects that depend on context, and effects that depend on the arrival time of words. Consider the network shown in Figure 3, which shows three snapshots taken during the processing of a sentence that induces a "cognitive doubletake":

(S4) The astronomer married a star.

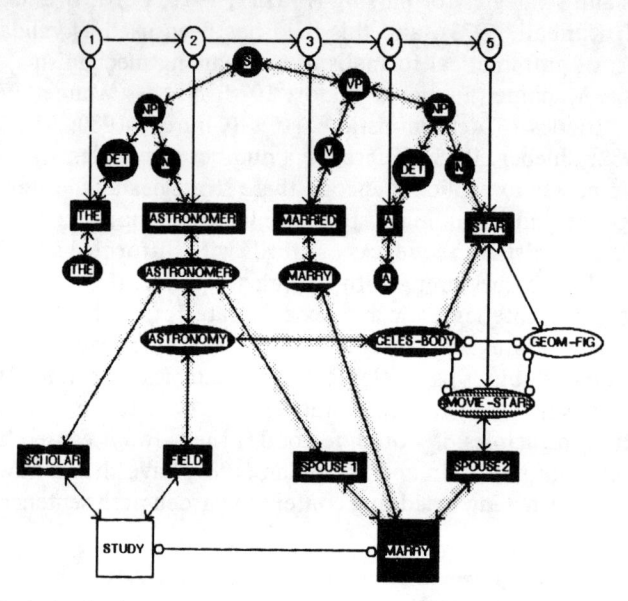

Figure 3a. Cycle 27 of "The astronomer married the star"; CELES-BODY seems to have won.

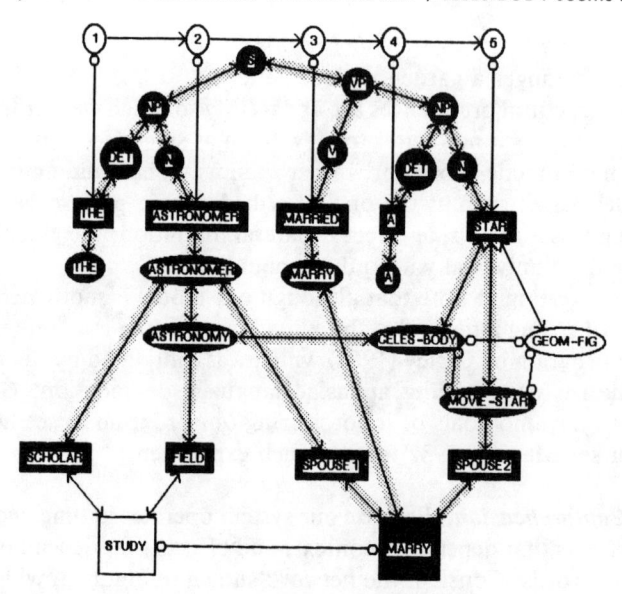

Figure 3b. Cycle 42; MOVIE-STAR has caught up.

Figure 3 includes three possible meanings for "star," namely (1) MOVIE-STAR—the featured player in dramatic acting, or (2) CELES-BODY—a celestial body, or 3) GEOM-FIG—a pentagram. We presume that "astronomer" primes CELES-BODY by the path of strong links: astronomer→ASTRONOMER→ASTRONOMY→CELES-BODY, but that MOVIE-STAR would be primed very little, if at all, because the its activation via the distributed context model would be very small. When the word "star" is encountered, the meaning CELES-BODY is initially highly preferred and seems to have won the competition (Figure 3a), but eventually, since CELES-

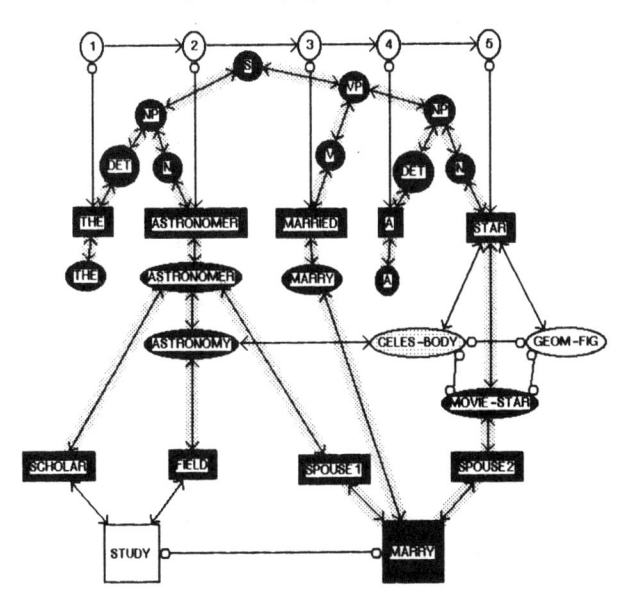

Figure 3c. Cycle 85; CELES-BODY has lost to MOVIE-STAR; consistency reigns.

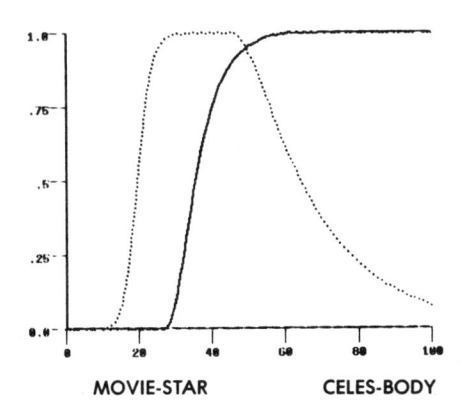

Figure 3d.

BODY is inanimate, whereas the object of MARRY should be human and animate, the MOVIE-STAR meaning of "star" catches up (Figure 3b) and wins out (Figure 3c).

In Figure 3d we show the activation levels for CELES-BODY and MOVIE-STAR as functions of time. One can see that the activation of CELES-BODY is initially very high, and that only later does MOVIE-STAR catch up to and eventually dominate it. We argue that, if activation level is taken as a prime determinant of the contexts of consciousness, then this model captures a common experience of people when hearing this sentence. This phenomenon is often reported as being humorous, and could be considered a kind of "semantic garden path." It should be emphasized that this behavior falls out of this model, and is *not* the result of juggling the weights until it works. In fact, the examples shown in this paper work in an essentially similar way over a broad range of link weightings.

*Case Frames and Selectional Restrictions.* Figure 3 includes some large square nodes at the bottom. These large boxes, representing case frames activated by "ASTRONOMER" and "MARRY", actually correspond to substantial structures of nodes and links, in basic agreement with the "exploded case frame" of Cottrell and Small (1983). Specifically, each case frame includes role slots, specific to the case frame: "MARRY" is attached via activation links to "MARRY-AGENT" and "MARRY-OBJECT" nodes, and these are each attached to semantic marker nodes for "HUMAN."

A scheme also is required to attach specific words, e.g., "astronomer," to roles as well as to the contexts of long-term memory. Finally, if the sentence's meaning is to be remembered, a scheme is necessary for dynamically connecting the active semantic and pragmatic nodes to long-term memory. All these schemes, in our view, require that there be (a) a way of collecting all active nodes, (b) a way of attaching the nodes to some other node (or nodes) unique to the set of active nodes, such that (c) the set of active nodes could be reactivated by activity of subsets of the set of nodes. Two methods for accomplishing this are Minsky's use of "K-lines" (Minsky, 1980) and Hinton's use of "microfeatures" (Hinton, 1981). The basic idea in both cases is that some node ("agent" for Minsky) or portion of a state vector (Hinton) is associated with specific activated nodes, and either bidirectional links (Minsky) or autoassociative hardware (Hinton) is used to recover the whole from any sufficiently large part.

Furthermore, we believe that each basic meaning node for a verb should itself be a composite structure: "eat," for instance, can be decomposed into schemas for moving food to a mouth, chewing, swallowing, and so on. Some ideas along these lines are explored in the section of DeJong and Waltz (1983) on "event shape diagrams." These diagrams can be viewed as plotting activation levels of such "microschemas." Ultimately, schemas should connect to even more detailed mechanisms for producing the experience of mental images (Waltz, 1979).

**Context: Introduction**

Earlier (in Figure 1) we used "context-setting" nodes such as "HUNT" and "GAMBLE" to prime particular word and phrase senses, in order to force appropriate interpretations of a noun phrase. There are, however, major problems that preclude the use of such context setting nodes as a solution to the problem of context-directed interpretation of language. A particular context-setting word, e.g., "hunting," may never have been explicitly mentioned earlier in the text or discourse, but may nonetheless be easily inferred by a reader or hearer. For example, preceding (S1) with:

(S12) John spent his weekend in the woods.

should suffice to induce the "hunting" context. Mention of such words or items as "outdoors," "hike," "campfire," "duck blind," "marksman," and so on, ought to also prime a hearer appropriately, even though some of these words (e.g., "outdoors" and "hike") are more closely related to many other concepts than to "hunting." We are thus apparently faced with either (a) the need to infer the special context-setting concept "hunting," given any of the words or items above; or (b) the need to provide connections between each of the words or items and *all* the various word senses they prime. There is, however, a better alternative.

We propose that each concept should be represented not merely as a unitary node, but should in addition be associated with a set of "microfeatures" that serve both (a) to define the concepts, at least partially, and (b) to associate the concept with others that share its microfeatures. We propose a large set of microfeatures (on the order of a thousand), each of which is potentially connected to every concept node in the system (potentially on the order of hundreds of thousands). Each concept is in fact connected to only some subset of the total set, via either bidirectional activation or bidirectional inhibition links. Closely related concepts have many microfeatures in common. The microfeatures are intended to be part of a module that can be driven by perception, language input, and memory.

We suggest that microfeatures should be chosen on the basis of first principles to correspond to the major distinctions humans make about situations in the world, that is, distinctions we must make to survive and thrive, and major divisions of history, geography, and topic. For example, some important microfeatures correspond to distinctions such as threatening/safe, animate/inanimate, edible/inedible, indoors/outdoors, good outcome/ neutral outcome/bad outcome, moving/still, intentional/unintentional, characteristic lengths of events (e.g., whether events require milliseconds, hours, or years), locations (particular cities, countries, continents, etc.) and historical times (dates or periods). As in Hinton's (1981) model, hierarchies arise naturally, based on subsets of shared microfeatures, but are not the fundamental basis for organizing concepts in a semantic network, as in most AI models.

*Microfeatures as a Priming Context—An Example.* Let us see how microfeatures could help solve the problems presented by the example from Figure 1. Figure 4 shows a partial set of microfeatures, corresponding to temporal event length or location type (setting) running horizontally. A small set of concepts relevant to our example is listed across the top. Solid circles denote strong connection of concepts to microfeatures, open circles, a weak connection, and crosses, a negative connection. A simple scoring scheme allows "weekend" and "outdoors" to appropriately prime concepts related to "fire at" and "deer" relative to "waste money" and "dollar," as well as the ability of "casino" or "video game" to induce an opposite priming effect, as shown in Figure 4b. It is interesting to compare these effects with the effects of priming with "hunting" or "gambling" directly. No relaxation was used, though it obviously could be (i.e., a concept could activate microfeatures, priming other concepts, and then the primed concepts could change the activation of the microfeatures, in turn activating new concepts and eventually settling down.[9] We have been experimenting with a number of possible weighting and propagation schemes, and have built up a much larger matrix than the one shown in Figure 4.

*Microfeatures and Context.* Ideally, the particular set of microfeatures associated with a concept should serve two purposes: (1) it should be sufficient to distinguish the concept from all others, and (2) it should have shared microfeatures with all the concepts that should be associated with the given concept, but that are not related to in the ways that we would generally class as common "free associations" or *n*-ary relations. The set of microfeatures is thus partially definitional, but there is strictly speaking, no such thing as a complete definition for a concept in our model. In this regard the microfeatures resemble the primitives used by Wilks in his "preference semantics" (1975). This view is also similar to that expressed in Minsky (1977), that is, that concepts are defined by their positions in a network, i.e., the things to which they are connected.

We have represented all possible microfeatures as a vector, where each position of the vector corresponds to an independent microfeature, and the numerical value at that position corresponds to the level of activation of

---

[9] We have tried hard to be fair in contructing Figure 4a, for example priming with "outdoor" rather than "woods," and including links between "casino" and "desert" to acknowledge Las Vegas. Time periods characterize event lengths. Locations are to be taken as settings or surroundings, *not* objects. All links are clearly culturally dependent though, we think, roughly in accord with current middle-class American language usage.

that feature. One might think that some microfeatures ought to be directly connected to each other by mutually inhibitory link or mutually activating links; for example "outdoor" and "indoor" microfeatures tend to be mutually inhibitory. However, both the features "indoor" and "outdoor"

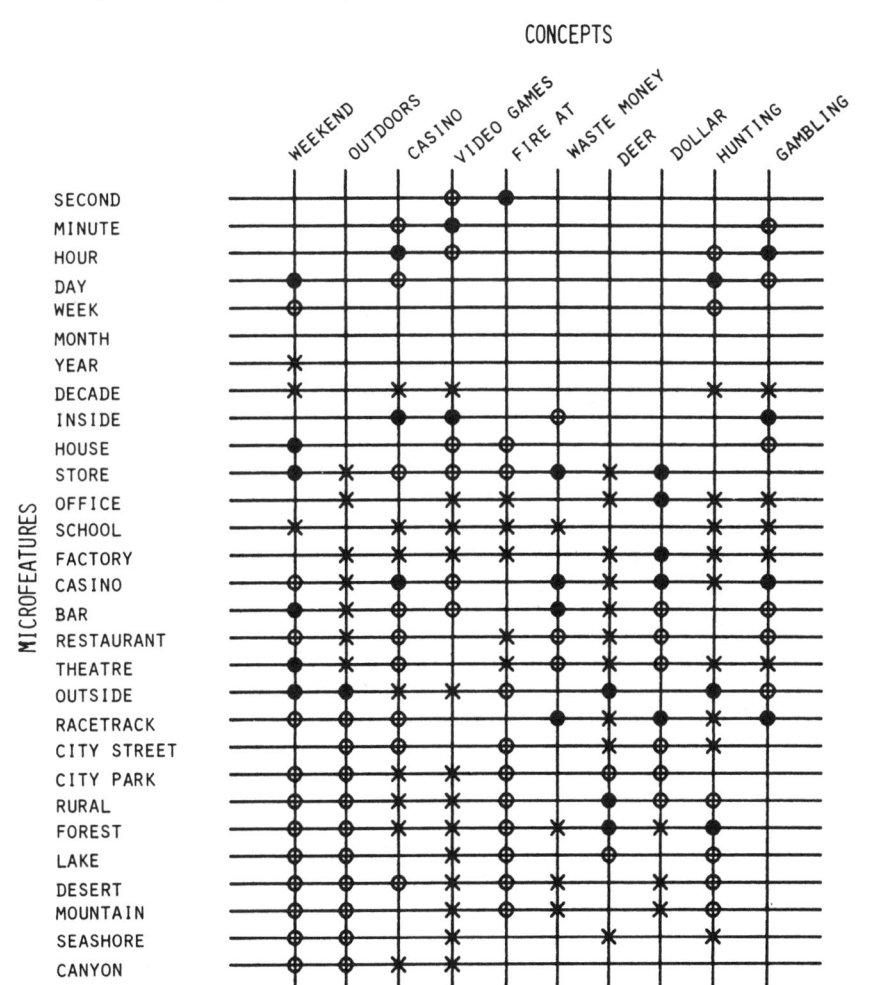

CONCEPTS: WEEKEND, OUTDOORS, CASINO, VIDEO GAMES, FIRE AT, WASTE MONEY, DEER, DOLLAR, HUNTING, GAMBLING

MICROFEATURES: SECOND, MINUTE, HOUR, DAY, WEEK, MONTH, YEAR, DECADE, INSIDE, HOUSE, STORE, OFFICE, SCHOOL, FACTORY, CASINO, BAR, RESTAURANT, THEATRE, OUTSIDE, RACETRACK, CITY STREET, CITY PARK, RURAL, FOREST, LAKE, DESERT, MOUNTAIN, SEASHORE, CANYON

- ● CONCEPT AND MICROFEATURE CHARACTERISTICALLY ASSOCIATED; WEIGHTING = 1
- ⊕ MILD ASSOCIATION; WEIGHTING = .5
- | COULD BE ASSOCIATED, BUT CHARACTERISTICALLY UNRELATED; WEIGHTING = 0
- ✱ NEGATIVELY ASSOCIATED; CONCEPT AND MICROFEATURE TEND TO BE MUTUALLY EXCLUSIVE; WEIGHTING = -.5

Figure 4a. This figure illustrates the use of microfeatures to provide contextual priming. At any given time, microfeatures will display some pattern of activation. Each concept has an induced activation level as a result of the microfeature activation values. The microfeature activations are modified whenever a concept is primed.

Fraction of Maximum Possible Score

Instantaneous priming effects on concepts; numbers show fraction of maximum possible score induced by the priming concepts. Microfeatures start at 0, and undergo a single priming cycle

| Priming Concepts | Primed Concepts | | | |
|---|---|---|---|---|
| | Fire-at | Waste | Deer | Dollar |
| Weekend | .41 | .55 | 0 | .46 |
| Outdoors | .41 | 0 | .44 | .08 |
| Casino | .05 | .59 | 0 | .42 |
| Video Games | .18 | .36 | 0 | .19 |
| Weekend+ Outdoors | .41 | .07 | .25 | .12 |
| Hunting | .36 | 0 | .50 | 0 |
| Gambling | .09 | .59 | 0 | .38 |

Figure 4b. For our example, assume "weekend" is primed, with all microfeatures initially at 0. The top line of Figure 4b shows the activation levels of concepts where the number represents a fraction of the maximum possible activation for that concept. These values prime various word sense nodes differentially.

in the sense we wish to use microfeatures may be present in varying degrees: for example, a room with big picture windows would have a high value for indoors, but a nonzero value for outdoors, while a dense forest or space under an umbrella would have a high value for outdoors, but a nonzero value for indoors, due to perceptual enclosure and partial protection above.

Only a subset of the possible combinations of microfeature values can ever occur as contexts; although a perceptual system could in principle induce any values for the microfeatures in a full system, the real world in fact behaves in an orderly manner, so that only certain value combinations would actually be observed. Characteristic constellations of microfeature values may occur frequently or persist through time.[10] Such constellations divide the world into classes of background situations that correspond to context-setting concepts, such as "hunting," "gambling," "working in an office," or "bargaining."

The microfeature vector could be primarily driven by a memory system rather than by the perceptual system, as in vivid remembering, or in planning. While the microfeatures are primarily determined by the need to form subsets of possible situations and actions, in practice the microfea-

[10] Contexts ought to be much more persistent than individual sentence meanings, which in turn ought to be much more persistent than syntactic constructs. For example, a single noun phrase-recognizing mechanism may need to be reused several times in processing a sentence, and thus would have to be rapidly deactivated as soon as its results (e.g., a case role entry) were stored or passed to other mechanisms. For related ideas, see Woods' paper on cascaded ATN's (Woods, 1980).

tures would probably become the organizing principle of all other memories as well.

Because of memory and perceptual constraints, it would not be possible to have more than a small number of context-settings represented at once in the vector. Most typically, one context-setting at a time is represented: at least three would be needed in a situation where one is, say, planning in an imagined future, while living in the present, using a remembered context-setting for help in planning or in coping with the present. Two or three contexts would be needed for understanding my interaction with another person, one for my own world view, and one for the other person's view of the world including me. If the other person's view of me is well-enough known and sufficiently different from my own view of myself, I would need at least three contexts: (1) myself, (2) the other person, and (3) the other person's model of me.[11]

Deeper embeddings would thus of necessity be hard: e.g., "my representation of the model of the other person that the other person believes I hold." Interactions of three or more people would also be hard for us to model in this view, in keeping with the observation that with larger groups we tend not to model each individual, but to divide up the larger groups into subgroups, "us" and "them," the "good guys" and the "bad guys," "allies," who "see eye-to-eye," that is, who tend to share a world view/context, and "enemies," whose views/contexts differ (Wilks & Bien, 1983).

## THE NEED FOR NEW ARCHITECTURE

Besides the evidence and phenomena which suggest a strongly integrated model for natural language, there is another reason we are looking towards a parallel model. Computer scientists, like cognitive scientists, tend to be limited by the conceptual framework of serial processing, the 30-year-old framework of the "von Neumann" machine, with its Central Processing Unit connected to its passive array of memory by a small bundle of wires.

---

[11] Presumably, as infants we can support only a single context, the egocentric one. Development of the ability to simultaneously support more than one context comes later, and may be the result of dividing the context vector into two subsets, each of which is reasonably complete, but which (1) can be separately activated as a whole, and (2) can support different activation patterns. Mechanisms for this could arise gradually, by processes such as reifying groups of microfeatures that frequently occur simultaneously. If we assume that activated microfeatures are sparsely scattered through the microfeature vector, it would be possible to support two or more separate contexts (e.g., me playing, mother reading). Alternatively, we may learn to "identify with" others, and use a single set of egocentric microfeatures to simulate their contexts. We hope our views may eventually have interesting connections to the ideas of Freud, Piaget, and others, but more than a footnote is premature at this point.

Backus addresses this problem in his 1977 Turing Lecture,[12] partially titled "Can Programming be Liberated from the von Neumann Style?":

> Conventional programming languages are basically high-level, complex versions of the von Neumann computer. Our thirty year old belief that there is only one kind of computer is the basis of our belief that there is only one kind of programming language, the conventional—von Neumann—language...(Backus, 1978, p. 615).
>
> Our fixation on von Neumann languages has continued the primacy of the von Neumann computer, and our dependency on it has made non-von Neumann languages uneconomical and has limited their development. The absence of full scale, effective programming styles founded on non-von Neumann principles has deprived designers of an intellectual foundation for new computer architectures (Backus, 1978, p. 616).

Backus's challenge, then, is to devise methods of computing which overcome the intellectual bottleneck in which both cognitive and computer scientists are stuck.

We have been consciously trying to answer Backus' challenge, by always informing our model with the constraints of "realizable parallelism."[13] Following in the footsteps of Fahlman, who devised a special purpose parallel processor for intersection search on a semantic network (Fahlman, 1979), and Hillis, who designed and is now building the Connection Machine (Hillis, 1981), we have designed two parallel communications architectures for modeling activation networks (Debrunner, 1983; Pollack, 1982).

## CONCLUSION

We have only scratched the surface of dynamic connectionist models for language interpretation. The ideas have a long history in psychology, but a very sparse history of computer implementation. Spreading activation and lateral inhibition provide a good framework for embedding comprehension phenomena which cannot even be approached with binary serial models. We have shown that disparate knowledge sources can be smoothly integrated and can be brought to bear simultaneously on the natural language processing task. While it is clearly very crude as a cognitive model, our microfeature and concept array is a beginning toward a system which has the (correct) property of understanding an input utterance as "a 'perturbation' to an

---

[12] (Backus, 1978, pp. 615–616). The Turing Award is the highest honor given each year by the Association for Computing Machinery.

[13] By "realizable parallelism," we mean the communication and computation constraints from parallel computer architecture. It makes little sense to design a parallel machine in which a million processors can execute concurrently, but all have to queue up to access a central memory bank. Similarly, one cannot posit a communications network based on a tremendous crossbar switch, or use parallelism to solve an NP-hard problem in unit time!

ongoing cognitive system that is trying to make sense of things" (Winograd, 1981, p. 245). We have also shown that structural preferences such as Minimal Attachment (Frazier, 1979) can be understood as side-effects of, rather than as strategies for, a syntactic processor, and that hypotheses about lexical disambiguation in context (Seidenberg, et al., 1980; Swinney, 1979) can nicely fit into a model with lateral inhibition while it could not be accounted for by activation alone. Garden-paths at different levels of processing can be explained by the breakdown of a common approximate consistent labeling algorithm—Lateral Inhibition—the "Universal Will to Disambiguate."

Questions that still need answering include:

1. What representation system underlies (and causes) activation and inhibition links? We believe at this time that they may be based on discriminators among a distributed set of "microfeatures" of much finer grain than those used in this paper (Hinton, 1981).
2. How can a network be dynamically generated without expanding the system's power to an unreasonable degree? Obviously, a production system could be used in the manner of the READER system (Thibadeau, Just, & Carpenter, 1982) or ACT* (Anderson, 1983), but we feel that the shared "blackboard" is a bottleneck to massive parallelism (See, for example the analysis of HEARSAY II by Fennell and Lesser, 1977.) Clearly, some dynamic generation is needed to deal with the problems of embedding and crosstalk. The only approaches we know of for dynamic construction of activation networks are Feldman's Dynamic Connections (1982) and McClelland's CIP (1984, pp. 113-146), but for pragmatic reasons in our experiments we have used "normal" computer programs to generate and connect up pieces of our networks.

Generally, we are very excited about the distributed approaches to cognitive modeling which have been on the rise in the past few years, and hope that this paper contributes to their ultimate ascendency.

## REFERENCES

Anderson, J. R. (1983). *The architecture of cognition*. Cambridge: Harvard University Press.
Backus, J. (1978). Can programming be liberated from the von Neumann style? A functional style and its algebra of programs. *Communications of the ACM, 21,* 613-641.
Berwick, R. C. (1983). Transformational grammar and artificial intelligence: A contemporary view. *Cognition and Brain Theory, 6,* 383-416.
Charniak, E. (1983). Passing markers: A theory of contextual influence in language comprehension. *Cognitive Science, 7,* 171-190.
Chomsky, N. (1957). *Syntactic structures*. The Hague: Mouton.
Church, K. W. (1980). On memory limitations in natural language processing. Master's thesis, MIT, Cambridge, MA.
Collins, A. M., & Loftus, E. F. (1975). A spreading activation theory of semantic processing. *Psychological Review, 82,* 407-428.

Cottrell, G. W., & Small, S. L. (1983). A connectionist scheme for modelling word sense disambiguation. *Cognition and Brain Theory, 6,* 89-120.
Crain, S., & Steedman, M. (1981, March). On not being led up the garden path: The use of context by the psychological parser. Paper presented to the Sloan Conference on Modelling Human Parsing, Austin, TX.
Debrunner, C. (1983). A two-dimensional activation cell. Working paper 41, Advanced Automation Research Group, Coordinated Science Laboratory, Urbana, IL.
DeJong, G. F., & Waltz, D. L. (1983). Understanding novel langauge. In N. Cercone (Ed.), *Computational linguistics*. Elmsford, NY: Pergamon Press.
Fahlman, S. E. (1979). *NETL: A system for representing and using real-world knowledge*. Cambridge: MIT Press.
Feldman, J. A. (1981). A connectionist model of visual memory. In G. E. Hinton & J. A. Anderson (Eds.), *Parallel models of associative memory*. Hillsdale, NJ: Erlbaum.
Feldman, J. A. (1982). Dynamic connections in neural networks. *Biological Cybernetics, 46,* 27-39.
Feldman, J. A., & Ballard, D. H. (1982). Connectionist models and their properties. *Cognitive Science, 6,* 205-254.
Fennel, R. D., & Lesser, V. R. (1977). Parallelism in AI problem-solving: A case study of HEARSAY II. *IEEE Transactions on Computers, C-26,* 98-111.
Fodor, J. (1982). *The modularity of mind*. Cambridge: MIT Press.
Ford, M., Bresnan, J., & Kaplan, R. (1982). A competence-based theory of syntactic closure. In J. Bresnan (Ed.), *The mental representation of grammatical relations*. Cambridge: MIT Press.
Frazier, L. (1979). *On comprehending sentences: Syntactic parsing strategies*. Indiana University Linguistics Club. Bloomington, IN.
Frazier, L., & Fodor, J. D. (1978). The sausage machine: A new two-stage parsing model. *Cognition, 6,* 291-326.
Gazdar, G., & Pullum, G. (1982, August). *GPSG: A theoretical synopsis*. Indiana University Linguistics Club. Bloomington, IN.
Gigley, H. M. (1982, March). A computational neurolinguistic approach to processing models of sentence comprehension. (COINS Tech. Rep. 82-9). Amherst: CIS Dept., University of Massachusetts.
Goodman, B. (1984). Communication and miscommunication. Unpublished doctoral dissertation. Department of Computer Science, University of Illinois.
Hewitt, C. (1976). Viewing control structures as patterns of passing messages. (AI Memo 410). MIT AI Lab, Cambridge, MA.
Hillis, W. D. (1981). The connection machine (computer architecture for the new wave). (AI MEMO 646). MIT AI Lab, Cambridge, MA.
Hinton, G. E. (1981). Implementing semantic networks in parallel hardware. In G. E. Hinton & J. A. Anderson (Eds.), *Parallel models of associative memory*. Hillsdale: Erlbaum.
Kay, M. (1973). The MIND system. In Rustin (Ed.), *Natural language processing*. New York: Algorithmics Press.
Kimball, J. (1973). Seven principles of surface structure parsing in natural language. *Cognition, 2,* 15-47.
Kwasny, S. C., & Sondheimer, N. K. (1981). Relaxation techniques for parsing ill-formed input. *American Journal of Computational Linguistics, 7,* 99-108.
MacKay, D. G. (1966). To end ambiguous sentences. *Perception and Psychophysics, 1,* 426-436.
Marcus, M. P. (1980). *A theory of syntactic recognition for natural language*. Cambridge: MIT Press.

Marcus, M. P. (1982). Consequences of functional deficits in a parsing model: Implications for Broca's aphasia. In M. A. Arbib, D. Caplan, & J. C. Marshall (Eds.), *Neural models of language processes.* New York: Academic.

Marslen-Wilson, W., & Tyler, L. K. (1980). The temporal structure of spoken language understanding. *Cognition, 8,* 1-72.

McClelland, J. L., & Rumelhart, D. E. (1980). An interactive activation model of the effect of context in perception. (Tech. Rep. No 91). Center for Human Information Processing, University of California at San Diego.

Milne, R. W. (1982). An explanation for minimal attachment and right association. *Proc. AAAI-82,* Carnegie-Mellon University, Pittsburgh, PA. 88-90.

Minsky, M. L. (1977, August). Plain talk about neurodevelopmental epistemology. *Proceedings of the Second International Joint Conference on Artificial Intelligence-77,* MIT, Cambridge, MA. 1083-1092.

Minsky, M. L. (1980). K-lines: A theory of memory. *Cognitive Science, 4,* 117-133.

Ortony, A., & Radin, D. (1983). Sapien: Spreading activation processor for information encoded network structures. (Tech. Rep. No. 296) Center for the Study of Reading, Champaign, IL.

Pollack, J. (1982). An activation/inhibition network VLSI cell. Working Paper 31, Advanced Automation Research Group, Coordinated Science Laboratory, Urbana, IL.

Pollack, J., & Waltz, D. (1982, August). NLP using spreading activation and lateral inhibition. *Proceedings of the 1982 Cognitive Science Conference,* Ann Arbor, MI. pp. 50-53.

Pylyshyn, Z. W. (1980). Computation and cognition: Issues in the foundations of cognitive science. *The Behavioral and Brain Sciences, 3,* 111-169.

Quillian, M. R. (1968). Semantic memory. In M. Minsky (Ed.), *Semantic information processing.* Cambridge: MIT Press.

Riesbeck, C., & Schank, R. C. (1976). Comprehension by computer: Expectation-based analysis of sentences in context. (Research Report 78), Computer Science Department, Yale University, New Haven, CT.

Sager, N. (1981). *Natural language information processing.* New York: Addison-Wesley.

Schank, R. C., Goldman, N., Rieger, C., & Riesbeck, C. (1973). MARGIE: Memory, analysis, response generation and inference in English. *Proceedings of the Second International Joint Conference on Artificial Intelligence,* Stanford University, Stanford, CA., pp. 255-262.

Seidenberg, M. S., Tanenhaus, M. K., & Leiman, J. M. (1980, March). *The time course of lexical ambiguity resolution in context.* (Tech. Rep. No. 164). Center for the Study of Reading, University of Illinois, Urbana.

Shieber, S. M. (1983). Sentence disambiguation by a shift-reduce parsing technique. *Proceedings of 21st Association of Computational Linguistics Conference,* Cambridge, MA.

Small, S. (1980). Word expert parsing: A theory of distributed word-based natural language understanding. (Tech. Rep. No. 954). Department of Computer Science, University of Maryland.

Swinney, D. A. (1979). Lexical access during sentence comprehension: (Re)consideration of context effects. *Journal of Verbal Learning and Verbal Behavior, 18,* 645-659.

Thibadeau, R., Just, M. A., & Carpenter, P. A. (1982). A model of the time course and content of reading. *Cognitive Science, 6,* 157-203.

Waltz, D. L. (1978). An English language question answering system for a large relational database. *Communications of the ACM, 21,* 526-539.

Waltz, D. L. (1979). On the function of mental imagery. *The Behavioral and Brain Sciences, 2,* 567-570.

Wanner, E. (1980). The ATN and the sausage machine: Which one is baloney? *Cognition, 8,* 209-225.

Wilks, Y. (1975). A preferential, pattern-seeking, semantics for natural language inference. *Artificial Intelligence, 6,* 53-74.
Wilks, Y., & Bien, J. (1983). Beliefs, points of view, and multiple environments. *Cognitive Science, 7,* 95-119.
Winograd, T. (1972). *Understanding natural language.* New York: Academic.
Winograd, T. (1981). What does it mean to understand language? In D. Norman (Ed.), *Perspectives on cognitive science.* Norwood, NJ: Ablex.
Woods, W. A. (1970). Transition network grammars for natural language analysis. *Communications of the ACM, 13,* 591-606.
Woods, W. A. (1980). Cascaded ATNs. *Journal of Association for Computational Linguistics, 6,* 1-12.

CHAPTER 8

# Feature Discovery by Competitive Learning*

DAVID E. RUMELHART
DAVID ZIPSER
*University of California at San Diego*

This paper reports the results of our studies with an unsupervised learning paradigm which we have called "Competitive Learning." We have examined competitive learning using both computer simulation and formal analysis and have found that when it is applied to parallel networks of neuron-like elements, many potentially useful learning tasks can be accomplished. We were attracted to competitive learning because it seems to provide a way to discover the salient, general features which can be used to classify a set of patterns. We show how a very simply competitive mechanism can discover a set of feature detectors which capture important aspects of the set of stimulus input patterns. We also show how these feature detectors can form the basis of a multilayer system that can serve to learn categorizations of stimulus sets which are not linearly separable. We show how the use of correlated stimuli can serve as a kind of "teaching" input to the system to allow the development of feature detectors which would not develop otherwise. Although we find the competitive learning mechanism a very interesting and powerful learning principle, we do not, of course, imagine that it is the only learning principle. Competitive learning is an essentially nonassociative statistical learning scheme. We certainly imagine that other kinds of learning mechanisms will be involved in the building of associations among patterns of activation in a more complete neural network. We offer this analysis of these competitive learning mechanisms to further our understanding of how simple adaptive networks can discover features important in the description of the stimulus environment in which the system finds itself.

This paper reports the results of our studies with an unsupervised learning paradigm which we have called "Competitive Learning." We have examined competitive learning using both computer simulation and formal analysis

* This research was supported by grants from the System Development Foundation and by Contract No. N00014-79-C-0323, NR 667-437 with the Personnel and Training Research Programs of the Office of Naval Research.

and have found that when it is applied to parallel networks of neuron-like elements, many potentially useful learning tasks can be accomplished. We were attracted to competitive learning because it seems to provide a way to discover the salient, general features which can be used to classify a set of patterns. The basic components of the competitive learning scheme are:

1. Start with a set of units that are all the same except for some randomly distributed parameter which makes each of them respond slightly differently to a set of input patterns.
2. Limit the "strength" of each unit.
3. Allow the units to compete in some way for the right to respond to a given subset of inputs.

The net result of correctly applying these three components to a learning paradigm is that individual units learn to specialize on sets of similar patterns and thus become "feature detectors" or "pattern classifiers." In addition to Frank Rosenblatt, whose work will be discussed, several others have exploited competitive learning in one form or another over the years. These include Christoph Von der Malsburg (1973), Stephen Grossberg (1976), Kunihiko Fukushima (1975), and Tuevo Kohonen (1982). Our analyses differ from many of these in that we focus on the development of feature detectors rather than pattern classification. We address these issues further below.

One of the central issues in the study of the processing capacities of neuron-like elements concerns the limitations inherent in a one-level system and the difficulty of developing learning schemes for multilayered systems. Competitive learning is a scheme in which important features can be discovered at one level that a multilayer system can use to classify pattern sets which cannot be classified with a single-level system.

Thirty-five years of experience have shown that getting neuron-like elements to learn some easy things is often quite straightforward, but designing systems with powerful general learning properties is a difficult problem, and the competitive learning paradigm does not change this fact. What we hope to show is that competitive learning is a powerful strategy which, when used in a variety of situations, greatly expedites some difficult tasks. Since the competitive learning paradigm has roots which go back to the very beginnings of the study of artificial learning devices, it seems reasonable to put the whole issue into historical perspective. This is even more to the point, since one of the first simple learning devices, the perceptron, caused a great furor and debate, the reverberations of which are still with us.

In the beginning, 35 or 40 years ago, it was very hard to see how anything resembling a neural network could learn at all, so any example of learning was immensely interesting. Learning was elevated to a status of great importance in those days because it was somehow uniquely associated with the properties of animal brains. After McCulloch and Pitts (1943) showed how neural-like networks could compute, the main problem then facing workers in this area was to understand how such networks could learn.

The first set of ideas that really got the enterprise going were contained in Donald Hebb's *Organization of Behavior* (1949). Before Hebb's work, it was believed that some physical change must occur in a network to support learning, but it was not clear what this change could be. Hebb proposed that a reasonable and biologically plausible change would be to strengthen the connections between elements of the network only when both the pre- and postsynaptic units were active simultaneously. The essence of Hebb's ideas still persist today in many learning paradigms. The details of the rules for changing weight may be different, but the essential notion that the strength of connections between the units must change in response to some function of the correlated activity of the connected units still dominates learning models.

Hebb's ideas remained untested speculations about the nervous system until it became possible to build some form of simulated network to test learning theories. Probably the first such attempt occurred in 1951 when Dean Edmonds and Marvin Minsky built their learning machine. The flavor of this machine and the milieu in which it operated is captured in Minsky's own words which appeared in a wonderful *New Yorker* profile of him by Jeremy Bernstein (1981):

> in the summer of 1951 Dean Edmonds and I went up to Harvard and built our machine. It had three hundred tubes and a lot of motors. It needed some automatic electric clutches, which we machined ourselves. The memory of the machine was stored in the positions of its control knobs, 40 of them, and when the machine was learning, it used the clutches to adjust its own knobs. We used a surplus gyropilot from a B24 bomber to move the clutches (p. 69).

This machine actually worked and was so fascinating to watch that Minsky remembers:

> We sort of quit science for awhile to watch the machine. We were amazed that it could have several activities going on at once in this little nervous system. Because of the random wiring it had a sort of fail safe characteristic. If one of the neurons wasn't working, it wouldn't make much difference and with nearly three hundred tubes, and the thousands of connections we had soldered there would usually be something wrong somewhere.... I don't think we ever debugged our machine completely, but that didn't matter. By having this crazy random design it was almost sure to work no matter how you built it (p. 69).

In fact the functioning of this machine apparently stimulated Minsky sufficiently to write his PhD thesis on a problem related to learning (Minsky, 1954). The whole idea must have generated rather wide interest; von Neumann for example, was on Minsky's PhD committee and gave him encouragement. Although Minsky was perhaps the first on the scene with a learning machine, the real beginnings of meaningful neuron-like network learning can probably be traced to the work of Frank Rosenblatt, a Bronx High School of Science classmate of Minsky's. Rosenblatt invented a class of simple neuron-like learning networks which he called perceptrons. In his book, *Principles of Neurodynamics* (1962), Rosenblatt brought together all of his results on perceptrons. In that book he gives a particularly clear description of what he thought he was doing:

> Perceptrons are not intended to serve as detailed copies of any actual nervous system. They're simplified networks, designed to permit the study of lawful relationships between the organization of a nerve net, the organization of its environment, and the "psychological" performances of which it is capable. Perceptrons might actually correspond to parts of more extended networks and biological systems; in this case, the results obtained will be directly applicable. More likely they represent extreme simplifications of the central nervous system, in which some properties are exaggerated and others suppressed. In this case, successive perturbations and refinements of the system may yield a closer approximation.
>
> The main strength of this approach is that it permits meaningful questions to be asked and answered about particular types of organizations, hypothetical memory mechanisms, and neural models. When exact analytical answers are unobtainable, experimental methods, either with digital simulation or hardware models, are employed. The model is not the terminal result, but a starting point for exploratory analysis of its behavior (p. 28).

Rosenblatt pioneered two techniques of fundamental importance to the study of learning in neural-like networks: digital computer simulation, and formal mathematical analysis, although he was not the first to simulate neural networks that could learn on digital computers (cf., Farley & Clark, 1954).

Since the paradigm of competitive learning uses concepts which appear in the work of Rosenblatt, it is worthwhile reviewing some of his ideas in this area. His most influential result was the "perceptron learning theorem" which boldly asserts:

> Given an elementary $\alpha$ - perceptron, a stimulus world W, and any classification C(W) for which a solution exits; let all stimuli in W occur in any sequence, provided that each stimulus must re-occur in finite time; then beginning from an arbitrary initial state, an error correction procedure will always yield a solution to C(W) in finite time,....(p. 596).

As it turned out, the real problem arose out of the phrase "for which a solution exists"—more about this later.

Less widely known is Rosenblatt's work on what he called "spontaneous learning." All network learning models require rules, which tell how to present the stimuli and change the values of the weights in accordance with the response of the model. These rules can be characterized as forming a spectrum, at one end of which is learning with an error-correcting teacher, and at the other is completely spontaneous, unsupervised discovery. In between is a continuum of rules which depend on manipulating the content of the input stimulus stream to bring about learning. These intermediate rules are often referred to as "forced learning." Here we are concerned primarily with attempts to design a perceptron that would discover something interesting without a teacher because this is similar to what happens in the competitive learning case. In fact, Rosenblatt was able to build a perceptron that was able to spontaneously dichotomize a random sequence of input patterns into classes such that the members of a single class were similar to each other, and different from the members of the other class. Rosenblatt realized that any randomly initialized perceptron would have to dichotomize an arbitrary input pattern stream into a "1-set" consisting of those patterns that happened to produce a response of 1, and a "0-set" consisting of those that produced a response of 0. Of course one of these sets could be empty by chance and neither would be of much interest in general. He reasoned that if a perceptron could reinforce these sets by an appropriate rule based only on the perceptron's spontaneous response and not on a teacher's error correction, it might eventually end up with a dichotomization in which the members of each set were more like each other than like the members of the opposite set. What was the appropriate rule to use to achieve the desired dicotomization? The first rule he tried for these perceptrons, which he called $C$ type, was to increment weights on lines active with patterns in the 1-set, and decrement weights on lines active with patterns in the 0-set. The idea was to force a dichotomization into sets whose members were similar in the sense that they activated overlapping subsets of lines. The results were disastrous. Sooner or later all the input patterns were classified in one set. There was no dichotomy but there was stability. Once one of the sets won, it remained the victor forever.

Not to be daunted, he examined why this undesirable result occurred and realized that the problem lay in the fact that since the weights could grow without limit, the set that initially had a majority of the patterns would receive the majority of the reinforcement. This meant that weights on lines which could be activated by patterns in both sets would grow to infinite magnitudes in favor of the majority set, which in turn would lead to the capture of minority patterns by the majority set, and ultimate total victory for the majority. Even where there was initial equality between the sets, inevitable fluctuations in the random presentation of patterns would create a majority set that would then go on to win. Rosenblatt overcame this problem by introducing mechanisms to limit weight growth in such a way that the set that was to be positively reinforced at active lines would compensate the other set by giving up some weight from all its lines. He called the modified perceptrons $C'$. An example of a $C'$ rule is to lower the magnitude of all weights by a fixed fraction of their current value before specifically incrementing the magnitude of some of the weights on the basis of the response to an input pattern. This type of rule had the desired result of making an equal dichotomy of patterns a stable rather than an unstable state. Patterns in each of the sets were similar to each other in the sense that they depended on similar sets of input lines to produce a response. In Rosenblatt's initial experiment, the main feature of similarity was not so much the shape of the patterns involved, but their location on the retina. That is, his system was able to spontaneously learn something about the geometry of its input line arrangement. Later, we will examine this important property of spontaneous geometry learning in considerable detail. Depending on the desired learning task, it can be either a boon or a nuisance.

Rosenblatt was extremely enthusiastic about his spontaneous learning results. In fact, his response can be described as sheer ecstasy. To see what he thought about his achievements, consider his claim (Rosenblatt, 1959):

> It seems clear that the class $C'$ perceptron introduces a new kind of information processing automaton: For the first time, we have a machine which is capable of having original ideas. As an analogue of the biological brain, the perceptron, more precisely, the theory of statistical separability, seems to come closer to meeting the requirements of a functional explanation of the nervous system than any system previously proposed (p. 449).

Although Rosenblatt's results were both interesting and significant, the claims implied in the above quote struck his contemporaries as unfounded. What was also significant was that Rosenblatt appeared to be saying that the type of spontaneous learning he had demonstrated was a property of perceptrons, which could not be replicated by ordinary computers. Consider the following quote from the same source:

> As a concept, it would seem that the perceptron has established, beyond doubt, the feasibility and principle of non-human systems which may embody human cognitive functions at a level far beyond that which can be achieved through present day automatons. The future of information processing devices which operate on statistical, rather than logical principles seems to be clearly indicated (p. 449).

It is this notion of Rosenblatt's—that perceptrons are in some way superior to computers—which ignited a debate in artificial intelligence (AI) that had significant effects on the development of neural-like network models for both learning and other cognitive processes. Elements of the debate are still with us today in arguments about what the brain can do that computers can't do. There is no doubt that this was an important issue in Rosenblatt's mind, and almost certainly contributed to the acrimonious debate at that time. Consider the following statement by Rosenblatt made at the important conference on Mechanization of Thought Processes back in 1959:

> Computers seem to share two main functions with the brain. a) Decision making, based on logical rule, and b) control, again based on logical rules. The human brain performs these functions, together with a third: interpretation of the environment. Why do we hold interpretation of the environment to be so important? The answer, I think, is to be found in the laws of thermodynamics. A system with a completely self contained logic can never spontaneously improve its ability to organize, and to draw valid conclusions from information (Rosenblatt, 1959, p. 423).

Clearly in some sense, Rosenblatt was saying that there were things that the brain and perceptrons, because of their statistical properties, could do which computers could not do. Now this may seem strange since Rosenblatt knew that a computer program could be written which would simulate the behavior of statistical perceptrons to any arbitrary degree of accuracy. Indeed, he was one of the pioneers in the application of digital simulation to this type of problem. What he was actually referring to is made clear when we examine the comments of other participants at the conference, such as Minsky (1959) and McCarthy (1959), who were using the symbol manipulating capabilities of the computer to directly simulate the logical processes involved in decision-making, theorem proving, and other intellectual activities of this sort. Rosenblatt believed the computer used in this way would be inadequate to mimic the brain's true intellectual powers. This task, he thought, could not only be accomplished if the computer or other electronic devices were used to simulate perceptrons. We can summarize these divergent points of view by saying that Rosenblatt was concerned not only with what the brain did, but with how it did it; whereas others, such as Minsky and McCarthy, were concerned with simulating what the brain did, and didn't really care how it was done. The subsequent history of AI has shown both the successes and failures of the standard AI approach. We still have the problems today, and it's still not clear to what degree computational strategies similar to the ones used by the brain must be employed in order to stimulate its performance.

In addition to producing fertilizer, as all debates do, this one also stimulated the growth of some new results on perceptrons, some of which came from Minsky. Rosenblatt had shown that a two layer perceptron could carry out any of the $2^{2^N}$ possible classifications of $N$ binary inputs; that is, a solution to the classification problem had always existed in principle. This result was of no practical value however, because $2^N$ units were required to accomplish the task in the completely general case. Rosenblatt's approach to this problem was to use a much smaller number of units in the first layer with each unit connected to a small subset of the $N$ inputs at random. His hope was that this would give the perceptron a high probability of learning to carry out classifications of interest. Experiments and formal analysis showed that these random devices could learn to recognize patterns to a significant degree but that they had severe limitations. Rosenblatt (1962) characterized his random perceptron as follows:

> It does not generalize well to similar forms occurring in new positions in the retinal field, and its performance in detection experiments, where a familiar figure appears against an unfamiliar background, is apt to be weak. More sophisticated psychological capabilities, which depend on the recognition of topological properties of the stimulus field, or on abstract relations between the components of a complex image, are lacking (pp. 191–192).

Minsky and Papert worked through most of the 60s on a mathematical analysis of the computing powers of perceptrons with the goal of understanding these limitations. The results of their work are available in a book called *Perceptrons* (Minsky & Papert, 1969). The central theme of this work is that parallel recognizing elements, such as perceptrons are beset by the same problems of scale as serial pattern recognizers. Combinatorial explosion catches you sooner or later, although sometimes in different ways in parallel than in serial. Minsky and Papert's book had a very dampening effect on the study of neuron-like networks as computational devices. Minsky has recently come to reconsider this negative effect:

> I now believe the book was overkill....So after being irritated with Rosenblatt for overclaiming and diverting all those people along a false path, I started to realize that for what you get out of it—the kind of recognition it can do—it is such a simple machine that it would be astonishing if nature did not make use of it somewhere. (Bernstein, 1981, p. 103).

Perhaps the real lesson from all this is that it really is worthwhile trying to put things in perspective.

Once the problem of scale has been understood, networks of neuron-like elements are often very useful in practical problems of recognition and classification. These networks are somewhat analogous to computers, in that they won't do much unless programmed by a clever person; networks, of course, are not so much programmed as designed. The problem of finding networks of practical size to solve a particular problem is challenging because relatively small changes in network design can have very large effects on the scale of a problem. Consider networks of neuron-like units that determine the parity of their $N$ binary inputs (see Figure 1). In the simple perceptrons studied by Minsky and Papert, units in the first layer output 1 only if all their inputs are 1 and output 0 otherwise. This takes $2^N$ units in the first layer, and a single linear threshold unit with a fan-in of $2^N$ in the second

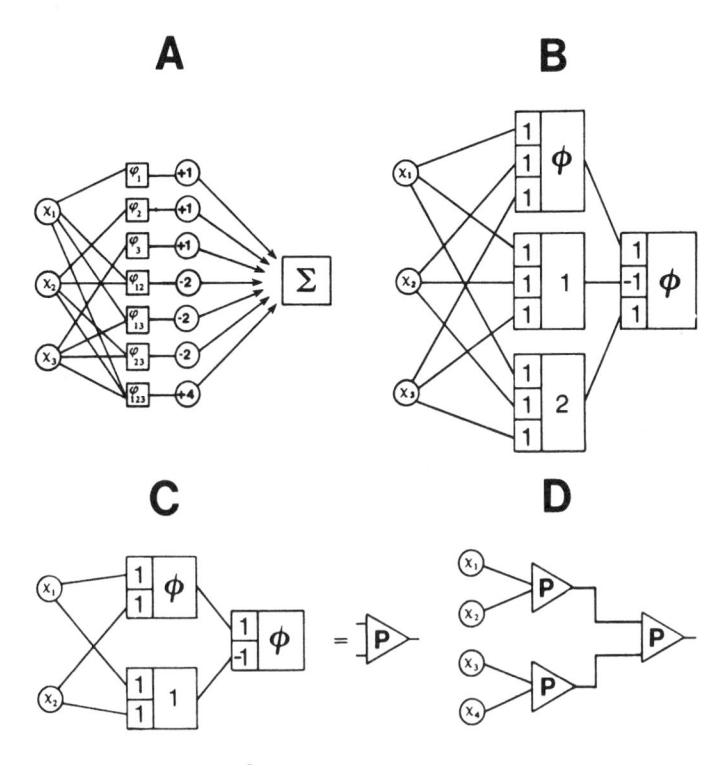

Figure 1. (A) Parity network from Minsky and Papert (1969). Each $\phi$ unit has an output of 1 only if all of its inputs are 1. $\Sigma$ is a linear threshold unit with threshold of 0, i.e., like all the other linear threshold units in the figure, it fires only when the sum of its weighted inputs is greater than the threshold. This and all the other networks signal odd parity with a 1 in the rightmost unit of the network. (B) Parity network made from two layers of linear threshold units. (C) Three unit network for determining the parity of a pair of inputs. (D) Two layer network using the subnetwork described in C. In general, the number of P-units is of order $N$ and the number of layers is of order $log_2 N$.

layer, to determine parity. If the units in the first layer are changed to linear threshold elements, then only $N$ of them are required, but all must have a fan-in of $N$. If we allow a multilayer network to do the job, then about $3N$ units are needed, but none needs a fan-in of more than 2. The number of layers is of order $log_2 N$. The importance of all this to the competitive learning paradigm, or any other for that matter, is that no network can learn what it is not capable of doing in principle. What any particular network can do is dependent on its structure and the computational properties of its component elements. Unfortunately there is no canonical way to find the best network or to determine what it will learn, so the whole enterprise still has much of the flavor of an experimental science.

# THE COMPETITIVE LEARNING MECHANISM

## Paradigms of Learning

It is possible to classify learning mechanisms in several ways. One useful classification is in terms of the learning paradigm in which the model is supposed to work. There are at least four common learning paradigms in neural-like processing systems:

*1. Auto Associator.* In this paradigm a set of patterns are repeatedly presented and the system is supposed to "store" the patterns. Then, later, parts of one of the original patterns or possibly a pattern similar to one of the original patterns is presented, and the task is to "retrieve" the original pattern through a kind of pattern completion procedure. This is an auto-association process in which a pattern is associated with itself so that a degraded version of the original pattern can act as a retrieval cue.

*2. Pattern Associator.* This paradigm is really a variant on the auto association paradigm. A set of *pairs* of patterns are repeatedly presented. The system is to learn that when one member of the pair is presented it is supposed to produce the other. In this paradigm one seeks a mechanism in which an essentially arbitrary set of input patterns can be paired with an arbitrary set of output patterns.

*3. Classification Paradigm.* The classification paradigm also can be considered as a variant on the previous learning paradigms, although the goals are sufficiently different and it is sufficiently common that it deserves separate mention. In this case, there are a fixed set of categories into which the stimulus patterns are to be classified. There is a training session in which the system is presented with the stimulus patterns along with the categories to which each stimulus belongs. The goal is to learn to correctly classify the stimuli so that in the future when a particular stimulus or a slightly distorted version of one of the stimuli is presented the system will classify it properly. This is the typical paradigm in which the perceptron is designed to operate and in which the perceptron convergence theorem is proved.

*4. Regularity Detector.* In this paradigm there is a population of stimulus patterns and each stimulus pattern, $S_k$, is presented with some probability $p_k$. The system is supposed to *discover* statistically salient features of the input *population*. Unlike the classification paradigm, there is no *a priori* set of categories into which the patterns are to be classified; rather, the system must develop its own featural representation of the input stimuli which captures the most salient features of the population of input patterns.

Competitive learning is a mechanism well-suited for regularity detection, as in the environment described in No. 4.

## Competitive Learning

The architecture of a competitive learning system (illustrated in Figure 2) is a common one. It consists of a set of hierarchically layered units in which each layer connects, via excitatory connections, with the layer immediately above it. In the most general case, each unit of a layer receives an input from each unit of the layer immediately below and projects to each unit in the layer immediately above. Moreover, within a layer, the units are broken into a set of inhibitory clusters in which all elements within a cluster inhibit all other elements in the cluster. Thus, the elements within a cluster at one level compete with one another to respond to the pattern appearing on the

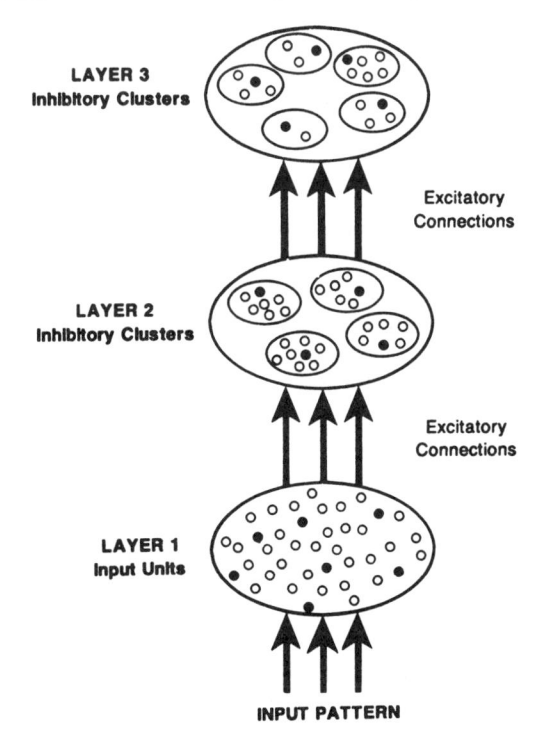

Figure 2. The architecture of the competitive learning mechanism. Competitive learning takes place in a context of sets of hierarchically layered units. Units are represented in the diagram as dots. Units may be active or inactive. Active units are represented by filled dots, inactive ones by open dots. In general, a unit in a given layer can receive inputs from all of the units in the next lower layer and can project outputs to all of the units in the next higher layer. Connections between layers are excitatory and connections within layers are inhibitory. Each layer consists of a set of clusters of mutually inhibitory units. The units within a cluster inhibit one another in such a way that only one unit per cluster may be active. We think of the configuration of active units on any given layer as representing the input pattern for the next higher level. There can be an arbitrary number of such layers. A given cluster contains a fixed number of units, but different clusters can have different numbers of units.

layer below. The more strongly any particular unit responds to an incoming stimulus, the more it shuts down the other members of its cluster.

There are many variations on the competitive learning theme. A number of researchers have developed variants of competitive learning mechanisms and a number of results already exist in the literature. We have already mentioned the pioneering work of Rosenblatt. In addition, Von der Malsburg (1973), Fukushima (1975), and Grossberg (1976), among others, have developed models which are competitive learning models, or which have many properties in common with competitive learning. We believe that the essential properties of the competitive learning mechanism are quite general. However, for the sake of concreteness, in this paper we have chosen to study, in some detail, the simplest of the systems which seem to be representative of the essential characteristics of competitive learning. Thus, the system we have analyzed has much in common with the previous work, but wherever possible we have simplified our assumptions. The system that we have studied most is described as follows:

1. The units in a given layer are broken into a set of nonoverlapping clusters. Each unit within a cluster inhibits every other unit within a cluster. The clusters are winner-take-all, such that the unit receiving the largest input achieves its maximum value while all other units in the cluster are pushed to the minimum value.[1] We have arbitrarily set the maximum value to 1 and the minimum value to 0.
2. Every element in every cluster receives inputs from the same lines.
3. A unit learns if and only if it wins the competition with other units in its cluster.
4. A stimulus pattern $S_j$ consists of a binary pattern in which each element of the pattern is either *active* or *inactive*. An active element is assigned the value 1 and an inactive element is assigned the value 0.
5. Each unit has a fixed amount of weight (all weights are positive) which is distributed among its input lines. The weight on the line connecting unit $i$ on the lower (or input) layer to unit $j$ on the upper layer, is designated $\omega_{ij}$. The fixed total amount of weight for unit $j$ is designated $\sum_i \omega_{ij} = 1$. A unit learns by shifting weight from its inactive to its active input lines. If a unit does not respond to a particular pattern no learning takes place in that unit. If a unit wins the competition, then each of its input lines give up some proportion $g$ of its weight and that weight is then distributed equally among the

---

[1] A simple circuit for achieving this result is attained by having each unit activate itself and inhibit its neighbors. Grossberg (1976) employs just such a network to *choose* the maximum value of a set of units.

active input lines.[2] More formally, the learning rule we have studied is:

$$\Delta\omega_{ij} = \begin{cases} 0 & \text{if } unit\ j\ loses\ on\ stimulus\ k \\ g\dfrac{c_{ik}}{n_k} - g\omega_{ij} & \text{if } unit\ j\ wins\ on\ stimulus\ k \end{cases}$$

where $c_{ik}$ is equal to 1 if in stimulus pattern $S_k$ unit $i$ in the lower layer is active and zero otherwise, and $n_k$ is the number of active units in pattern $S_k$ (thus

$$n_k = \sum_i c_{ik}).$$

Figure 3 illustrates a useful geometric analogy to this system. We can consider each stimulus pattern as a vector. If all patterns contain the same number of active lines, then all vectors are the same length and each can be viewed as a point on an $N$ dimensional hypersphere, where $N$ is the number of units in the lower level, and therefore, also the number of input lines received by each unit in the upper level. Each $x$ in Figure 3a represents a particular pattern. Those patterns which are very similar are near one another on the sphere, those which are very different will be far from one another on the sphere. Now, note that since there are $N$ input lines to each unit in the upper layer, its weights can also be considered a vector in $N$ dimensional space. Since all units have the same total quantity of weight, we have $N$ dimensional vectors of approximately fixed length for each unit in the cluster.[3] Thus, properly scaled, the weights themselves form a set of vectors which (approximately) fall on the surface of the same hypersphere. In figure 3b, the o's represent the weights of two units superimposed on the same sphere with the stimulus patterns. Now, whenever a stimulus pattern is presented, the unit which responds most strongly is simply the one whose weight vector is nearest that for the stimulus. The learning rule specifies that whenever a unit wins a competition for a stimulus pattern, it moves a percentage $g$ of the way from its current location toward the location of the stimulus pattern on the hypersphere. Now, suppose that the input patterns fell into

[2] This learning rule was proposed by Von der Malsburg (1973). As Grossberg (1976) points out, renormalization of the weights is not necessary. The same result can be obtained by normalizing the input patterns and then assuming that the weights approach the values on the input lines. Normalizing weights is simpler to implement than normalizing patterns, so we chose that option. For most of our experiments, however, it does not matter which of these two rules we chose since all patterns were of the same magnitude.

[3] It should be noted that this geometric interpretation is only approximate. We have used the constraint that $\sum_i \omega_{ij} = 1$ rather than the constraint that $\sum_i \omega_{ij}^2 = 1$. This latter constraint would ensure that all vectors are in fact the same length. Our assumption only assures that they will be approximately the same length.

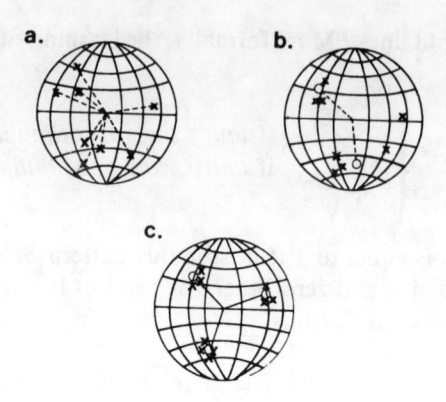

Figure 3. A geometric interpretation of competitive learning. (a) It is useful to conceptualize stimulus patterns as vectors who tips all lie on the surface of a hypersphere. We can then directly see the similarity among stimulus patterns as distance between the points on the sphere. In the figure, a stimulus pattern is represented as an x. The figure represents a population of eight stimulus patterns. There are two clusters of three patterns and two stimulus patterns which are rather distinct from the others. (b) It is also useful to represent the weights of units as vectors falling on the surface of the same hypersphere. Weight vectors are represented in the figure as 0's. The figure illustrates the weights of two units falling on rather different parts of the sphere. The response rule of this model is equivalent to the rule that whenever a stimulus pattern is presented, the unit whose weight vector is closest to that stimulus pattern on the sphere wins the competition. In the figure, one unit would respond to the cluster in the northern hemisphere and the other unit would respond to the rest of the stimulus patterns. (c) The learning rule of this model is roughly equivalent to the rule that whenever a unit wins the competition (i.e., is closest to the stimulus pattern), that weight vector is moved toward the presented stimulus. The figure shows a case in which there are three units in the cluster and three natural groupings of the stimulus patterns. In this case, the weight vectors for the three units will each migrate toward one of the stimulus groups.

some number, $M$, "natural" groupings. Further, suppose that an inhibitory cluster receiving inputs from these stimuli contained exactly $M$ units (as in Figure 3c). After sufficient training, and assuming that the stimulus groupings are sufficiently distinct, we expect to find one of the vectors for the $M$ units placed roughly in the center of each of the stimulus groupings. In this case, the units have come to detect the grouping to which the input patterns belong. In this sense, they have "discovered" the structure of the input pattern sets.

## Some Features of Competitive Learning

There are several characteristics of a competitive learning mechanism which make it an interesting candidate for further study, for example:

1. Each cluster classifies the stimulus set into $M$ groups, one for each unit in the cluster. Each of the units captures roughly an equal number of stimulus patterns. It is possible to consider a cluster as forming an $M$-ary feature in which every stimulus pattern is classified as having exactly one of the $M$ possible values of this feature. Thus, a cluster containing two units acts as a binary feature detector. One element of the cluster responds when a particular feature is present in the stimulus pattern, otherwise the other element responds.
2. If there is *structure* in the stimulus patterns, the units will break up the patterns along structurally relevant lines. Roughly speaking, this means that the system will find clusters if they are there. (A key problem, which we address below, is specifying the *nature* of the structure that this system discovers.)
3. If the stimuli are highly structured the classifications are highly stable. If the stimuli are less well-structured, the classifications are more variable, and a given stimulus pattern will be responded to first by one and then by another member of the cluster. In our experiments, we started the weight vectors in random directions and presented the stimuli randomly. In this case, there is rapid movement as the system reaches a relatively stable configuration (such as one with a unit roughly in the center of each cluster of stimulus patterns). These configurations can be more or less stable. For example, if the stimulus points don't actually fall into nice clusters, then the configurations will be relatively unstable and the presentation of each stimulus will modify the pattern of responding so that the system will undergo continual evolution. On the other hand, if the stimulus patterns fall rather nicely into clusters, then the system will become very stable in the sense that the same units will always respond to the same stimuli.[4]
4. The particular grouping done by a particular cluster depends on the starting value of the weights and the sequence of stimulus patterns actually presented. A large number of clusters, each receiving inputs from the same input lines can, in general, classify the inputs into a larger number of different groupings, or alternatively, discover a variety of independent features present in the stimulus

---

[4] Grossberg (1976) has addressed this problem in his very similar system. He has proved that if the patterns are sufficiently sparse, and/or when there are enough units in the cluster, then a system such as this will find a perfectly stable classification. He also points out that when these conditions don't hold, the classification can be unstable. Most of our work is with cases in which there is *no* perfectly stable classification and the number of patterns is *much* larger than the number of units in the inhibitory clusters.

population. This can provide a kind of coarse coding of the stimulus patterns.[5]

## Formal Analysis

Perhaps the simplest mathematical analysis that can be given of the competitive learning model under discussion involves the determination of the sets of *equilibrium states* of the system—that is, states in which the average inflow of weight to a particular line is equal to the average outflow of weight on that line. Let $p_k$ be the probability that stimulus $S_k$ is presented on any trial. Let $v_{jk}$ be the probability that unit $j$ wins when stimulus $S_k$ is presented. Now we want to consider the case in which $\sum_k \Delta \omega_{ij} V_{jk} p_k = 0$, that is, the case in which the average change in the weights is zero. We refer to such states as *equilibrium states*. Thus, using the learning rule and averaging over stimulus patterns we can write

$$0 = g\sum_k \frac{c_{ik}}{n_k} p_k v_{jk} - g\sum_k \omega_{ij} p_k v_{jk}$$

which implies that at equilibrium

$$\omega_{ij} \sum_k p_k v_{jk} = \sum_k \frac{p_k c_{ik} v_{jk}}{n_k}$$

and thus

$$\omega_{ij} = \frac{\sum_k \dfrac{p_k c_{ik} v_{jk}}{n_k}}{\sum_k p_k v_{jk}}$$

There are a number of important observations to note about this equation. First, note that $\sum_k p_k v_{ik}$ is simply the probability that unit $k$ wins averaged over all stimulus patterns. Note further that $\sum_k p_k c_{ik} v_{jk}$ is the probability that that input line $i$ is active and unit $j$ wins. Thus, the ratio

$$\frac{\sum_k p_k c_{ik} v_{jk}}{\sum_k p_k v_{jk}}$$

---

[5] There is a problem in that one can't assure that the different clusters will discover different features. A slight modification of the system in which clusters "repel" one another can insure that different clusters find different features. We shall not pursue that further in this paper.

is the conditional probability that line $i$ is active given unit $j$ wins, $p(line_i = 1 | unit_j \ wins)$. Thus, if all patterns are of the same size, i.e., $n_k = n$ for all $k$, then the weight $\omega_{ij}$ becomes proportional to the probability that line $i$ is active given unit $j$ wins. That is,

$$\omega_{ij} \rightarrow np(line_i = 1 | unit_j \ wins).$$

We are now in a position to specify the response, at equilibrium, of unit $j$ when stimulus $S_l$ is presented. Let $\alpha_{jl}$ be the input to unit $j$ in the face of stimulus $S_l$. This is simply the sum of weights on the active input lines. This can be written

$$\alpha_{jl} \rightarrow \sum_i \omega_{ij} c_{il} = \sum_i c_{il} \frac{\sum_k \frac{p_k c_{ik} v_{jk}}{n_k}}{\sum_k p_k v_{jk}}$$

which implies that at equilibrium

$$\alpha_{jl} = \frac{\sum_i p_i r_{li} v_{ji}}{\sum_i p_i v_{ji}}$$

where $r_{li}$ represents the overlap between stimulus $l$ and stimulus $i$,

$$r_{li} = \sum_k \frac{c_{ki} c_{kl}}{n_i}$$

Thus, at equilibrium a unit responds most strongly to patterns that overlap other patterns to which the unit responds, and responds most weakly to patterns that are far from patterns to which it responds. Finally, it should be noted that there is another set of restrictions on the value of $v_{jk}$—the probability that stimulus unit $j$ responds to stimulus $S_k$. In fact, the competitive learning rule we have studied has the further restriction that

$$v_{jk} = \begin{cases} 1 \leftarrow \rightarrow \alpha_{jk} > \alpha_{ik} \text{ for } all \ i \neq j \\ 0 \qquad otherwise \end{cases}$$

Thus, in general, there are many solutions to the equilibrium equations described above. The competitive learning mechanisms can only reach those equilibrium states in which the above-stated relationships between the $v_{jk}$ and the $\alpha_{jk}$ also hold.

Whenever the system is in a state in which, on average, the weights are not changing, we say that the system has reached an *equilibrium state*. In such a state the values of $\alpha_{jk}$ become relatively stable, and therefore, the values of $v_{ik}$ become stable. When this happens, the system always responds the same way to a particular stimulus pattern. However, it is possible that the weights will be pushed out of equilibrium by an unfortunate sequence of stimuli. In this case, the system can move toward a new equilibrium state (or possibly back to a previous one). Some equilibrium states are more stable than others in the sense that the $v_{ik}$ is very unlikely to change values for long periods of time. In particular, this will happen whenever the largest $\alpha_{jk}$ is much larger than any other $\alpha_{ik}$ for all stimulus patterns $S_k$. In this case, small movements in the weight vector of one of the units is very unlikely to change which unit responds to which stimulus pattern. Such equilibrium states are said to be highly *stable*. We should expect, then, that after it has been learning for a period of time, the system will spend most of its time in the most highly stable of the equilibrium states. One good measure of the stability of an equilibrium state is given by the average amount by which the input to the winning units is greater than the response of all of the other units averaged over all patterns and all units in a cluster. This measure is given by $T$ below.

$$T = \sum_k p_k \sum_{j,i} v_{jk}(\alpha_{jk} - \alpha_{ik})$$

The larger the value of $T$, the more stable the system can be expected to be, and the more time we can expect the system to spend in that state. Roughly, if we assume that the system moves into states which maximize $T$, we can show that this amounts to maximizing the overlap among patterns within a group while minimizing the overlap among patterns between groups. In the geometric analogy above, this will occur when the weight vectors point toward maximally compact stimulus regions which are as distant as possible from other such regions.

## SOME EXPERIMENTAL RESULTS

### Dipole Experiments

The essential structure that a competitive learning mechanism can discover is represented in the overlap of stimulus patterns. The simplest stimulus population in which stimulus patterns can overlap with one another is one constructed out of *dipoles*—stimulus patterns consisting of exactly two active elements and the rest inactive. If we have a total of $N$ input units there are $\frac{N(N-1)}{2}$ possible dipole stimuli. Of course, if the actual stimulus population consists of all $\frac{N(N-1)}{2}$ possibilities, there is no structure to be discovered. There are no clusters for our units to point at (unless we have one unit for each of the possible stimuli, in which case we can point a weight vector at each of the possible input stimuli). If, however, we restrict the possible dipole stimuli in certain ways, then there can be meaningful groupings of the stimulus patterns that the system can find. Consider, as an example, a case in which the stimulus lines could be thought of as forming a two dimensional grid in which the only possible stimulus patterns were those which formed adjacent pairs in the grid. If we have an $N \times M$ grid, there are $N(M-1) + M(N-1)$ possible stimuli. Figure 4 shows one of the 24 possible adjacent dipole patterns defined on a $4 \times 4$ grid. We carried out a number of experiments employing stimulus sets of this kind. In most of these experiments we employed a two layer system with a single inhibitory cluster of size two. Figure 5 illustrates the architecture of one of our experiments. The results of three runs with this architecture are illustrated in Figure 6, which shows the relative values of the weights for the two units. The

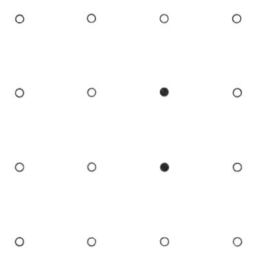

Figure 4. A dipole stimulus defined on a 4×4 matrix of input units. The rule for generating such stimuli is simply that any two adjacent units may be simultaneously active. Nonadjacent units may not be active and more than two units may not be simultaneously active. Active units are indicated by filled circles.

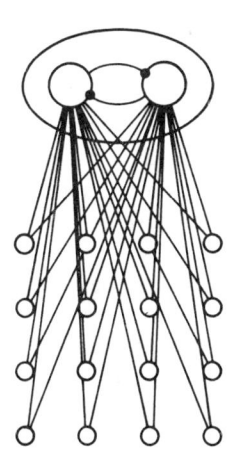

Figure 5. The architecture of a competitive learning system with 16 input units and one cluster of size two in the second layer.

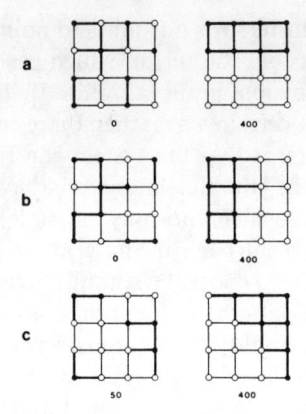

Figure 6. Relative weight values for the two members of the inhibitory cluster. (a) The results for one run with the dipole stimuli defined over a two-dimensional grid. The left-hand grid shows the relative values of the weights initially and the right-hand grid shows the relative values of the weights after 400 trials. A filled circle means that unit 1 had the larger weight on the corresponding input. An unfilled circle means that unit 2 had the larger weight. A heavy line connecting two circles means that unit 1 responded to the stimulus pattern consisting of the activation of the two circles, and a light line means that unit 2 won the corresponding pattern. In this case the system has divided the grid horizontally. (b) The results for a second run under the same conditions. In this case the system has divided the grid horizontally. (c) The results for a third run. In this case the left-hand grid represents the state of the system after 50 trials. Here the grid was divided diagonally.

values are shown laid out on a $4 \times 4$ grid so that weights are next to one another if the units with which they connect are next to one another. The relative values of the weights are indicated by the filling of the circles. If a circle is filled, that indicates that unit 1 had the largest weight on that line. If the circle is unfilled, that means that unit 2 had the largest weight on that line. The grids on the left indicate the initial configurations of the weights. The grids on the right indicate the final configurations of weights. The lines connecting the circles represent the possible stimuli. For example, the dipole stimulus pattern consisting of the upper left input line and the one immediately to the right of it is represented by the line connecting the upper-left circle in the grid with its right neighbor. The unit that wins when this stimulus is presented is indicated by the width of the line connecting the two circles. The wide line indicates that unit 1 was the winner, the narrow line indicates that unit 2 was the winner. It should be noted, therefore, that two unfilled circles must always be joined by a narrow line and two filled circles must always be joined by a wide line. The reason for this is that if a particular unit has more weight on both of the active lines then that unit *must* win the competition. The results clearly show that the weights move from a rather chaotic initial arrangement to an arrangement in which essentially all of those on one side of the grid are filled and all on the other side are unfilled.

The border separating the two halves of the grid may be at any orientation, but most often, it is oriented vertically and horizontally, as shown in the upper two examples. Only rarely is the orientation diagonal as in the example in the lower right-hand grid. Thus, we have a case in which each unit has chosen a coherent half of the grid to which they respond. It is important to realize that as far as the competitive learning mechanism is concerned the 16 input lines are unordered. The two dimensional grid-like arrangement exists only in the statistics of the population of stimulus patterns. Thus, the system has *discovered* the dimensional structure inherent in the stimulus population and has devised binary feature detectors that tell which half of the grid contains the stimulus pattern. Note, each unit responds to roughly half of the stimulus patterns. Note, also, that while some units break the grid vertically, some break the grid horizontally, and some break it diagonally; a combination of several clusters offers a rather more precise classification of a stimulus pattern.

In other experiments, we tried clusters of other sizes. For example, Figure 7 shows the results for a cluster of size four. It shows the initial configuration and its sequence of evolution after 100, 200, 400, 800, and after 4000 training trials. Again, initially the regions are chaotic. After training, however, the system settles into a state in which stimuli in compact regions of the grid are responded to by the same units. It can be seen, in this case, that the trend is toward a given unit responding to a maximally compact group of stimuli. In this experiment, three of the units settled on compact square regions while the remaining one settled on two nonconnected stimulus regions. It can be shown that the state into which the system settled does not quite maximize the value $T$, but does represent a relatively stable equilibrium state.

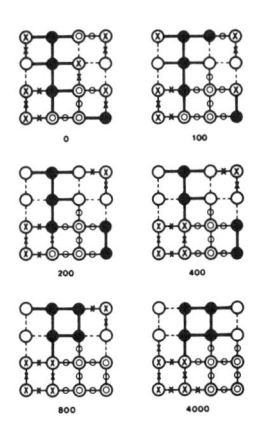

Figure 7. The relative weights of each of the four elements of the cluster after 0, 100, 200, 400, 800, and 4000 stimulus presentations.

In the examples discussed thus far, the system, to a first approximation, settled on a highly compact representation of the input patterns in which all patterns in a region are captured by one of the units. The grids discussed above have all been two-dimensional. There is no need to restrict the analysis to a two-dimensional grid. In fact, a two unit cluster will, essentially, pass a plane through a space of any dimensionality. There is a preference for planes perpendicular to the axes of the spaces. Figure 8 shows a typical result for the system learning a three-dimensional space. In the case of three dimensions, there are three equally good planes which can be passed through the space and, depending on the starting directions of the weight vectors and on the sequence of stimuli, different clusters will choose different ones of these planes. Thus, a system which receives input from a set of such clusters will be given information as to which *quadrant* of the space the pattern appears in. It is important to emphasize that the coherence of the space is *entirely* in the choice of input stimuli, *not* in the architecture of the competitive learning mechanism. The system *discovers* the spatial structure in the input lines.

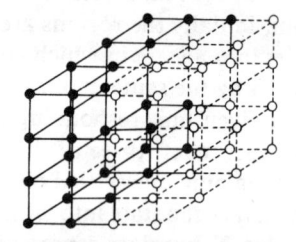

Figure 8. The relative weights for a system in which the stimulus patterns were chosen from a three-dimensional grid after 4000 presentations.

*Formal Analysis.* For the dipole examples described above, it is possible to develop a rather precise characterization of the behavior of the competitive learning system. Recall our argument that the most stable equilibrium state (and therefore the one the system is *most* likely to end up in) is the one that maximizes the function

$$T = \sum_k p_k \sum_{j,j} v_{jk}(\alpha_{jk} - \alpha_{ik}).$$

Now, in the dipole examples, all stimulus patterns of the stimulus population were equally likely (i.e., $p_k = \frac{1}{N}$), all stimulus patterns involve two active lines, and for every stimulus pattern in the population of patterns there are a fixed number of other stimulus patterns in the population which overlap it.[6] This implies that $\sum_{k}^{N} r_{kj} = R$ for all $j$. With these assumptions, it is possible to show that maximizing $T$ is equivalent to minimizing the function

$$\sum_{i}^{M} \frac{B_i}{N_i}$$

(see appendix for derivation), where $N_i$ is the number of patterns on which unit $i$ wins, $M$ is the number of units in the cluster, and $B_i$ is the number of cases in which $i$ responds to a particular pattern and does not respond to a pattern which overlaps it. This is the number of *border* patterns to which unit $i$ responds. Formally, we have

$$B_i = \sum_{j}^{N} \sum_{k}^{N} v_{ij}(1 - v_{ik}) \, r_{jk} > 0.$$

From this analysis, it is clear that the most stable states are ones in which the size of the border is minimized. Since, total border region is minimized when regions are spherical, we can conclude that in a situation in which stimulus pairs are drawn from adjacent points in a high-dimensional hyperspace, our competitive learning mechanism will form essentially spherical regions that partition the space into one such spherical region for each element of the cluster.

Another result of our simulations which can be explained by these equations is the tendency for each element of the cluster to capture roughly equally-sized regions. This results from the interconnectedness of the stimulus population. The result is easiest in the case in which $M = 2$. In this case, the function we want to minimize is given by

$$\frac{B_1}{N_1} + \frac{B_2}{N_2}.$$

Now, in the case of $M = 2$, we have $B_1 = B_2$, since the two regions must border on one another. Moreover, we have $N_1 + N_2 = N$, since every pattern is either responded to by unit 1 or unit 2. Thus, we want to minimize the function

$$B \left( \frac{1}{N_1} + \frac{1}{N - N_1} \right).$$

[6] Note that this latter condition does not quite hold for the examples presented above due to edge effects. It is possible to eliminate edge effects by the use of a torus. We have carried out experiments on tori as well, and the results are essentially the same.

This function is minimized when $N_1 = \frac{N}{2}$. Thus, there are two pressures which determine the performance of the system in these cases:

1. There is a pressure to reduce the number of border stimuli to a minimum.
2. There is a pressure to divide the stimulus patterns among the units in a way that depends on the total amount of weight that unit has. If two units have the same amount of weight, they will capture roughly equal numbers of equally-like stimulus patterns.

## Learning Words and Letters

It is common practice to handcraft networks to carry out particular tasks. Whenever one creates such a network that performs a task rather successfully, the question arises as to how such a network might have evolved. The word perception model developed in McClelland and Rumelhart (1981) and Rumelhart and McClelland (1982) is one such case in point. That model offers rather detailed accounts of a variety of word perception experiments, but it was crafted to do its job. How could it have evolved naturally? Could a competitive learning mechanism create such a network?

Let's begin with the fact that the word perception model required a set of position-specific letter detectors. Suppose that a competitive learning mechanism is faced with a set of words—to what features would the system learn to respond? Would it create position-specific letter detectors or their equivalent? We proceeded to answer this question by again viewing the lower level units as forming a two-dimensional grid. Letters and words could then be presented by activating those units on the grid corresponding to the points of a standard CRT font. Figure 9 gives examples of some of the stimuli used in our experiments. The grid we used was a $7 \times 14$ grid. Each letter occurred in a $7 \times 5$ rectangular region on the grid. There was room for two letters with some space in between, as shown in the figure. We then carried out a series of experiments in which we presented a set of word and/or letter stimuli to the system allowing it to extract relevant features.

Before proceeding with a description of our experiments, it should be mentioned that these experiments required a slight addition to the competitive learning mechanism. The problem was that, unlike the dipole stimuli, the letter stimuli only sparsely covered the grid and many of the units in the lower level never became active at all. Therefore, there was a possibility that, by chance, one of the units would have most of its weight on input lines that were never active, whereas another unit may have had most of its weight on lines common to all of the stimulus patterns. Since a unit never

a

b

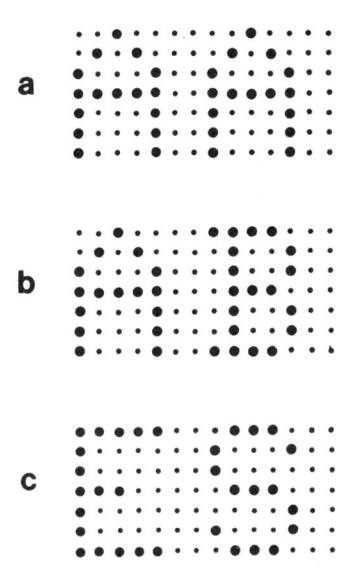

c

Figure 9. Example stimulus for the word and letter experiments.

learns unless it wins, it is possible that one of the units will never win, and therefore never learn. This, of course, takes the competition out of competitive learning. This situation is analogous to the situation in the geometric analogy in which all of the stimulus points are relatively close together on the hypersphere, and one of the weight vectors, by chance, points near the cluster while the other one points far from the stimuli. (See Figure 10). It is clear that the more distant vector is not closest to any stimulus and thus can never move toward the collection. We have investigated two modifications to the system which deal with the problem. One, which we call the leaky learning model, modifies the learning rule to state that *both* the winning *and*

Figure 10. A geometric interpretation of changes in stimulus sensitivity. The larger the circle around the head of the weight vector the more sensitive the unit. The decision as to which unit wins is made on the basis of the distance from the circle rather than from the head of the weight vector. In the example, the stimulus pattern indicated by the y is actually closer to the head of one vector 0, but since it is closer to the circle surrounding vector p, unit p would win the competition.

the losing units move toward the presented stimulus, the close vector simply moves much further. In symbols this suggests that

$$\Delta \omega_{ij} = \begin{cases} g_l \frac{c_{ik}}{n_k} - g_w \omega_{ij} & \text{if unit } j \text{ loses on stimulus } k \\ g_w \frac{c_{ik}}{n_k} - g_w \omega_{ij} & \text{if unit } j \text{ wins on stimulus } k \end{cases}$$

where $g_l$ is the learning rate for the losing units, $g_w$ is the learning rate for the winning unit, and where $g_l \ll g_w$. In our experiments we made $g_l$ an order of magnitude smaller than $g_w$. This change has the property that it slowly moves the losing units into the region where the actual stimuli lie, at which point they begin to capture some units and the ordinary dynamics of competitive learning take over.

The second method is similar to that employed by Bienenstock, Cooper, and Munro (1982), in which a unit modulates its own sensitivity so that when it is not receiving enough inputs, it becomes increasingly sensitive. When it is receiving too many inputs, it decreases its sensitivity. This mechanism can be implemented in the present context by assuming that there is a threshold and that the relevant activation is the degree to which the unit exceeds its threshold. If, whenever a unit fails to win it decreases its threshold, and whenever it does win it increases its threshold, then this method will also make all of the units eventually respond, thereby engaging the mechanism of competitive learning. This second method can be understood in terms of the geometric analogy that the weight vectors have a circle surrounding the end of the vector. The relevant measure is not the distance to the vector itself, but the distance to the circle surrounding the vector. Every time a unit loses, it increases the radius of the circle; every time it wins, it decreases the radius of the circle. Eventually, the circle on the losing unit will be large enough to be closer to some stimulus pattern than the other units.

We have used both of these mechanisms in our experiments and they appear to result in essentially similar behavior. The former, the leaky learning method, does not alter the formal analysis as long as the ratio $\frac{g_l}{g_w}$ is sufficiently small. The varying threshold method is more difficult to analyze and may, under some circumstances, distort the competitive learning process somewhat. After this diversion, we can now return to our experiments on the development of word/position specific letter detectors and other feature detectors.

***Position Specific Letter Detectors.*** In our first experiment, we presented letter pairs drawn from the set: *AA AB BA* and *BB*. We began with clusters of size two. The results were unequivocal. The system developed position-specific letter detectors. In some experimental runs, one of the units responded whenever *AA* or *AB* was presented, and the other responded whenever *BA* or *BB* was presented. In this case, unit 1 represents an *A* detector in position 1 and unit 2 represents a *B* detector for position 1. Moreover, as in the word perception model, the letter detectors are, of course, in a mutually inhibitory pool. On other experimental runs, the pattern was reversed. One of the units responded whenever there was an *A* in the second position and the other unit responded whenever there was a *B* in the second position. Figure 11 shows the final configuration of weights for one of our experimental runs. Note, that although the units illustrated here respond *only* to the letter in the first position, there is still weight on the active lines in the second position. It is just that the weights on the first position differentiate between *A* and *B*, whereas those on the second position respond equally to the two letters. In particular, as suggested by our formal analysis, asymptotically the weights on a given line are proportional to the probability that that line is active when the unit wins. That is, $\omega_{ij} \rightarrow P(units_i = 1 | unit_j \ wins)$. Since the lower level units unique to *A* occur equally as often as those unique to *B*, the weights on those lines are roughly equal. The input lines common to the two letters are on twice as often as those unique to either letter, and hence, they have twice as much weight. Those lines that never come on reach zero weight.

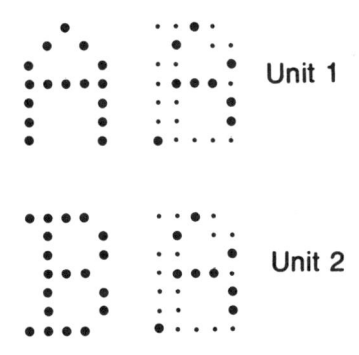

Figure 11. The final configuration of weights for a system trained on the stimulus patterns A B C D.

*Word Detection Units.* In another experiment, we presented the same stimulus patterns, but increased the elements in the cluster from 2 to 4. In this case, each of the four level two units came to respond to one of the four input patterns—in short, the system developed *word detectors*. Thus, if layer two were to consist of a number of clusters of various sizes, large clusters with approximately one unit per word pattern will develop into word detectors, while smaller clusters with approximately the number of letters per spatial position, will develop into position-specific letter detectors. As we shall see below, if the number of elements of a cluster is substantially less than the number of letters per position, then the cluster will come to detect position-specific letter features.

*Effects of Number of Elements Per Serial Position.* In another experiment, we varied the number of elements in a cluster and the number of letters per serial position. We presented stimulus patterns drawn from the set *AA AB AC AD BA BB BC BD.* In this case, we found that with clusters of size two, one unit responded to the patterns beginning with *A* and the other responded to those beginning with *B*. In our previous experiment, when we had the same number of letters in each position, we found that the clusters were indifferent as to which serial position they responded. Some responded to position 1 and others to position 2. In this experiment, we found that a two-element cluster always becomes a letter detector specific to serial position in which two letters vary. Similarly, in the case of clusters of size four we found that they always became letter detectors for the position in which four letters varied. Thus, in this case one responded to an *A* in the second position, one responded to a *B* in the second position, one responded to a *C* in the second position, and one responded to a *D* in the second position. Clearly, there are two natural ways to cluster the stimulus patterns—two levels of structure. If the patterns are to be put in two categories, then the binary feature *A* or *B* in the first position is the relevant distinction. On the other hand, if the stimuli are to be grouped into four groups, the four value feature determining the second letter is the relevant distinction. The competitive learning algorithm can discover either of the levels of structure—depending on the number of elements in a cluster.

*Letter Similarity Effects.* In another experiment, we studied the effects of letter similarity to look for units which detect letter features. We presented letter patterns consisting of a letter in the first position only. We chose the patterns so they formed two natural clusters based on the similarity of the letters to one another. We presented the letters *A, B, S,* and *E.* The letters were chosen so that they fell naturally into two classes. In our font, the letters *A* and *E* are quite similar and the letters *B* and *S* are very similar. We used a cluster of size two. Naturally, one of the units responded to the *A* or the *E* while the other unit responded to the *B* or the *S*. The weights were largest on those features of the stimulus pairs which were common among each of these similar pairs. Thus, the system developed subletter size feature detectors for the features relevant to the discrimination.

*Correlated Teaching Inputs.* We carried out one other set of experiments with the word/letter patterns. In this case, we used clusters of size two and presented simuli drawn from the set *AA BA SB EB*. Note that on the left hand side, we have the same four letters as we had in the previous experiment, but on the right hand side we have only two patterns; these two patterns are correlated with the letter in the first position. An *A* in the second position means that the first position contains either an *S* or an *E*. Note further, that those correlations between the first and second positions are in opposition to the "natural" similarity of the letters in the first serial position. In this experiment, we first trained the system on the four stimuli already described. Since the second serial position had only two letters in it, the size two cluster became a position-specific letter detector for the second serial position. One unit responded to the *A* and one to the *B* in the second position. Notice that the units are also responding to the letters in the first serial position as well. One unit is responding to an *A* or a *B* in the first position while the other responds to an *E* or an *S*. Figure 12 shows the patterns of weights developed by the two units. After training, the system was then presented patterns containing only the first letter of the pair and, as expected, the system had learned the "unnatural" classification of the letters in the first position. Here the strong correlation between the first and second position led the competitive learning mechanism to override the strong correlation between the highly similar stimulus patterns in the first serial position. This suggests that even though the competitive learning system is an "unsupervised" learning mechanism, one can control what it learns by controlling the statistical structure of the stimulus patterns being presented to it. In this sense, we can think of the right-hand letter in this experiment as being a kind of *teaching* stimulus aimed at determining the classification learned for other aspects of the stimulus. It should also be noted that this teaching mechanism is essentially the same as the so-called error-less learning procedure used by Terrace (1963) in training pigeons to peck a certain color key by associating that color with a response situation where their pecking is determined by other factors. As we shall see below, this correlational teaching mechanism is useful in allowing the competitive learning mechanism to discover features which it otherwise would be unable to discover.

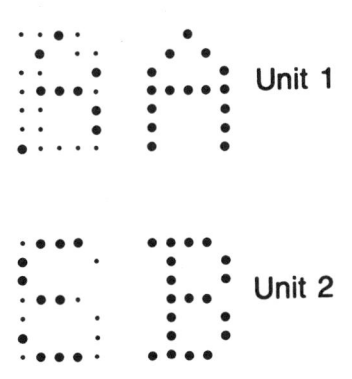

Figure 12. The pattern of weights developed in the correlated learning experiment.

## Horizontal and Vertical Lines

One of the classically difficult problems for a linear threshold device like a perceptron is to distinguish between horizontal and vertical lines. In general horizontal and vertical lines are not linearly separable and require a multilayer perceptron system to distinguish them. One of the goals of the competitive learning device is for it to discover features which at a higher level of analysis might be useful for discriminating patterns with a linear threshold type device which might not otherwise be discriminable. It is therefore of some interest to see what kinds of features the competitive learning mechanism discovers when presented with a set of vertical and horizontal lines. In the following discussion, we chronicle a series of experiments on this problem. Several of the experiments ended in failure, but we were able to discover a way in which competitive learning systems can be put together to build a hierarchical feature detection system capable of discriminating vertical and horizontal lines. We proceed by sketching several of our failures as well as our successes because the way in which the system fails is elucidating. It should be noted at the outset that our goal is not so much to present a model of how the human learns to distinguish between vertical and horizontal lines (indeed, such a distinction is probably pre-wired in the human system), but rather to show how competitive learning can discover features which allow for the system to learn distinctions with multiple layers of units that cannot be learned by single-layered systems. Learning to distinguish vertical and horizontal lines is simply a paradigm case.

In this set of experiments, we represented the lower level of units as if they were on a $6 \times 6$ grid. We then had a total of 12 stimulus patterns, each consisting of turning on six level one units in a row on the grid. Figure 13 illustrates the grid and several of the stimulus patterns. Ideally, one might hope that one of the units would respond whenever a vertical line is presented, the other would respond whenever a horizontal line is presented. Unfortunately, a little thought indicates that this is impossible. Since every input unit participates in exactly one vertical and one horizontal line, there is no configuration of weights which will distinguish vertical from horizontal. This is exactly why no linear threshold device can distinguish between vertical and horizontal lines in one level. Since that must fail, we might hope that some clusters in the competitive learning device will respond to vertical lines by assigning weights as illustrated in Figure 14. In this case, one unit of the pair would respond whenever the first, second, or fourth vertical line was presented, and another would respond whenever the third, fifth, or sixth vertical line was presented; since both units would receive about the same input in the face of a horizontal line, we might expect that sometimes one and sometimes the other would win the competition, but that the primary response would be to vertical lines. If other clusters settled down similarly to horizontal lines, then a unit at the third level looking at the output of the

# FEATURE DISCOVERY BY COMPETITIVE LEARNING

a

b

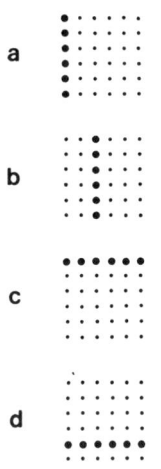

c

d

Figure 13. Stimulus patterns for the horizontal/vertical discrimination experiments.

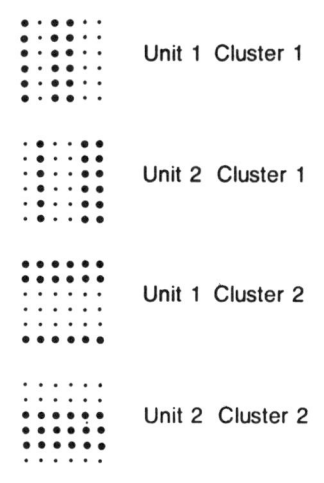

Unit 1 Cluster 1

Unit 2 Cluster 1

Unit 1 Cluster 2

Unit 2 Cluster 2

Figure 14. A possible weight configuration which could distinguish vertical from horizontal.

various clusters could distinguish vertical and horizontal. Unfortunately, that is not the pattern of weights discovered by the competitive learning mechanism. Rather, a typical pattern of weights is illustrated in Figure 15. In this arrangement, each cluster responds to exactly three horizontal and three vertical lines. Such a cluster has lost all information that might distinguish vertical from horizontal. We have discovered a feature of absolutely no use in this distinction. In fact, such features systematically throw away the information relevant to horizontal versus vertical. Some further thought indicates why such a result occurred. Note, in particular, that two horizontal lines have exactly *nothing* in common. The grid that we show in the dia-

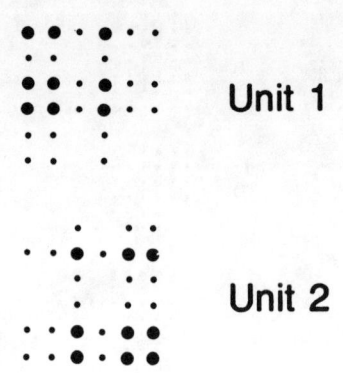

Figure 15. A typical configuration of weights for the vertical/horizontal discrimination.

grams is merely for our convenience. As far as the units are concerned there are 36 unordered input units; sometimes some of those units are active. Pattern similarity is determined entirely by pattern overlap. Since horizontal lines don't intersect, they have no units in common, thus they are not seen as similar at all. However, every horizontal line intersects with every vertical line and thus has much more in common with vertical lines than with other horizontal ones. It is this similarity that the competitive learning mechanism has discovered.

Now, suppose that we change the system somewhat. Suppose that we "teach" the system the difference between vertical and horizontal (as we did in the previous experiments with letter strings). In this experiment we used a $12 \times 6$ grid. On the right-hand side of the grid we presented either a vertical or a horizontal line, as we did before. On the left-hand side of the grid we always presented the uppermost horizontal line whenever any horizontal line was presented on the right-hand grid, and we always presented the vertical line furthest to the left on the left-hand grid whenever we presented any vertical line on the right-hand side of the grid. We then had a cluster of two units receiving inputs from all $12 \times 6 = 72$ lower level units. (Figure 16 shows several of the stimulus patterns.)

As expected, the two units soon learned to discriminate between vertical and horizontal lines. One of the units responded whenever a vertical line was presented and the other responded whenever a horizontal line was presented. They were responding, however, to the pattern on the left-hand side rather than to the vertical and horizontal pattern on the right. This too should be expected. Recall that the value of the $\omega_{ij}$ approaches a value which is proportional to the probability that input unit $i$ is active, given that unit $j$ won the competition. Now, in the case of the unit that responds to vertical line for example, every unit on the right hand grid occurs equally often so that all of the weights connecting to units in that grid have equal weights. The same is true for the unit responding to the horizontal line. The weights

# FEATURE DISCOVERY BY COMPETITIVE LEARNING 237

a
```
• · · · · · · · • · · ·
• · · · · · · · • · · ·
• · · · · · · · • · · ·
• · · · · · · · • · · ·
• · · · · · · · • · · ·
```

b
```
• · · · · · · • · · · ·
• · · · · · · • · · · ·
• · · · · · · • · · · ·
• · · · · · · • · · · ·
• · · · · · · • · · · ·
```

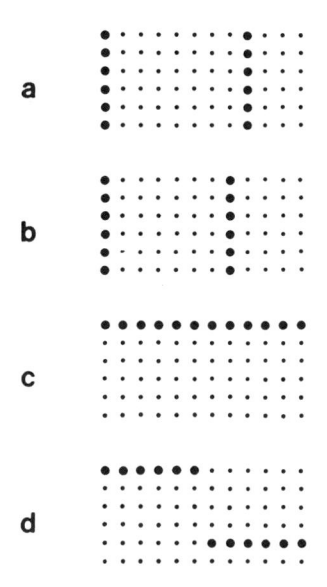

c
```
• • • • • • • • • • • •
· · · · · · · · · · · ·
· · · · · · · · · · · ·
· · · · · · · · · · · ·
· · · · · · · · · · · ·
```

d
```
• • • • • • · · · · · ·
· · · · · · · · · · · ·
· · · · · · · · · · · ·
· · · · · · · · · · · ·
· · · · · · • • • • • •
```

Figure 16. Stimulus patterns for the vertical/horizontal discrimination experiments with a correlated "teaching" input on the right-hand side.

on the right-hand grid are identical for the two cluster members. Thus, when the "teacher" is turned off, and only the right-hand figure is presented, the two units respond randomly and show no evidence of having learned the horizontal/vertical distinction.

Suppose, however, that we have four, rather than two, units in the level two clusters. We ran this experiment and found that of the four units, two of them divided up the vertical patterns and two of them divided up the horizontal patterns. Figure 17 illustrates the weight values for one of our runs. One of the units took three of the vertical line patterns, another unit took three other vertical patterns. A third unit responded to three of the horizontal line patterns and the last unit responded to the remaining three horizontal lines. Moreover, after we took away the "teaching" pattern, the system continued to classify the vertical and horizontal lines just as it did when the left-hand "teaching" pattern was present.

In one final experiment with vertical and horizontal lines, we developed a three level system in which we used the same stimulus patterns as in the previous experiment, the only difference was that we had *two* clusters of four units at the second level and one cluster of two units at the third level. Figure 18 shows the architecture employed. In this case, the two four element clusters each learned to respond to subsets of the vertical and horizontal lines as in the previous experiment. The two clusters generally responded to different subsets, however. Thus, when the upper horizontal line was presented, unit 1 of the first cluster responded and unit 3 of the second cluster

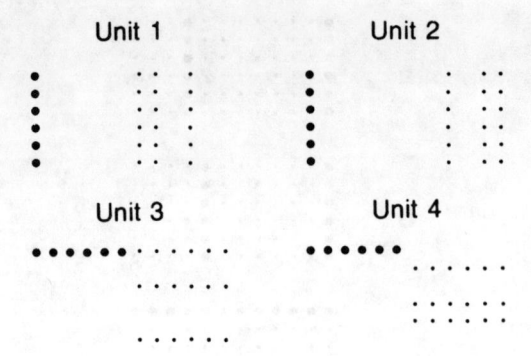

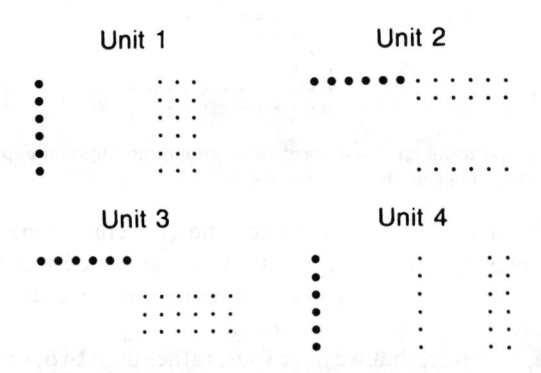

Figure 17. The weight values for the two clusters of size four for the vertical/horizontal discrimination experiment with a correlated "teaching" stimulus.

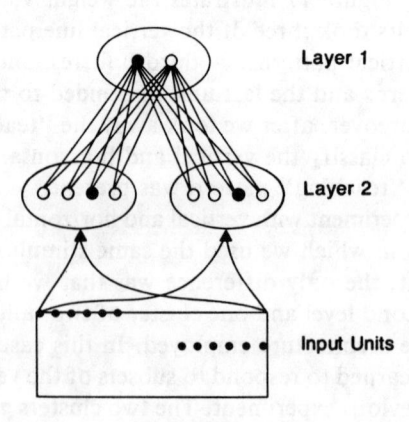

Figure 18. The architecture for the three level horizontal/vertical discrimination experiment.

responded. When the bottom horizontal line was presented, unit 1 of the first cluster responded again, but unit 4 of the second cluster also responded. Thus, the cluster of size two at the highest level was receiving a kind of dipole stimulus. It has four inputs and on any trial, two of them are active. As with our analysis of dipole stimuli, we know that stimuli that overlap are always put in the same category. Note that when a vertical line is presented, one of the two units in each of the middle layers of clusters that responds to vertical lines will become active, and that none of the units which respond to horizontal lines will ever be active; thus, this means that there are two units in each middle layer cluster which respond to vertical lines. Whenever a vertical line is presented, one of the units in each cluster will become active. None of the horizontal units will ever be active in the face of a vertical stimulus. Thus, one of the units at the highest level learns to respond whenever a vertical line is presented and the other unit responds whenever a horizontal line is presented. Once the system has been trained, this occurs despite the absence of the "teaching" stimulus. Thus, what we have shown is that the competitive learning mechanism can, under certain conditions, develop feature detectors which allow the system to distinguish among patterns which are not differentiable by a simple linear unit in one level.

## CONCLUSION

We have shown how a very simple competitive mechanism can discover a set of feature detectors which capture important aspects of the set of stimulus input patterns. We have also shown how these feature detectors can form the basis of a multilayer system that can serve to learn categorizations of stimulus sets which are not linearly separable. We have shown how the use of correlated stimuli can serve as a kind of "teaching" input to the system to allow the development of feature detectors which would not develop otherwise. Although we find the competitive learning mechanism a very interesting and powerful learning principle, we do not, of course, imagine that it is the only learning principle. Competitive learning is an essentially non-associative statistical learning scheme. We certainly imagine that other kinds of learning mechanisms will be involved in the building of associations among patterns of activation in a more complete neural network. We offer this analysis of these competitive learning mechanisms to further our understanding of how simple adaptive networks can discover features important in the description of the stimulus environment in which the system finds itself.

# APPENDIX

For the case of homogeneous *dipole* stimulus patterns, it is possible to derive an expression for the most *stable* equilibrium state of the system. We say that a set of dipole stimulus patterns is homogeneous if (1) they are equally likely and (2) for every input pattern in the set there are a fixed number of other input patterns which overlap them. These conditions were met in our simulations. Our measure of stability is given by

$$T = \sum_k p_k \sum_j \sum_i v_{jk}(\alpha_{jk} - \alpha_{ik}).$$

Since $p_k = \frac{1}{N}$, we can write

$$T = \frac{1}{N}\sum_i\sum_j\sum_k v_{jk}\alpha_{jk} - \frac{1}{N}\sum_i\sum_j\sum_k v_{jk}\alpha_{ik}.$$

Summing the first portion of the equation over $i$ and the second over $j$ we have

$$T = \frac{M}{N}\sum_j\sum_k v_{jk}\alpha_{jk} - \frac{1}{N}\sum_i\sum_k \alpha_{ik}\sum_j v_{jk}.$$

Now note that when $p_k = \frac{1}{N}$, we have $\alpha_{ik} = \dfrac{\sum_j r_{kj}v_{ij}}{\sum_l v_{kl}}$. Furthermore, $\sum_l v_{lk} = 1$ and $\sum_k v_{lk} = N_l$, where $N_l$ is the number of patterns captured by unit $l$. Thus, we have

$$T = \frac{M}{N}\sum_j\sum_k v_{jk}\alpha_{jk} - \frac{1}{N}\sum_i\sum_k \frac{\sum_l r_{kl}v_{il}}{N_i}.$$

Now, since all stimuli are the same size, we have $r_{ij} = r_{ji}$. Moreover, since all stimuli have the same number of neighbors, we have $\sum_i r_{ij} = \sum_j r_{ij} = R$, where $R$ is a constant determined by the dimensionality of the stimulus space from which the dipole stimuli are drawn. Thus, we have

$$T = \frac{M}{N}\sum_j\sum_k v_{jk}\alpha_{jk} - \frac{R}{N}\sum_i \frac{\sum_l v_{il}}{N_i}$$

and we have

$$T = \frac{M}{N}\sum_j\sum_k v_{jk}\alpha_{jk} - \frac{RM}{N}.$$

Since $R$, $M$, and $N$ are constants, we have that $T$ is maximum whenever $T' = \sum_j\sum_k v_{jk}\alpha_{jk}$ is maximum. Now substituting for $\alpha_{jk}$, we can write

$$T' = \sum_j \frac{1}{N_j}\sum_k\sum_l r_{kl}v_{jk}v_{jl}.$$

We can now substitute for the product $v_{jk}v_{jl}$ the term $v_{jk} - v_{jk}(1 - v_{jl})$. We then can write

$$T' = \sum_j \frac{1}{N_j}\sum_k\sum_l r_{kl}v_{jk} - \sum_j \frac{1}{N_j}\sum_k\sum_l r_{kl}v_{jk}(1 - v_{jl}).$$

Summing the first term of the equation first over $l$, then over $k$, and then over $j$, gives us

$$T' = MR - \sum_j \frac{1}{N_j}\sum_k\sum_l r_{kl}v_{jk}(1 - v_{jl}).$$

Now, recall that $r_{kl}$ is given by the degree of stimulus overlap between stimulus $i$ and stimulus $k$. In the case of dipoles there are only three possible values of $r_{kl}$.

$$r_{kl} = \begin{cases} 0 & \text{no overlap} \\ 1 & k = l \\ 1/2 & \text{otherwise} \end{cases}$$

Now, the second term of the equation for $T'$ is 0 if either $r_{kl} = 0$ or if $v_{jk}(1 - v_{jl}) = 0$. Since $v_{ik}$ is either $1$ or $0$, this will be zero whenever $j = l$. Thus, for all non-zero cases in the second term we have $r_{kl} = \frac{1}{2}$. Thus we have

$$T' = MR - \frac{1}{2}\sum_j \frac{1}{N_j}\sum_k\sum_l v_{jk}(1 - v_{jl}).$$

Finally, note that $\sum_k\sum_l v_{jk}(1 - v_{jl})$ is $1$ and $r_{kl}$ is $\frac{1}{2}$ in each case in which different units capture neighboring patterns. We refer to this as a case of *bad neighbors* and let $B_j$ designate the number of bad neighbors for unit $j$. Thus, we have

$$T' = MR - \frac{1}{2}\sum_j \frac{B_j}{N_j}.$$

Finally, we can see that $T'$ will be a maximum whenever $T'' = \sum_j \frac{B_j}{N_j}$ is minimum. Thus, minimizing $T''$ leads to the maximally stable solution in this case.

## REFERENCES

Bernstein, J. (1981). Profiles: AI, Marvin Minsky. *The New Yorker, 57,* 50-126.

Bienenstock, E. L., Cooper, L. N., & Munro, P. W. (1982). Theory for the development of neuron selectivity; Orientation specificity and binocular interaction in visual cortex. *Journal of Neuroscience, 2,* 32-48.

Farley, B. G., & Clark, W. A. (1954). Simulation of self-organizing systems by digital computer. *I.R.E. Transactions of Information Theory, Vol. PGIT-4,* 76-84.

Fukushima, K. (1975). Cognitron: A self-organizing multilayered neural network. *Biological Cybernetics, 20,* 121-136.

Grossberg, S. (1976). Adaptive pattern classification and universal recoding, I: Parallel development and coding of neural feature detectors. *Biological Cybernetics, 23,* 121-134.

Hebb, D. O. (1949). *The organization of behavior.* New York: Wiley.

Kohonen, T. (1982). Clustering, taxonomy, and topological maps of patterns. In M. Lang (Ed.), *Proceedings of the Sixth International Conference on Pattern Recognition* (pp. 114-125). Silver Spring, MD: IEEE Computer Society Press.

McCarthy, J. (1959). Comments. In *Mechanisation of thought processes: Proceedings of a Symposium held at the National Physical Laboratory, November 1958. Vol. 1* (p. 464). London: Her Majesty's Stationery Office.

McClelland, J. L., & Rumelhart, D. E. (1981). An interactive activation model of context effects in letter perception: Part 1. An account of basic findings. *Psychological Review, 88,* 375-407.

McCulloch, W. S., & Pitts, W. (1943). A logical calculus of the ideas immanent in neural nets. *Bulletin of Mathematical Biophysics, 5,* 115-137.

Minsky, M. (1954). Neural nets and the brain-model problem. Unpublished doctoral dissertation, Princeton University.

Minsky, M. L. (1959). Some methods of artificial intelligence and heuristic programming. In *Mechanisation of thought processes: Proceedings of a Symposium held at the National Physical Laboratory, November, 1958. Vol. 1* (pp. 3-28). London: Her Majesty's Stationery Office.

Minsky, M., & Papert, S. (1969). *Perceptrons.* Cambridge, MA: MIT Press.

Rosenblatt, F. (1959). Two theorems of statistical separability in the perceptron. In *Mechanisations of thought processes: Proceedings of a Symposium held at the National Physical Laboratory, November 1958. Vol. 1* (pp. 421-456). London: Her Majesty's Stationery Office.

Rosenblatt, F. (1962). *Principles of neurodynamics.* New York: Spartan.

Rumelhart, D. E., & McClelland, J. L. (1982). An interactive activation model of context effects in letter perception: Part 2: The contextual enhancement effect and some tests and extensions of the model. *Psychological Review, 89,* 60-94.

Terrace, H. S. (1963). Discrimination learning with and without errors. *Journal of the Experimental Analysis of Behavior, 6,* 1-27.

Von der Malsburg, C. (1973). Self-organizing of orientation sensitive cells in the striate cortex. *Kybernetik, 14,* 85-100.

CHAPTER 9

# Competitive Learning: From Interactive Activation to Adaptive Resonance

STEPHEN GROSSBERG
*Boston University*

Functional and mechanistic comparisons are made between several network models of cognitive processing: competitive learning, interactive activation, adaptive resonance, and back propagation. The starting point of this comparison is the article of Rumelhart and Zipser (1985) on feature discovery through competitive learning. All the models which Rumelhart and Zipser (1985) have described were shown in Grossberg (1976b) to exhibit a type of learning which is temporally unstable. Competitive learning mechanisms can be stabilized in response to an arbitrary input environment by being supplemented with mechanisms for learning top-down expectancies, or templates; for matching bottom-up input patterns with the top-down expectancies; and for releasing orienting reactions in a mismatch situation, thereby updating short-term memory and searching for another internal representation. Network architectures which embody all of these mechanisms were called adaptive resonance models by Grossberg (1976c). Self-stabilizing learning models are candidates for use in real-world applications where unpredictable changes can occur in complex input environments. Competitive learning postulates are inconsistent with the postulates of the interactive activation model of McClelland and Rumelhart (1981), and suggest different levels of processing and interaction rules for the analysis of word recognition. Adaptive resonance models use these alternative levels and interaction rules. The self-organizing learning of an adaptive resonance model is compared and contrasted with the teacher-directed learning of a back propagation model. A number of criteria for evaluating real-time network models of cognitive processing are described and applied.

## 1. INTRODUCTION

Many cognitive scientists are now rapidly translating their intuitions about human intelligence into real-time network models. As each research group

---

Supported in part by the Air Force Office of Scientific Research (AFOSR 85-0149 and AFOSR F49620-86-C-0037) and the National Science Foundation (NSF IST-8417756).
Acknowledgements: Thanks to Cynthia Suchta for her valuable assistance in the preparation of the manuscript.

injects a stream of new models into this sprawling literature, it becomes ever more essential to penetrate behind the many ephemeral differences between models to the deeper architectural level on which a formal model lives. What are the key issues, principles, properties, mechanisms, and data that may be used to distinguish one model from another? How may we decide whether two seemingly different models are really formally equivalent, or are probing profoundly different aspects of cognitive processing?

This article outlines a comparative analysis of network models within a focused conceptual domain. Its starting point is the recent article by Rumelhart and Zipser (1985) on competitive learning models that was published in this journal as part of a special issue on connectionist models. The discussions raised by this article lead naturally to a comparison of several distinct models, notably competitive learning, interactive activation, adaptive resonance, and back propagation models. Before considering these models, I briefly discuss why real-time network models are so important, and why their very promise makes them difficult to understand.

## 2. EMERGENT PROPERTIES OF NETWORK INTERACTIONS: FUNCTION VERSUS MECHANISM

A key issue leading to network models concerns how the behavior of individuals adapts successfully in real-time to constraints imposed by their environments. In order to analyse this issue, one needs to identify the functional level on which an individual's behavioral success is defined. Much theoretical and experimental evidence suggests that this is the level of neural networks, rather than the level of individual nerve cells. Key behavioral properties are often emergent properties due to interactions among many cells in a neural network. Thus, the study of real-time networks is important because behavior can best be understood on the level of a network analysis.

Often a network's emergent properties are much more complex than the network components from which they arise. In a good network model, the whole is far greater than the sum of its parts. In addition, the formal relationships among those emergent properties may be quite subtle, and may reflect the delicate interplay of behavioral properties that are characteristic of living organisms. Thus network models can excite our interest by showing us how subtle and complex functional properties can emerge from interactions among simple components.

The very fact, however, that simple network laws can generate complex behaviors makes network models difficult to understand. In order to effectively analyse a network model, one needs powerful analytic and computational methods to derive the complex emergent properties of the network from a description of its simple components. A network model cannot, in principle, be understood merely as a list of processing rules or as a computer program. In order to adequately describe the dynamism of such a network, it is necessary to use a mathematical formalism that can naturally analyse interactions which may occur in a nonlinear fashion across thousands or even millions of components. The need for such a formal analysis is especially great when the network can learn, since the same laws define the network at all times, but its functional properties may be radically different before and after learning occurs.

The distinction between a network's emergent functional properties and its simple mechanistic laws also clarifies why the controversy surrounding the relationship of an intelligent system's abstract properties to its mechanistic instantiation has been so enduring. Without a linkage to mechanism, a network's functional properties cannot be formally or physically explained. On the other hand, how do we decide which mechanisms are crucial for generating desirable functional properties and which mechanisms are adventitious? Two seemingly different models can be equivalent from a functional viewpoint if they both generate similar sets of emergent properties. An analysis which proves such a functional equivalence between models does not, however, minimize the importance of their mechanistic descriptions. Rather, such an analysis identifies mechanistic variations which are not likely to be differentiated by evolutionary pressures which select for these functional properties on the basis of behavioral success.

Another side of such an evolutionary analysis concerns the identification of the fundamental network modules which are specialized by the evolutionary process for use in a variety of behavioral tasks. How do evolutionary variations of a single network module, or blueprint, generate behavioral properties which, on the level of raw experience, seem to be phenomenally different and even functionally unrelated? Although each specialized network may generate a characteristic bundle of emergent properties, parametric changes of these specialized networks within the framework of a single network may generate bundles of emergent properties that are qualitatively different. In order to identify the mechanistic unity behind this phenomenal diversity, appropriate analytic methods are once again indispensable.

In summary, the relationship between the emergent functional properties that govern behavioral success and the mechanisms that generate these properties is far from obvious. A single network module may generate qualitatively different functional properties when its parameters are changed. Conversely, two mechanisms which are mechanistically different may generate formally homologous functional properties. The intellectual difficulties caused by these possibilities are only compounded by the fact that we are designed by evolution to be serenely ignorant of our own mechanistic substrates. The very cognitive and learning mechanisms which enable us to group, or chunk, ever more complex information into phenomenally simple unitized representations act to hide from us the myriad interactions that subserve these representations during every moment of experience. Thus we cannot turn to our daily intuitions or to our lay language for secure guidance in discovering or analysing network models. The simple lesson that the whole is greater than the sum of its parts forces us to use an abstract mathematical language that is capable of analysing interactive emergence and functional equivalence.

## 3. PROCESSING LEVELS AND INTERACTIONS: MODELS OR METAPHORS?

A network model is usually easy to define using just a few equations. These equations specify the dynamical laws governing the model's nodes, or cells, including the processing *levels* in which these nodes are embedded. The equations also specify the *interactions* between these nodes, including which nodes are connected by pathways and the types of signals or other processes that go on in these pathways. Inputs to the network, outputs from the network, parameter choices within the network, and initial values of network variables often complete the model description. Such components are common to essentially all real-time network models. Thus, to merely say that a model has such components is scientifically vacuous.

How, then, can we decide when a network model is wrong? Such a task is deceptively simple. If the model's levels are incorrectly chosen, then it is wrong. If its interactions are incorrectly chosen, then it is wrong. And so on. The only escape from such a critique would be to demonstrate that a different set of levels and interactions can be correctly chosen, and shares similar functional properties with the original model. The new choice of levels and interactions would, however, constitute a new model. The old model would still be wrong. Such an analysis would show that the shared model properties are essentially model-independent, yet that there exist finer tests to distinguish between models. I will describe several such tests below.

In the absence of such a literal process of model selection and rejection, it would soon become impossible to criticize a model at all, since its authors could claim that they really did not intend their model to be literally interpreted. To avoid criticism or disconfirmation, the model could be turned into a metaphor of itself, or even into a vaguely outlined framework that could be broadly enough defined to include all future modeling possibilities, including possibilities that flatly contradicted the original model.

McClelland (1985, p. 144) essentially advocated this position when he wrote "we would not view the interactive activation model as a description of a mechanism at all...it allows us to study interactive activation models of a wide range of phenomena at a psychological or functional level without necessarily worrying about the plausibility of assuming that they provide an adequate description of the actual implementation." As noted above, dissociation of a functional description from a mechanistic description is impossible in a network model. The possibility that a particular mechanistic instantiation may be functionally equivalent to a different instantiation does not in the least free us from committing ourselves to particular classes of mechanisms. McClelland's usage would become acceptable only if we agreed to use the term "interactive activation" model to mean any real-time network model. Such a usage would, however, make the term scientifically vacuous. In addition, the interactive activation model is a relatively recent member of the family of real-time network models in psychology. It has added no new qualitative concepts to this class of models *as a framework,* hence it needs to be analysed *as a model* in order to appreciate its contribution to the network modeling literature. All models discussed in this article will be treated literally as models, rather than as metaphors or frameworks of models.

## 4. FEATURE DISCOVERY BY COMPETITIVE LEARNING

I will use the Rumelhart and Zipser (1985) article to motivate my analysis of a number of issues which promise to play a central role in evaluating the strengths and weaknesses of various network models. Rumelhart and Zipser (1985) analyse a type of learning model which is called a *competitive learning* model. They acknowledge that competitive learning models have been intensively studied for some time and thus conclude that "it seems reasonable to put the whole issue into historical perspective" (p. 76). They also note that "It is a common practice to handcraft networks to carry out particular tasks. Whenever one creates such a network that performs a task rather successfully, the question arises as to how such a network might have evolved. The word perception model developed in McClelland and Rumelhart (1981) and Rumelhart and McClelland (1982) is one such case in point. That model offers rather detailed accounts of a variety of word perception experiments, but it was crafted to do its job. How could it have evolved naturally? Could a competitive learning mechanism create such a network?" (p. 98). Thus, these authors ask how a competitive learning network can be joined to an interactive activation network in order to endow the latter type of network with a learning capability.

Their discussion does not, however, acknowledge that both the levels and the interactions of a competitive learning model are incompatible with those of an interactive activation model (Grossberg, 1984). The authors likewise do not state that the particular competitive learning model which they have primarily analysed is identical to the model introduced and analysed in

Grossberg (1976a, 1976b), nor that this model was consistently embedded into an adaptive resonance model in Grossberg (1976c) and later developed in Grossberg (1978) to articulate the key functional properties which McClelland and Rumelhart described when they introduced the interactive activation model in McClelland and Rumelhart (1981). In summary, the stated goal of Rumelhart and Zipser (1985)—to join a competitive learning model with a model capable of generating functional properties that are shared with the interactive activation model—was carried out using an adaptive resonance model in Grossberg (1978). In addition, the interactive activation model *as a model* is incapable of participating in such a synthesis.

The Rumelhart and Zipser (1985) article thus raises a number of issues which make real-time network models so difficult to understand and to differentiate. How can an adaptive resonance model share functional properties with an interactive activation model, yet be mechanistically consistent with a competitive learning model with which the interactive activation model is mechanistically inconsistent? What design principles are realized by an adaptive resonance model but not by an interactive activation model which can be used to distinguish these models on a deep computational level? The analysis of Rumelhart and Zipser (1985) provides no light into these matters. Indeed, these authors also stated that "our analyses differ from many of these [former analyses] in that we focus on the development of feature detectors rather than pattern classification" (p. 76). A glance at such titles as Malsburg (1973) and Grossberg (1976a, 1976b) shows that this observation is also inaccurate.

## 5. THE PROBLEM OF TEMPORALLY UNSTABLE LEARNING

Analysis of the competitive learning model revealed a fundamental problem which is shared by most other learning models that are now being developed and which was overcome by the adaptive resonance theory. I will now illustrate this general problem using a competitive learning model, before indicating that adaptive variants of the interactive activation model cannot solve it.

The particular competitive learning models described in Grossberg (1976b) and in Rumelhart and Zipser (1985) were used to show how a stream of input patterns to a network level $F_1$ can adaptively tune the weights, or long term memory ($LTM$) traces, in the pathways from $F_1$ to a coding level $F_2$. Although these $LTM$ traces may initially be randomly chosen, the presentation of inputs at $F_1$ can alter the $LTM$ traces through learning in such a way that $F_2$ eventually parses the input patterns into sets which activate distinct recognition categories. Appendix 1 describes this competitive learning scheme as well as the formal identity of the Grossberg (1976b) model with the model studied by Rumelhart and Zipser (1985).

In Grossberg (1976b), a theorem was proved which described input environments to which the model responds by learning a temporally stable recognition code. This theorem is described in Appendix 2. The theorem proved that, if not too many input patterns are presented to $F_1$, relative to the number of coding nodes in $F_2$, or if the input patterns form not too many clusters, then learning of the recognition code eventually stabilizes. In addition, the learning process elicits the best distribution of *LTM* traces that is consistent with the structure of the input environment. The computer simulations of Rumelhart and Zipser (1985) essentially confirm this theorem.

Despite the demonstration of input environments that can be stably coded, it was also shown, through explicit counterexamples, that a competitive learning model cannot learn a temporally stable code in response to arbitrary input environments. Moreover, these counterexamples included input environments that could easily occur in many important applications. In these counterexamples, as a list of input patterns perturbed level $F_1$ through time, the response of level $F_2$ to the *same* input pattern could be different on each successive presentation of that input pattern. Moreover, the $F_2$ response to a given input pattern might never settle down as learning proceeded.

Such unstable learning in response to a prescribed input is due to the learning that occurs in response to the other, intervening, inputs. In other words, the network's adaptability, or plasticity, enables prior learning to be washed away by more recent learning in response to a wide variety of input environments. Carpenter and Grossberg (1987a, 1987b) have extended this instability analysis by describing infinitely many input environments in which periodic presentation of just four input patterns can cause temporally unstable learning. This instability problem is not, moreover, peculiar to competitive learning models. As I shall indicate below, it is a problem of almost all learning models that are now being developed.

## 6. THE STABILITY-PLASTICITY DILEMMA: SELF-STABILIZED LEARNING IN A COMPLEX AND CHANGING ENVIRONMENT

This instability problem was too fundamental to be ignored. In addition to showing that learning could become unstable in response to a complex input environment, the analysis also showed that learning could all too easily become unstable due to simple changes in an input environment. Changes in the probabilities of inputs, or in the deterministic sequencing of inputs, could readily wash away prior learning.

The seriousness of this problem can be dramatized by imagining that you have grown up in Boston before moving to Los Angeles, but periodically return to Boston to visit your parents. Although you may need to learn many new things to enjoy life in Los Angeles, these new learning experiences do not prevent you from knowing how to find your parent's house or otherwise remembering Boston. A multitude of similar examples illustrate that we are designed to successfully adapt to environments whose rules may change—without necessarily forgetting our old skills. Moreover, we are designed to successfully adapt to environments whose rules may change *unpredictably,* and can do so even if no one tells us that the environment has changed. We can adapt, in short, without a teacher, and through a direct confrontation with our experiences. Such adaptation is called *self-organization* in the network modeling literature.

The instability of the competitive learning model thus emphasized the fundamental nature of the *stability-plasticity dilemma* (Grossberg, 1980, 1982a, 1982b): How can a learning system be designed to remain plastic in response to significant new events, yet also remain stable in response to irrelevant events? How does the system know how to switch between its stable and its plastic modes in order to prevent the relentless degradation of its learned codes by the "blooming buzzing confusion" of irrelevant experience? How can it do so without using a teacher? The problem addresses one of the key capabilities that makes a human cognitive system so remarkable: its ability to learn internal representations of awesome amounts of the widest possible variety of environmental stimuli in real-time and without a teacher. The stability-plasticity dilemma articulates one sense in which a cognitive system is *universal.* Unlike the individual senses, which are specialized to deal with particular classes of inputs, a cognitive system is designed to integrate unanticipated combinations of events from all the senses into coherent moments of resonant recognition.

Rumelhart and Zipser (1985) were able to ignore this fundamental issue by considering simple input environments whose probabilistic rules do not change through time. Other modelers, for example, Kohonen (1984), have stabilized learning in their applications of the competitive learning model by externally shutting off plasticity before the learned code can be erased. This approach creates the danger of shutting off plasticity too soon, in which case important information is not learned, or too late, in which case important learned information can be erased. The only way to overcome instability using this approach in an unpredictable input environment is to assume that the observer, or teacher, who shuts off plasticity is omniscient. If a model of an omniscient teacher is available, however, then you will not also need a model of a potentially unstable learning process.

Yet other modelers, such as Ackley, Hinton, and Sejnowski (1985), Hopfield (1982), Knapp and Anderson (1984), McClelland and Rumelhart (1985), Rumelhart, Hinton, and Williams (1986), and Sejnowski and Rosenberg (1986), have stabilized their models by externally restricting the input environment. They thereby recast the problem of model instability into one about model capacity: What sorts of restricted input environments can these models handle before their learned codes are washed away by the flux of input experience? None of these learning models has yet addressed the general instability problem that was articulated a decade ago.

## 7. THE INCOMPATIBILITY OF THE COMPETITIVE LEARNING AND INTERACTIVE ACTIVATION MODELS: LETTER AND WORD LEVELS DO NOT EXIST

Before outlining a solution of the stability-plasticity dilemma, I indicate the nature of the inconsistency between the competitive learning model and the interactive activation model. Both the levels and the interactions of the two models are incompatible.

In a competitive learning model, *all* interactions between levels are excitatory. The only inhibitory interactions occur within each level. By contrast, in the Rumelhart and McClelland (1982, p. 61) model "Each letter node is assumed to activate all of those word nodes consistent with it and inhibit all other word nodes." Thus, the two models postulate different types of interlevel interactions. The selective activations and inhibitions that are hypothesized to exist between consistent and inconsistent letter nodes and word nodes must obviously be learned. In Grossberg (1984), it was shown that such connections cannot be learned using competitive learning mechanisms. Thus the postulated connections from letter nodes to word nodes are inconsistent, in a fundamental way, with competitive learning mechanisms.

An equally serious issue concerns the fact that the letter level and the word level which are postulated in the interactive activation model do not exist either in a language learning model that is based upon competitive learning mechanisms or, I would claim, *in vivo*. Instead, these levels code what I have called *items* and *lists,* respectively (Cohen & Grossberg, 1986; Grossberg, 1978, 1982a, 1984; 1987b; Grossberg & Stone, 1986). This insight is hinted at in the simulations which Rumelhart and Zipser (1985) have performed on lists of letters. These simulations are inconsistent with the existence of a letter level and a word level because both letters and words can have representations on both levels $F_1$ and $F_2$.

The difference between levels built up from letters and words and levels built up from items and lists can begin to be appreciated through the following observations (Grossberg, 1984). McClelland and Rumelhart (1981) postulated that a stage of letter nodes precedes a stage of word nodes. They used these stages to discuss the processing of letters in 4-letter words. The hypothesis of separate stages for letter and word processing implies that letters are not also represented on the level of words of length four. In order to be of general applicability, these concepts should certainly be generalizable to words of length less than four, notably to 1-letter words such as *A* and *I*. A consistent extension of the McClelland and Rumelhart stages would require that those letters which are also words, such as *A* and *I,* are represented on both the letter level and the word level, whereas those letters which are not words, such as *E* and *F,* are represented only on the letter level. How this distinction could be learned by an unsupervised learning model remains unclear.

This problem of processing units is symptomatic of a more general difficulty. The letter and word levels contain only nodes that represent letters and words. What did these nodes represent before their respective letters and words were learned? Where will the nodes come from to represent the letters and words that the model individual has not yet learned? Are these nodes to be created *de novo*? They certainly cannot be created *de novo* within the five or six trials that enable a pseudoword to acquire many of the recognition characteristics of a word (Salasso, Shiffrin, & Feustel, 1984)? These concerns clarify the need to define a processing substrate that can represent the learned units of a subject's internal lexicon before, during, or after they are learned.

The assumption of separate letter and word levels also requires special assumptions to deal with various data, such as the data of Wheeler (1970) and Samuel, van Santen, and Johnston (1982, 1983) concerning the word superiority effect. If separate letter and word levels exist, then letters such as *A* and *I* which are also words should, as words, be able to prime their letter representations. In constrast, letters such as *D* and *E* which are not words should receive no significant priming from the word level. One might therefore expect easier recognition of *A* and *I* than of *D* and *E*. Wheeler (1970) showed that this is not the case.

The assumptions of separate letter and word levels could escape this contradiction by assuming that *all* letters can be recognized so much more quickly than words of length at least two that no priming whatsoever can be received from the word level before letter recognition is complete. This assumption seems to be incompatible with the word length data of Samuel, van Santen, and Johnston (1982, 1983). These authors showed that recognition improves if a letter is embedded in words of greater length. Thus a letter that is presented alone for a fixed time before a mask appears is recognized less well than a letter presented for the same amount of time in a word length 2, 3, or 4. These data cast doubt on any explanation based on speed of processing alone, since they suggest that priming of letters due to multiletter words is effective.

In contrast, within a model which uses a item level and a list level, *all* familiar letters possess both item and list representation, not just letters such as *A* and *I* that are also words. Thus a model which uses an item level and a list level can readily explain the Wheeler (1970) data. An analysis of how item and list representations are built up led, in fact, to the prediction of a word length effect for words of lengths 1, 2, 3, and 4 (Grossberg, 1978,

Section 41; reprinted in Grossberg, 1982a). Cohen and Grossberg (1986a, 1987b) describe computer simulations of how item and list levels interact.

## 8. ADAPTIVE RESONANCE THEORY: SELF-STABILIZATION OF CODE LEARNING IN AN ARBITRARY INPUT ENVIRONMENT

A formal analysis of how to overcome the learning instability experienced by a competitive learning model led to the introduction of an expanded theory, called adaptive resonance theory (*ART*), in Grossberg (1976c). This formal analysis showed that a certain type of top-down learned feedback and matching mechanism could significantly overcome the instability problem. It was also realized that top-down attentional mechanisms, which had earlier been discovered through an analysis of interactions between cognitive and reinforcement mechanisms (Grossberg, 1975), had the same properties as these code-stabilizing mechanisms. In other words, once it was recognized how to formally solve the instability problem, it also became clear that one did not need to invent any qualitatively new mechanisms to do so. One only needed to remember to include previously discovered attentional mechanisms! These additional mechanisms enable code learning to self-stabilize in response to an essentially arbitrary input environment. For a recent mathematical proof of this type of stability, see Carpenter and Grossberg (1987b).

The types of top-down effects, such as the "rich-get-richer" and "gang" effects, which McClelland and Rumelhart (1981) experimentally reported were predicted formal properties of the *ART* theory as developed in Grossberg (1978). Such properties are shared by many networks which undergo reciprocal bottom-up and top-down feedback exchanges. They arose in *ART* as predictions about the emergent properties of network architectures that were designed to guarantee self-stabilizing self-organization of cognitive recognition codes—the very properties that are absent from the interactive activation model. In addition to such properties, *ART* has by now been used to analyse and predict data about speech perception, word recognition and recall, visual perception, olfactory coding, classical and instrumental conditioning, decision making under risk, event related potentials, neural substrates of learning and memory, critical period termination, and amnesias (Banquet & Grossberg, 1987; Carpenter & Grossberg, 1987a, 1987b, 1987c; Cohen & Grossberg, ;1987a, 1987b; Grossberg, 1982b, 1984, 1987a, 1987b; Grossberg & Gutowski, 1987; Grossberg & Levine, 1987; Grossberg & Stone, 1986a, 1986b). Thus *ART* has already demonstrated an explanatory and predictive competence as an interdisciplinary physical theory. It is now being developed both to expand its predictive range and to implement it in real-time hardware. In Sections 9–15, some properties of *ART* are outlined as a basis for comparisons with other learning models in the literature, such as the back propagation model. These comparisons delineate issues that could just as easily be raised in the evaluation of any network learning model.

## 9. SOLVING THE STABILITY-PLASTICITY DILEMMA USING INTERACTING ATTENTIONAL AND ORIENTING SYSTEMS

In addition to the bottom-up mechanisms of a competitive learning model, an *ART* system includes processes for learning of top-down expectancies, or templates; for matching bottom-up input patterns with top-down expectancies; and for releasing orienting reactions in a mismatch situation, thereby leading to rapid updating, or reset, of short term memory as the network carries out a hypothesis testing scheme that searches for and, if necessary, leads to learning of a better representation of the input pattern (Figure 1).

Using these mechanisms, an *ART* system can generate recognition codes adaptively, and without a teacher, in response to a series of environmental inputs. As learning proceeds, interactions between the inputs and the system generate new steady states, or equilibrium points. The steady states are formed as the system discovers and learns *critical feature patterns,* or prototypes, that represent invariants of the set of all experienced input patterns. These learned codes are dynamically buffered, or stabilized, against relentless recoding by irrelevant inputs. The formation of steady states is internally controlled using mechanisms that suppress possible sources of system instability.

An *ART* system can adaptively switch between its stable and plastic modes. It is capable of plasticity in order to learn about significant new events, yet it can also remain stable in response to irrelevant events. In order to make this distinction, an *ART* system is sensitive to *novelty*. It is capable, without a teacher, of distinguishing between familiar and unfamiliar events, as well as between expected and unexpected events.

Multiple interacting memory systems are needed to monitor and adaptively react to the novelty of events without an external teacher. Within *ART*, interactions between two functionally complementary subsystems are used to process familiar and unfamiliar events. Familiar events are processed within an attentional subsystem, which is built up from a competitive learning network. The attentional subsystem establishes ever more precise internal representations of and responses to familiar events. It also learns the top-down expectations that help to stabilize the learned bottom-up codes of familiar events. As described above, however, the attentional subsystem is unable simultaneously to maintain stable representations of familiar categories and to create new categories for unfamiliar patterns in certain input environments. An isolated attentional subsystem can become either too rigid and incapable of creating new categories for unfamiliar patterns, or too unstable and capable of ceaselessly recoding the categories of familiar patterns as the statistics of the input environment change.

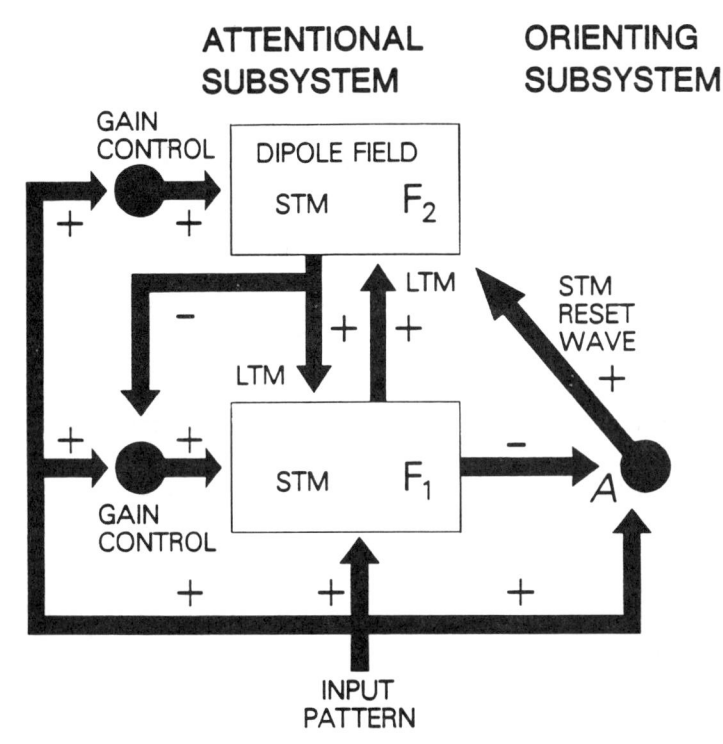

**Figure 1.** Anatomy of the attentional-orienting system: Two successive stages, $F_1$ and $F_2$, of the attentional subsystem encode patterns of activation in short term memory (STM). Bottom-up and top-down pathways between $F_1$ and $F_2$ contain adaptive long term memory (LTM) traces which multiply the signals in these pathways. The remainder of the circuit modulates these STM and LTM processes. Modulation by gain control enables $F_1$ to distinguish between bottom-up input patterns and top-down priming, or template, patterns, as well as to match these bottom-up and top-down patterns. Gain control signals also enable $F_2$ to react supraliminally to signals from $F_1$ while an input pattern is on. The orienting subsystem A generates a reset wave to $F_2$ when mismatches between bottom-up and top-down patterns occur at $F_1$. This reset wave selectively and enduringly inhibits active $F_2$ cells until the input is shut off. (Reprinted with permission from Carpenter and Grossberg, 1987b).

The second subsystem is an orienting subsystem that resets the attentional subsystem when an unfamiliar event occurs. Interactions between the attentional subsystem and the orienting subsystem help to express whether a novel pattern is familiar and well represented by an existing recognition code, or unfamiliar and in need of a new recognition code (Figure 1). Within an *ART* system, attentional mechanisms play a major role in self-stabilizing the learning of an emergent recognition code. A mechanistic analysis of the role of attention in learning has led Carpenter and Grossberg (1987a, 1987b) to distinguish between four types of attentional mechanisms attentional priming, attentional gain control, attentional vigilance, and intermodality competition (see Figure 4).

# 10. SELF-SCALING COMPUTATIONAL UNITS, SELF-ADJUSTING MEMORY SEARCH, DIRECT ACCESS, AND ATTENTIONAL VIGILANCE

Four properties are basic to the workings of an *ART* network. Violating any one of these properties prevents the network from learning well in certain input environments. Essentially all other learning models violate one or more of these properties.

## A. Self-Scaling Computational Units: Critical Feature Patterns

Properly defining signal and noise in a self-organizing system raises a number of subtle issues. Pattern context must enter the definition so that input features which are treated as irrelevant noise when they are embedded in a given input pattern may be treated as informative signals when they are embedded in a different input pattern. The system's unique learning history must also enter the definition so that portions of an input pattern which are treated as noise when they perturb a system at one stage of its self-organization may be treated as signals when they perturb the same system at a different stage of its self-organization. The present systems automatically self-scale their computational units to embody context- and learning-dependent definitions of signal and noise.

One property of these self-scaling computational units is illustrated in Figure 2. In Figure 2a, each of the two input patterns is composed of three features. The patterns agree at two of the three features, but disagree at the third feature. A mismatch of one out of three features may be designated as informative by the system. When this occurs, these mismatched features are treated as signals which can elicit learning of distinct recognition codes for the two patterns. Moreover, the mismatched features, being informative, are incorporated into these distinct recognition codes through the learning process.

In Figure 2b, each of the two input patterns is composed of 31 features. The patterns are constructed by adding identical subpatterns to the two patterns in Figure 2a. Thus the input patterns in Figure 2b disagree at the same features as the input patterns in Figure 2a. In the patterns of Figure 2b, however, this mismatch is less important, other things being equal, than in the patterns of Figure 2a. Consequently, the system may treat the mismatched features as noise. A single recognition code may be learned to represent both of the input patterns in Figure 2b. The mismatched features would not be learned as part of this recognition code because they are treated as noise.

The assertion that *critical feature patterns* are the computational units of the code learning process summarizes this self-scaling property. The term *critical feature* indicates that not all features are treated as signals by the system. The learned units are *patterns* of critical features because the perceptual context in which the features are embedded influences which

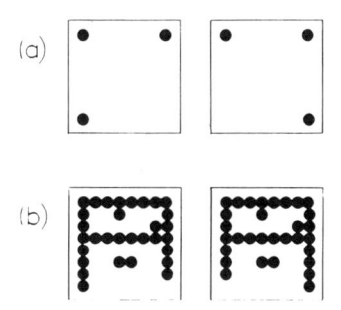

(a) [figure: two squares, left with dots at 3 corners plus bottom-left, right with dots at 3 corners plus bottom-right]

(b) [figure: two E-like dot patterns]

**Figure 2.** Self-scaling property discovers critical features in a context-sensitive way: (a) Two input patterns of 3 features mismatch at 1 feature. When this mismatch is sufficient to generate distinct recognition codes for the two patterns, the mismatched features are encoded in *LTM* as part of the critical feature patterns of these recognition codes. (b) Identical subpatterns are added to the two input patterns in (a). Although the new input patterns mismatch at the same one feature, this mismatch may be treated as noise due to the additional complexity of the two new patterns. Both patterns may thus learn to activate the same recognition code. When this occurs, the mismatched feature is deleted from *LTM* in the critical feature pattern of the code.

features will be processed as signals and which features will be processed as noise. Thus a feature may be a critical feature in one pattern (Figure 2a) and an irrelevant noise element in a different pattern (Figure 2b).

### B. Self-Adjusting Memory Search

No pre-wired search algorithm, such as a search tree, can maintain its efficiency as a knowledge structure evolves due to learning in a unique input environment. A search order that may be optimal in one knowledge domain may become extremely inefficient as that knowledge domain becomes more complex due to learning.

An *ART* system is capable of a parallel memory search that adaptively updates its search order to maintain efficiency as its recognition code becomes arbitrarily complex due to learning. This self-adjusting search mechanism is part of the network design whereby the learning process self-stabilizes by engaging the orienting subsystem.

None of these mechanisms is akin to the rules of a serial computer program. Instead, the circuit architecture as a whole generates a self-adjusting search order and self-stabilization as emergent properties that arise through system interactions. Once the *ART* architecture is in place, a little randomness in the initial values of its memory traces, rather than a carefully wired search tree, enables the search to carry on until the recognition code self-stabilizes.

### C. Direct Access to Learned Codes

A hallmark of human recognition performance is the remarkable rapidity with which familiar objects can be recognized. The existence of many learned recognition codes for alternative experiences does not necessarily interfere with rapid recognition of an unambiguous familiar event. This type of rapid recognition is very difficult to understand using models wherein trees or other serial algorithms need to be searched for longer and longer periods as a learned recognition code becomes larger and larger.

In an *ART* model, as the learned code becomes globally self-consistent and predictively accurate, the search mechanism is automatically disengaged. Subsequently, no matter how large and complex the learned code may become, familiar input patterns *directly access,* or activate, their learned code, or category. Unfamiliar patterns can also directly access a learned category if they share invariant properties with the critical feature pattern of the category. In this sense, the critical feature pattern acts as a prototype for the entire category. As in human pattern recognition experiments, a "prototype" input pattern that perfectly matches a learned critical feature pattern may be better recognized than any of the "exemplar" input patterns that gave rise to the critical feature pattern (Posner, 1973; Posner & Keele, 1968, 1970). Grossberg and Stone (1986a) have shown, moreover, that these direct access properties can be used to explain *RT* and error data from lexical decision and word familiarity experiments.

Unfamiliar input patterns which cannot stably access a learned category engage the self-adjusting search process in order to discover a network substrate for a new recognition category. After this new code is learned, the search process is automatically disengaged and direct access ensues.

We use the term critical feature pattern, rather than prototype, because critical feature patterns are learned, matched, and regulate future learning in a manner different from classical prototype models. Estes (1986) compared several types of category learning models in the light of recent data and showed that exemplar models, prototype models, and exemplar similarity models all have their merits. An *ART* model can also be sensitive to exemplars, prototypes, or similarity between exemplars, depending upon the experimental conditions. One factor that mediates between these alternatives is now summarized.

**D. Environment as a Teacher: Modulation of Attentional Vigilance**

Although an *ART* system self-organizes its recognition code, the environment can also modulate the learning process and thereby carry out a teaching role. This teaching role allows a system with a fixed set of feature detectors to function successfully in an environment which imposes variable performance demands. Different environments may demand either coarse discriminations or fine discriminations to be made among the same set of objects.

In an *ART* system, if an erroneous recognition is followed by an environmental disconfirmation, such as a punishment, then the system becomes more *vigilant.* This change in vigilance may be interpreted as a change in the system's attentional state which increases its sensitivity to mismatches between bottom-up input patterns and active top-down critical feature patterns. A vigilance change alters the size of a single parameter in the network. The interactions within the network respond to this parameter change by learning recognition codes that make finer distinctions. In other words, if the network erroneously groups together some input patterns, then negative reinforcement can help the network to learn the desired distinction by making the system more vigilant. The system then behaves *as if* it has a better set of feature detectors. Thus at a level of very high vigilance, a category may emerge that accepts only one exemplar. At lower levels of vigilance, similarity relationships among the accepted exemplars help to mold the category's emergent critical feature pattern. Different vigilance levels may, moreover, be imposed by environmental feedback in response to easy or difficult discriminations during the course of a single experiment or experience.

The ability of a vigilance change to alter the course of pattern recognition illustrates a theme that is common to a variety of neural processes: a one-dimensional parameter change that modulates a simple nonspecific neural process can have complex specific effects upon high-dimensional neural information processing.

## 11. BOTTOM-UP ADAPTIVE FILTERING AND CONTRAST-ENHANCEMENT IN SHORT-TERM MEMORY

The typical network reactions to a single input pattern I within a temporal stream of input patterns are now briefly summarized. Each input pattern may be the output pattern of a preprocessing stage. Different preprocessing is given, for example, to speech signals and to visual signals before the outcome of such modality-specific preprocessing ever reaches the attentional subsystem. The preprocessed input pattern I is received at the stage $F_1$ of an attentional subsystem. Pattern $I$ is transformed into a pattern $X$ of activation across the nodes, or abstract "feature detectors", of $F_1$ (Figure 3). The transformed pattern $X$ is said to represent I in short term memory (*STM*). In $F_1$ each node whose activity is sufficiently large generates excitatory signals along pathways to target nodes at the next processing stage $F_2$. A pattern $X$ of *STM* activities across $F_1$ hereby elicits a pattern $S$ of output signals from $F_1$. When a signal from a node in $F_1$ is carried along a pathway to $F_2$, the signal is multiplied, or *gated,* by the pathway's long term memory (*LTM*) trace. The *LTM* gated signal (i.e., signal times *LTM* trace), not the signal alone, reaches the target node. Each target node sums up all of its *LTM* gated signals. In this way, pattern $S$ generates a pattern $T$ of *LTM*-gated and summed input signals to $F_2$ (Figure 4a). The transformation from $S$ to $T$ is called an *adaptive filter*.

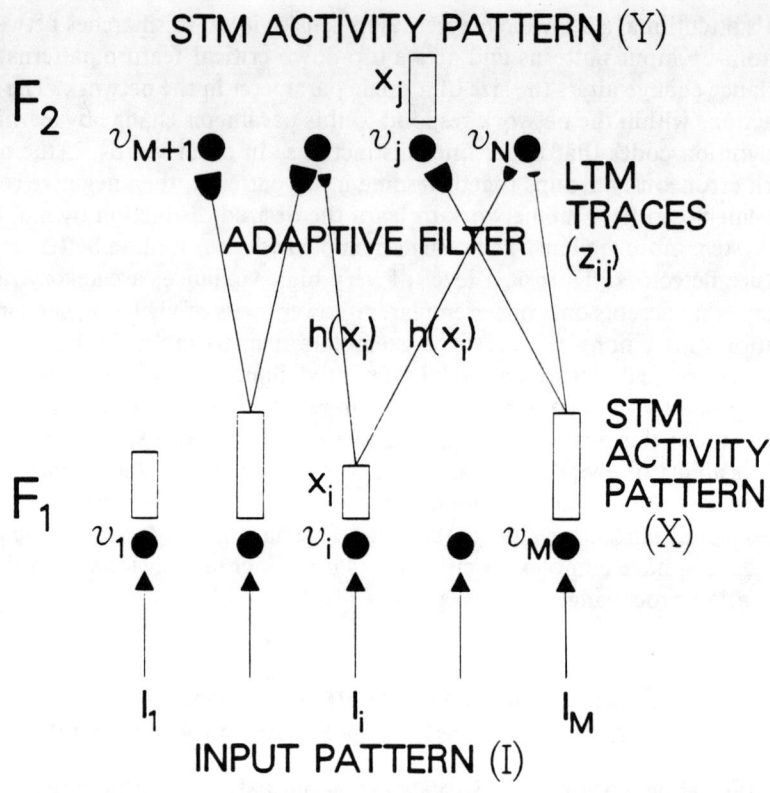

**Figure 3.** Stages of bottom-up activation: The input pattern $I$ generates a pattern of STM activation $X$ across $F_1$. Sufficiently active $F_1$ nodes emit bottom-up signals to $F_2$. This signal pattern $S$ is gated by long term memory (LTM) traces within the $F_1 \rightarrow F_2$ pathways. The LTM gated signals are summed before activating their target nodes in $F_2$. This LTM-gated and summed signal pattern $T$ generates a pattern of activation $Y$ across $F_2$. The nodes in $F_1$ are denoted by $v_1, v_2, \ldots, v_M$. The nodes in $F_2$ are denoted by $v_{M+1}, v_{M+2}, \ldots v_N$. The input to node $v_i$ is denoted by $I_i$. The STM activity of node $v_i$ is denoted by $x_i$. The LTM trace of the pathway from $v_i$ to $v_j$ is denoted by $z_{ij}$.

The input pattern $T$ to $F_2$ is quickly transformed by interactions among the nodes of $F_2$. These interactions contrast-enhance the input pattern $T$. The resulting pattern of activation across $F_2$ is a new pattern $Y$. The contrast-enhanced pattern $Y$, rather than the input pattern $T$, is stored in STM by $F_2$. These interactions also occur in a competitive learning model.

## 12. TOP-DOWN TEMPLATE MATCHING AND STABILIZATION OF CODE LEARNING

As soon as the bottom-up STM transformation $X \rightarrow Y$ takes place, the STM activities $Y$ in $F_2$ elicit a top-down excitatory signal pattern $U$ back to $F_1$

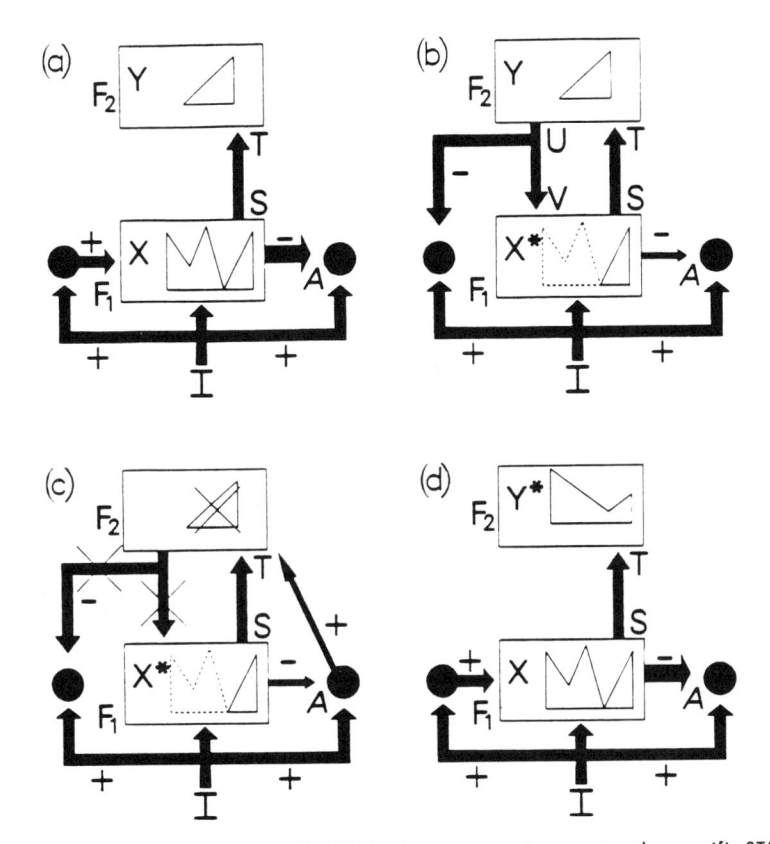

**Figure 4.** Search for a correct $F_2$ code: (a) the input pattern $I$ generates the specific STM activity pattern $X$ at $F_1$ as it nonspecifically activates $A$. Pattern $X$ both inhibits $A$ and generates the output signal pattern $S$. Signal pattern $S$ is transformed into the input pattern $T$, which activates the STM pattern $Y$ across $F_2$. (b) Pattern $Y$ generates the top-down signal pattern $U$ which is transformed into the template pattern $V$. If $V$ mismatches $I$ at $F_1$, then a new STM activity pattern $X^*$ is generated at $F_1$. The reduction in total STM activity which occurs when $X$ is transformed into $X^*$ causes a decrease in the total inhibition from $F_1$ to $A$. (c) Then the input-driven activation of $A$ can release a nonspecific arousal wave in $F_2$, which resets the STM pattern $Y$ at $F_2$. (d) After $Y$ is inhibited, its top-down template is eliminated, and $X$ can be reinstated at $F_1$. Now $X$ once again generates input pattern $T$ to $F_2$, but since $Y$ remains inhibited $T$ can activate a different STM pattern $Y^*$ at $F_2$. If the top-down template due to $Y^*$ also mismatches $I$ at $F_1$, then the rapid search for an appropriate $F_2$ code continues.

(Figure 4b). Only sufficiently large *STM* activities in $Y$ elicit signals in $U$ along the feedback pathways $F_2 \rightarrow F_1$. As in the bottom-up adaptive filter, the top-down signals $U$ are also gated by *LTM* traces and the *LTM*-gated signals are summed at $F_1$ nodes. The pattern $U$ of output signals from $F_2$ hereby generates a pattern $V$ of *LTM*-gated and summed input signals to $F_1$. The transformation from $U$ to $V$ is thus also an adaptive filter. The pattern $V$ is called *a top-down template,* or *learned expectation.*

Two sources of input now perturb $F_1$: the bottom-up input pattern $I$ which gave rise to the original activity pattern $X$, and the top-down template pattern $V$ that resulted from activating $X$. The activity pattern $X^*$ across $F_1$ that is induced by $I$ and $V$ taken together is typically different from the activity pattern $X$ that was previously induced by $I$ alone. In particular, $F_1$ acts to *match* $V$ against $I$. The result of this matching process determines the future course of learning and recognition by the network.

The entire activation sequence

$$I \to X \to S \to T \to Y \to U \to V \to X^* \tag{1}$$

takes place very quickly relative to the rate with which the *LTM* traces in either the bottom-up adaptive filter $S \to T$ or the top-down adaptive filter $U \to V$ can change. Even though none of the *LTM* traces changes during such a short time, their prior learning strongly influences the *STM* patterns $Y$ and $X^*$ that evolve within the network by determining the transformations $S \to T$ and $U \to V$. I now sketch how a match or mismatch of $I$ and $V$ at $F_1$ regulates the course of learning in response to the pattern $I$, and in particular solves the stability-plasticity dilemma.

## 13. INTERACTIONS BETWEEN ATTENTIONAL AND ORIENTING SUBSYSTEMS: STM RESET AND SEARCH

In Figure 4a, as input pattern $I$ generates an *STM* activity pattern $X$ across $F_1$. The input pattern $I$ also excites the orienting subsystem $A$, but pattern $X$ at $F_1$ inhibits $A$ before it can generate an output signal. Activity pattern $X$ also elicits an output pattern $S$ which, via the bottom-up adaptive filter, instates an *STM* activity pattern $Y$ across $F_2$. In Figure 4b, pattern $Y$ reads a top-down template pattern $V$ into $F_1$. Template $V$ mismatches input $I$, thereby significantly inhibiting *STM* activity across $F_1$. The amount by which activity in $X$ is attenuated to generate $X^*$ depends upon how much of the input pattern $I$ is encoded within the template pattern $V$.

When a mismatch attenuates *STM* activity across $F_1$, the total size of the inhibitory signal from $F_1$ to $A$ is also attenuated. If the attenuation is sufficiently great, inhibition from $F_1$ to $A$ can no longer prevent the arousal source $A$ from firing. Figure 4c depicts how disinhibition of $A$ releases an arousal burst to $F_2$ which equally, or nonspecifically, excites all the $F_2$ cells. The cell populations of $F_2$ react to such an arousal signal in a state-dependent fashion. In the special case that $F_2$ chooses a single population for *STM* storage, the arousal burst selectively inhibits, or resets, the active population in $F_2$. This inhibition is long-lasting. One physiological design for $F_2$ processing which has these properties is a *gated dipole field* (Grossberg, 1982a, 1987a). A gated dipole field consists of opponent processing channels which are gated by habituating chemical transmitters. A nonspecific arousal burst induces selective and enduring inhibition of active populations within a gated dipole field.

In Figure 4c, inhibition of $Y$ leads to removal of the top-down template $V$, and thereby terminates the mismatch between $I$ and $V$. Input pattern $I$ can thus reinstate the original activity pattern $X$ across $F_1$, which again generates the output pattern $S$ from $F_1$ and the input pattern $T$ to $F_2$. Due to the enduring inhibition at $F_2$, the input pattern $T$ can no longer activate the original pattern $Y$ at $F_2$. A new pattern $Y^*$ is thus generated at $F_2$ by $I$ (Figure 4d).

The new activity pattern $Y^*$ reads-out a new top-down template pattern $V^*$. If a mismatch again occurs at $F_1$, the orienting subsystem is again engaged, thereby leading to another arousal-mediated reset of $STM$ at $F_2$. In this way, a rapid series of $STM$ matching and reset events may occur. Such an $STM$ matching and reset series controls the system's hypothesis testing and search of $LTM$ by sequentially engaging the novelty-sensitive orienting subsystem. Although $STM$ is reset sequentially in time via this mismatch-mediated, self-terminating $LTM$ search process, the mechanisms which control the $LTM$ search are all parallel network interactions, rather than serial algorithms. Such a parallel search scheme continuously adjusts itself to the system's evolving $LTM$ codes. The $LTM$ code depends upon both the system's initial configuration and its unique learning history, and hence cannot be predicted *a priori* by a pre-wired search algorithm. Instead, the mismatch-mediated engagement of the orienting subsystem realizes the type of self-adjusting search that was described in Section 10B.

The mismatch-mediated search of $LTM$ ends when an $STM$ pattern across $F_2$ reads-out a top-down template which matches $I$, to the degree of accuracy required by the level of attentional vigilance (Section 10D), or which has not yet undergone any prior learning. In the latter case, a new recognition category is then established as a bottom-up code and top-down template are learned.

## 14. ATTENTIONAL GAIN CONTROL AND PATTERN MATCHING: THE 2/3 RULE

The $STM$ reset and search process described in Section 13 makes a paradoxical demand upon the processing dynamics of $F_1$: the *addition* of new excitatory top-down signals in the pattern $V$ to the bottom-up signals in the pattern $I$ causes a *decrease* in overall $F_1$ activity (Figures 4a and 4b). Some auxiliary mechanism must exist to distinguish between bottom-up and top-down inputs. This auxiliary mechanism is called *attentional gain control* to distinguish it from *attentional priming* by the top-down template $V$. While $F_2$ is active, the attentional priming mechanism delivers *excitatory specific learned* template patterns to $F_1$. Top-down attentional gain control has an *inhibitory nonspecific unlearned* effect on the sensitivity with which $F_1$ responds to the template pattern, as well as to other patterns received by $F_1$. The attentional gain control process enables $F_1$ to tell the difference between bottom-up and top-down signals.

In Figure 4a, during bottom-up processing, a suprathreshold node in $F_1$ is one which receives both a specific input from the input pattern $I$ and a nonspecific attentional gain control input. In Figure 4b, during the matching of simultaneous bottom-up and top-down patterns, attentional gain control signals to $F_1$ are inhibited by the top-down channel. Nodes of $F_1$ must then receive sufficiently large inputs from both the bottom-up and the top-down signal patterns to generate suprathreshold activities. Nodes which receive a bottom-up input or a top-down input, but not both, cannot become suprathreshold: mismatched inputs cannot generate suprathreshold activities. Attentional gain control thus leads to a matching process whereby the addition of top-down excitatory inputs to $F_1$ can lead to an overall decrease in $F_1$'s *STM* activity. Since, in each case, an $F_1$ node becomes active only if it receives large signals from two of the three input sources, we call this matching process the 2/3 Rule (Figure 5).

## 15. STABLE CODE LEARNING IN AN ARBITRARY INPUT ENVIRONMENT

If an *ART* system violates the 2/3 Rule, there are infinitely many input sequences, each containing only four distinct patterns, that cannot be stably encoded (Carpenter & Grossberg, 1987b). It has also been mathematically proved that, when the 2/3 Rule is reinstated, the *ART* architecture self-organizes, self-stabilizes, and self-scales its learning of a recognition code in response to an arbitrary ordering of arbitrarily many, arbitrarily chosen binary input patterns (Carpenter & Grossberg, 1987b). Moreover, each of the *LTM* traces oscillates at most once through time as learning proceeds in response to any such environment. Thus, learning in an *ART* architecture is remarkably stable. Figure 6 illustrates computer simulations of alphabet learning by an *ART* circuit. At two difference values of the vigilance parameter $\rho$, different numbers of recognition categories are learned. In both cases, code learning is complete and self-stabilizes in response to the 26 letters after only 3 trials.

Computer simulations of code learning using a coding level $F_2$ which carries out a multiple scale, distributed decomposition of its input patterns have also been carried out (Cohen & Grossberg, 1986, 1987). Such a design for $F_2$, and by extension for the higher coding levels $F_3, F_4, \ldots$ fed by $F_2$, is called a *masking field*. A masking field instantiates the *list level* that was described in Section 7. Such a network can simultaneously detect multiple groupings within its input patterns and assigns weights to the codes for

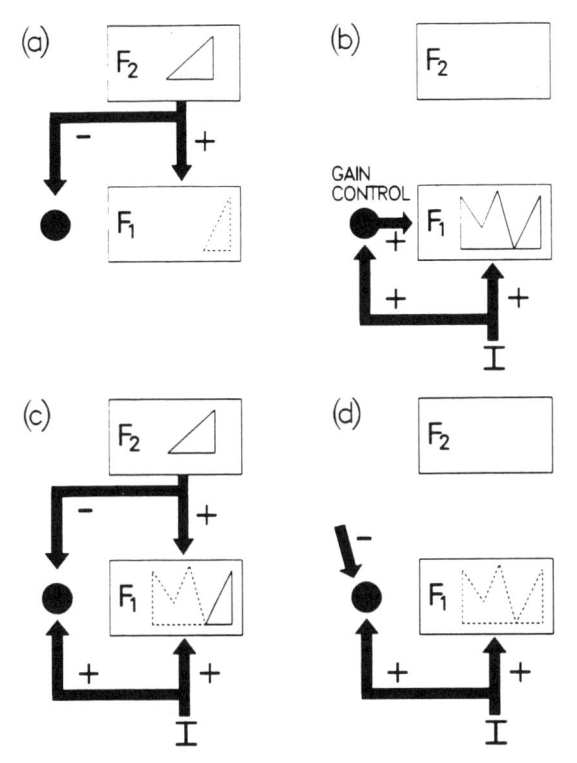

**Figure 5.** Matching by the 2/3 Rule: (a) A top-down template from $F_2$ inhibits the attentional gain control source as it subliminally primes target $F_1$ cells. (b) Only $F_1$ cells that receive bottom-up inputs and gain control signals can become supraliminally active. (c) When a bottom-up input pattern and a top-down template are simultaneously active, only those $F_1$ cells that receive inputs from both sources can become supraliminally active. (d) Intermodality inhibition can shut off the $F_1$ gain control source and thereby prevent a bottom-up input from supraliminally activating $F_1$. Similarly, disinhibition of the $F_1$ gain control source may cause a top-down prime to become supraliminal.

these groupings which are predictive with respect to the contextual information embedded within the patterns and the prior learning of the system. A masking field automatically rescales its sensitivity as the overall size of an input pattern changes, yet also remains sensitive to the microstructure within each input pattern. In this way, such a network distinguishes between codes for pattern wholes and for pattern parts, yet amplifies the code for a pattern part when it becomes a pattern whole in a new input context. This capability is useful in speech recognition, visual object recognition, and cognitive information processing.

To achieve these properties, a masking field $F_2$ performs a new type of multiple scale analysis in which unpredictive list codes are competitively masked, or inhibited, and predictive codes are amplified in direct response

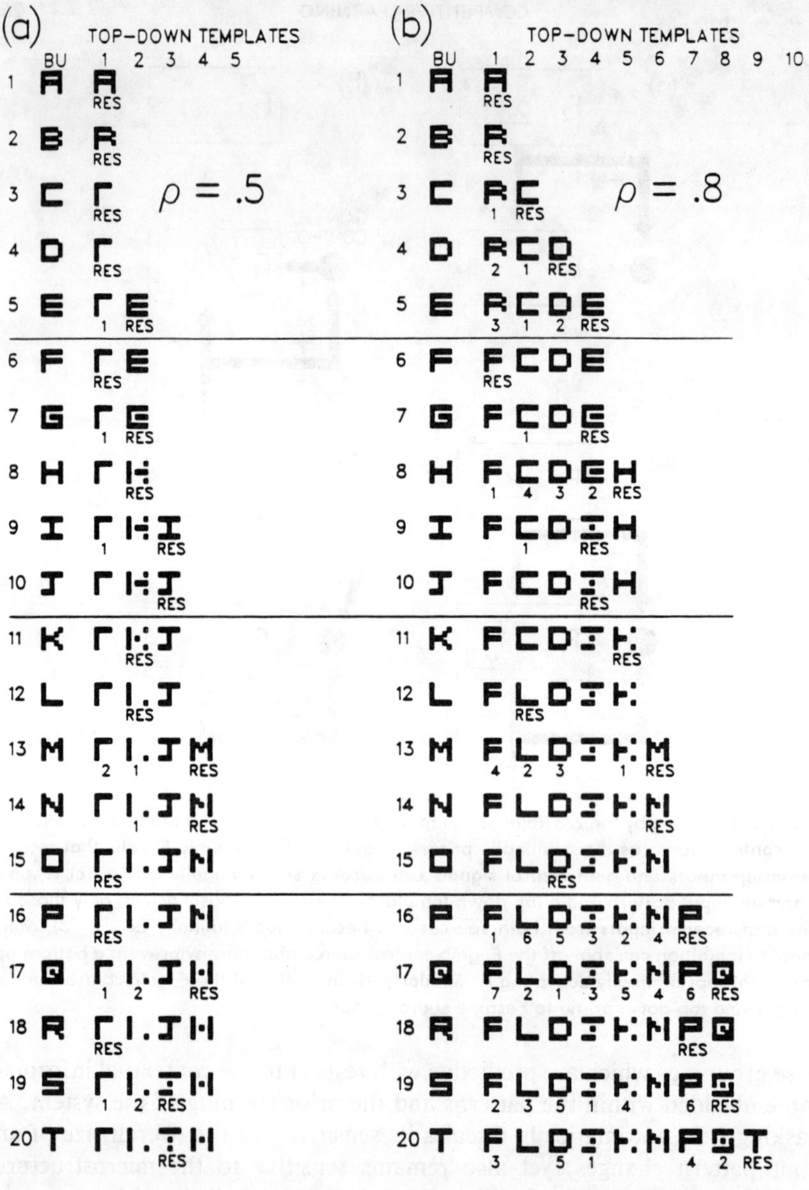

**Figure 6.** Alphabet learning: Code learning in response to the first presentation of the first 20 letters of the alphabet is shown. Two different vigilance levels were used, $\rho = .5$ and $\rho = .8$. Each row represents the total code that is learned after the letter at the left-hand column of the row is presented at $F_1$. Each column represents the critical feature pattern that is learned through time by the $F_2$ node listed at the top of the column. The critical feature patterns do not, in general, equal the pattern exemplars which change them through learning. Instead, each critical feature pattern acts like a prototype for the entire set of these exemplars, as well as for unfamiliar exemplars which share invariant properties with familiar exemplars. The simulation illustrates the "fast learning" case, in which the altered LTM traces reach a new equilibrium in response to each new stimulus. Slow learning is more gradual than this. (Reprinted with permission from Carpenter and Grossberg, 1987b).

to trainable signals from an adaptive filter $F_1 \rightarrow F_2$ that is activated by an input source $F_1$. An adaptive sharpening property obtains whereby a familiar input pattern causes a more focal spatial activation of its recognition code than an unfamiliar input pattern. The recognition code also becomes less distributed when an input pattern contains more information on which to base an unambiguous prediction of which input pattern is being processed. Thus, a masking field suggests a solution of the credit assignment problem by embodying a real-time code for the predictive evidence contained within its input patterns. Such a network processing level can be used to build up an *ART* system $F_1 \leftrightarrow F_2 \leftrightarrow F_3 \leftrightarrow \ldots$ with any number of processing levels.

## 16. THE BACK PROPAGATION AND NETtalk MODELS

The *ART* architecture may be usefully compared with the back propagation (*BP*) model of Rumelhart, Hinton, and Williams (1986). The similarities and differences of these models highlight many of the types of formal comparisons that can help to evaluate other network learning models.

The *BP* model is a steepest descent algorithm in which each *LTM* trace, or weight, in the network is adjusted to minimize its contribution to the total mean square error between the desired and actual system outputs. Although steepest descent algorithms have a long history in technology and the neural modelling literature, the *BP* model has attracted widespread interest, partly because of the demonstration of Sejnowski and Rosenberg (1986), in which the *BP* algorithm is part of a system that learns to convert printed text into spoken language. Despite the appeal of this demonstration, the *BP* model does not model a brain process, as will be shown below. This shortcoming does not limit the model's possible value in technological applications which can benefit from a steepest descent algorithm, but it undermines the model's usefulness in explaining behavioral or neural data.

The *BP* model is usually described as a three level model, with levels $F_1$, $F_2$, $F_3$, such that level $F_2$ is a level of "hidden units" between $F_1$ and $F_3$. The purpose of the model is to learn an associative map between the input level $F_1$ and the output level $F_3$. The map is designed to be sufficiently distributed to allow alterations in the inputs at $F_1$ to generate appropriate alterations in the outputs at $F_3$. Such a possibility depends upon general projection properties of distributed associated maps (Kohonen, 1984). The key property demonstrated by computer simulations of the *BP* model is that it can learn a distributed associative map.

Some of the claims for the *BP* model have been based on comparisons with the early Perceptron model (Rosenblatt, 1962). Sejnowski and Rosenfeld (1986) have written that "until recently, learning in multilayered networks was an unsolved problem and considered by some impossible... In a multilayered machine the internal, or hidden, units can be used as feature detectors which perform a mapping between input units and output units, and the difficult problem is to discover proper features." Carpenter and Grossberg (1986) note, in contrast, that learning an associative map using hidden units is an old problem with definite solutions in the neural modelling literature subsequent to the introduction of the Perceptron. Indeed, *ART* was developed in part to develop a theory of how learning of an associative map could proceed in a self-stabilizing fashion (Figure 7). A basic difference does, however, exist between models of associative map learning, such as *ART* and the *BP* model. The former model is self-organizing, whereas the *BP* model requires an external teacher.

The way in which this teacher works is what distinguishes the *BP* model from other types of steepest descent learning algorithms, such as the classical Adaline model (Widrow, 1962). The teaching algorithm is also what makes the *BP* model impossible as a model of a brain process. In addition to the levels $F_1$, $F_2$, and $F_3$ and the pathways $F_1 \rightarrow F_2 \rightarrow F_3$, the *BP* model also requires levels $F_4$, $F_5$, $F_6$, and $F_7$ as well as a complicated set of highly specific interactions between these levels and the rest of the network (Figure 8). These levels and interactions will now be described.

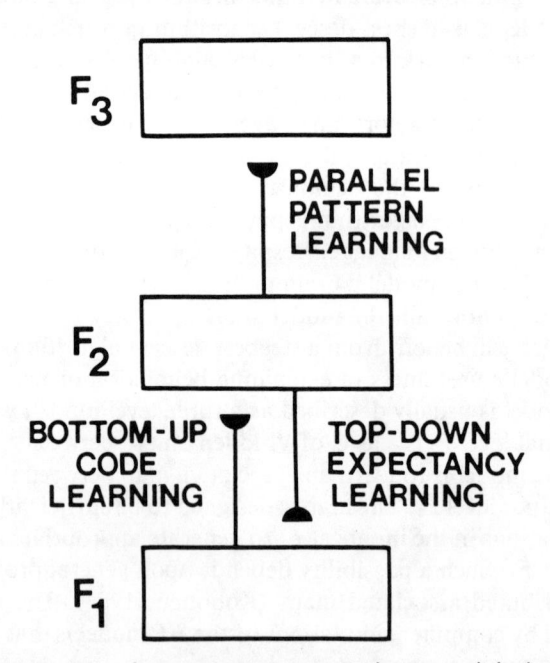

**Figure 7.** Self-organization of an associative map can be accomplished using a network with three levels $F_1$, $F_2$, and $F_3$. Levels $F_1$ and $F_2$ regulate learning within bottom-up pathways $F_1 \rightarrow F_2$ and top-down pathways $F_2 \rightarrow F_1$. This learning process discovers compressed recognition codes with invariant properties for the set of input patterns processed at $F_1$. Activation of these recognition codes at $F_2$ enables the activated sampling cells to learn output patterns at $F_3$. The total transformation $F_1 \rightarrow F_3$ defines the associative mapping.

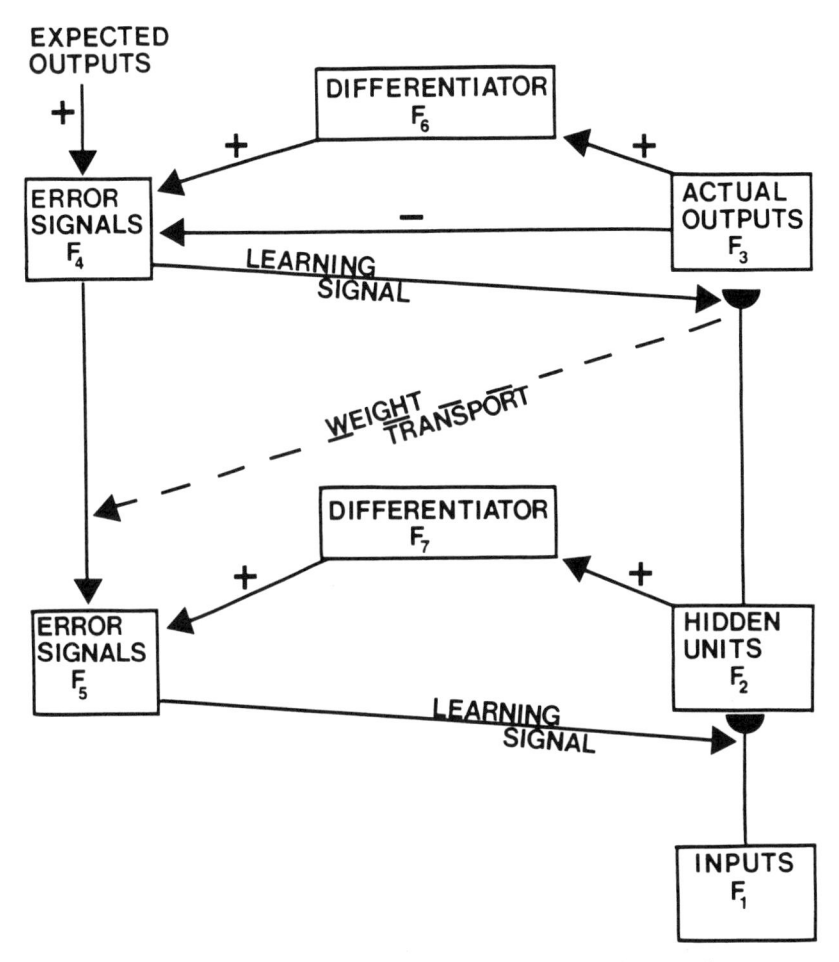

**Figure 8.** Circuit diagram of the back propagation model: In addition to the processing levels $F_1$, $F_2$, $F_3$, there are also levels $F_4$, $F_5$, $F_6$, and $F_7$ to carry out the computations which control the learning process. The transport of learned weights from the $F_2 \rightarrow F_3$ pathways to the $F_4 \rightarrow F_5$ pathways shows that this algorithm cannot represent a learning process in the brain.

Inputs delivered to $F_1$ propagate forward through $F_2$ to $F_3$, where they generate the actual outputs of the network. The desired, or expected, outputs are independently delivered to level $F_4$ by an external teacher on every learning trial. The actual outputs are subtracted from the expected outputs at $F_4$ to generate error signals. These error signals propagate from $F_4$ to the $F_2 \rightarrow F_3$ pathways, where they change the weights in the $F_2 \rightarrow F_3$ pathways.

Back propagation proceeds as follows. The weights computed in the bottom-up $F_2 \rightarrow F_3$ pathways are *transported* to the top-down $F_4 \rightarrow F_5$ pathways.

Once in these pathways, the differences between expected and real outputs at $F_4$ are multiplied by the transported weights within the $F_4 \to F_5$ pathways to generate weighted error signals that determine the inputs to $F_5$. These inputs activate $F_5$, which in turn generates output signals to the $F_1 \to F_2$ pathways. These output signals act as error signals which change the weights in the $F_1 \to F_2$ pathways.

Such a physical transport of weights has no plausible physical interpretation. The weights in the $F_2 \to F_3$ pathways must be computed *within* these pathways in order to multiply signals from $F_2$ to $F_3$. These weights cannot also exist *within* the pathways from $F_4$ to $F_5$ in order to multiply signals from $F_4$ to $F_5$ without being physically transported from $(F_2 \to F_3)$ to $(F_4 \to F_5)$ pathways, thereby violating basic properties of locality. Moreover, the levels $F_3$ and $F_4$ cannot be lumped together, because $F_3$ must record actual outputs, whereas $F_4$ must record differences between expected and actual outputs. The *BP* model is thus not a model of a brain process.

The computation of the error signal has an additional complexity. In addition to subtracting each actual output at $F_3$ from each expected output at $F_4$, the *derivative* of each actual output is also computed. The difference between each expected and actual output is multiplied by the corresponding derivative in addition to being multiplied by the corresponding transported weight. Thus, there exist additional levels $F_6$ and $F_7$ at each layer for converting outputs into derivatives of outputs before signalling these derivatives, with great positional specificity, to the correct transported weights (Figure 8). This complex interaction scheme must be replicated at every stage of hidden units that is used in a *BP* model.

## 17. COMPARING ADAPTIVE RESONANCE AND BACK PROPAGATION MODELS

Some *BP* mechanisms are evocative of *ART* mechanisms. The *BP* mechanisms do not, however, possess the key properties which endow an *ART* model with its computational power.

### A. Stability
The learned code of the *BP* model is unstable in a complex environment. It keeps tracking whatever expected outputs are imposed from outside. An omniscient teacher would be needed to decide if the model had learned enough in response to an unpredictable input environment. The learned code of an *ART* model is self-stabilizing in an arbitrary input environment.

### B. Expectations as Exemplars or as Prototypes
Within a *BP* model, an expected or template pattern is imposed on every trial by an external teacher. Errors are computed by comparing each component of the expected output pattern with the corresponding component of

the actual output pattern. There is no self-scaling property to alter the importance of each expected component when it is embedded in expected outputs of variable complexity. There is no concept of a critical feature pattern, or prototype. Instead, the expected pattern in a *BP* model is a particular exemplar at every stage of learning, rather than a prototype that gradually discovers invariant properties of all the exemplars that are ever experienced.

In contrast, an *ART* model learns its own expectations without a teacher. Because an *ART* model is self-scaling, it can learn critical feature patterns, or expected prototypes, by evaluating the predictive importance of particular features in input patterns of variable complexity at each stage of learning.

### C. Weight Transport or Top-Down Template Learning

In both a *BP* model and an *ART* model, both bottom-up and top-down *LTM* traces exist. In a *BP* model (Figure 8), the top-down *LTM* traces in $F_4 \rightarrow F_5$ pathways are formal transports of the learned $F_2 \rightarrow F_3$ *LTM* traces. In an *ART* model (Figure 1), the top-down *LTM* traces in $F_2 \rightarrow F_1$ pathways are directly learned by a real-time associative process. These top-down *LTM* weights are not transports of the learned *LTM* traces in the $F_1 \rightarrow F_2$ pathways, and they need not equal these bottom-up *LTM* traces. Thus, an *ART* model is designed so that both bottom-up learning and top-down learning are part of a single information processing hierarchy, which can be realized by a locally computable real-time process.

### D. Matching to Alter Information Processing and/or to Regulate Learning

In both the *BP* model and an *ART* model, there exists a concept of matching. Within an *ART* model, matching both alters information processing and regulates the learning process. In particular, the 2/3 Rule (Section 14) enables a top-down expectation to subliminally sensitize the network in preparation for any exemplar of an expected class of input patterns, and to coherently deform such an exemplar, when it occurs, towards the prototype of the class. This *STM* transformation, also helps to regulate any learning that may be necessary to generate a globally self-consistent recognition code.

In contrast, matching within the *BP* model only changes *LTM* weights. It does not have any effects on the fast information processing that occurs within each input trial.

### E. Learning an Associative Mapping

*BP* and *ART* provide different descriptions of how associative maps between seen language and spoken language are learned. Figure 9 describes a macrocircuit that schematizes our conception of this process (Cohen & Grossberg, 1986; Grossberg, 1978, 1986, 1987b; Grossberg & Stone, 1986a). The associative map $V^* \rightarrow \{A_4, A_5\}$ in Figure 9 joins seen language to spoken language.

Unlike a *BP* model, all the learning of recognition codes that is triggered by auditory, visual, or motor patterns in Figure 9 is regulated by self-organizing mechanisms in reciprocal bottom-up and top-down adaptive filters. Once these codes self-stabilize their invariant recognition properties, the learning of associative maps between these code invariants can also proceed in a self-organizing fashion.

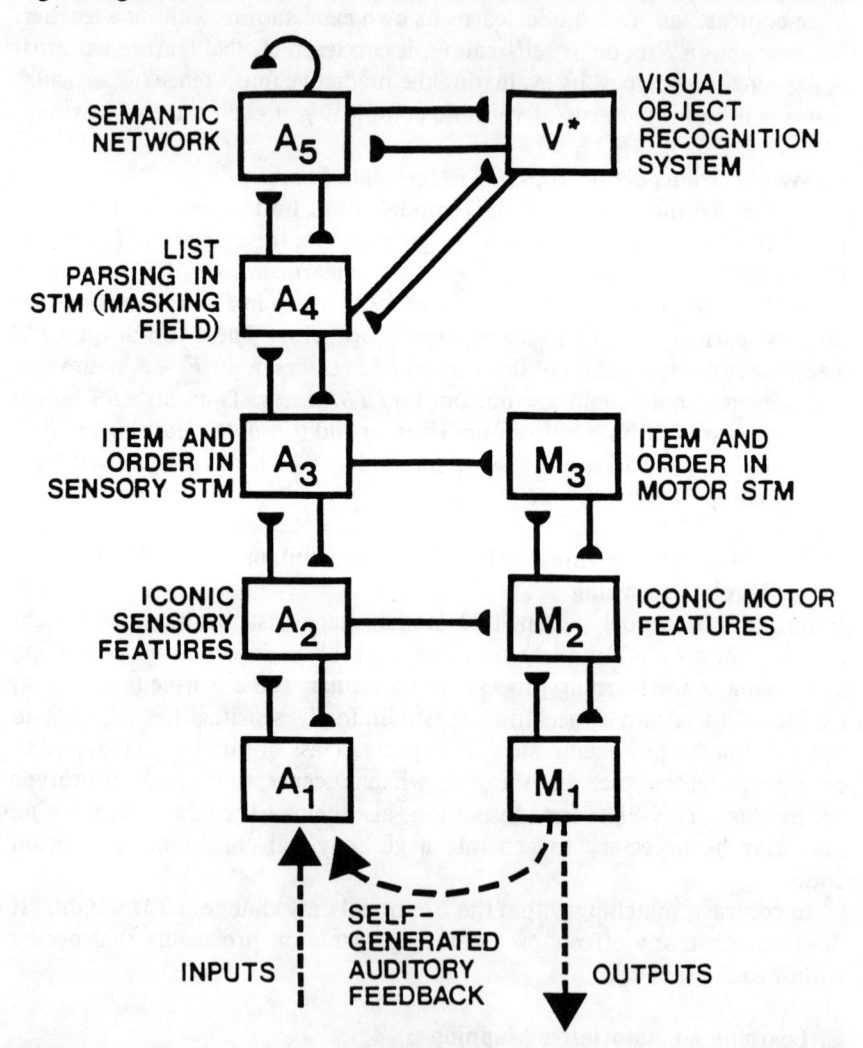

**Figure 9.** A macrocircuit governing self-organization of recognition and recall processes: Auditorily mediated language processes (the $A_i$), visual recognition processes ($V^*$), and motor control processes (the $M_j$) interact internally via conditionable pathways (black lines) and externally via environmental feedback (dotted lines) to self-organize the various processes which occur at the different network stages.

## F. Speech Invariants, Coherence, and Perception

The NETtalk application of the back propagation algorithm (Sejnowski & Rosenberg, 1986) uses a familiar associative learning device: the number of nodes in $F_1$ and $F_3$ is chosen to be large enough to separate features of the inputs and outputs, thereby avoiding too much cross-talk in the associative map, but small enough to enable some generalization to occur among the distributed $F_1 \to F_3$ projections. In particular, the time between letter scans is represented in NETtalk by leaving some coding slots in $F_1$ empty. This mechanism does not generalize to a model capable of computing the temporal invariances of reading or speech perception.

The NETtalk model also makes the strong assumptions that exactly seven letter slots in $F_1$ correspond to a single, isolated phoneme slot in $F_3$, and that this isolated phoneme slot corresponds to the entry in the middle letter slot. These assumptions prevent the model from attempting to solve the fundamental problem of how speech sounds are coherently grouped in real-time. Furthermore, it is not clear how a phoneme-by-phoneme match between actual output and expected output could be realized during a learning episode *in vivo*.

In addition to assuming the automatic isolation, scaling, and centering of information, NETtalk also postulates that each phoneme slot in $F_3$ contains 23 separate nodes. These nodes provide enough spatial dimensions to represent a large number of articulatory features, such as point of articulation, voicing, vowel height, etc. Extra nodes are introduced to encode stress and syllable boundaries. The model builds in the transformations from visual input to $F_1$ nodal representation and from $F_3$ nodal representation to phoneme sound. Because the model automates all of its $F_1$ and $F_3$ representations, all questions about visual and speech perception, as such, lie outside its scope.

The *ART* speech model in Figure 9 was derived from postulates concerning real-time constraints on speech learning and perception. In particular, the model includes mechanisms capable of learning some speech invariants (Grossberg, 1986, 1987b), and the top-down expectancies between its processing levels have coherent grouping properties. One of the primary function of such templates is to define and complete resonant contexts of features, no less than to generate error signals for self-regulating changes in associative weights.

In summary, the *BP* model suggests a new way to use steepest descent to learn associative maps between input and output environments which are statistically stationary and not too complex. Desirable properties of associative map learning are, however, shared by many associative learning models. Real-time network models must do more than learn an associative map or to store distributed codes for carefully controlled environments. Moreover, the use of an unphysical process such as weight transport in a model casts an unanswerable doubt over all empirical applications of the model.

## 18. CONCLUDING REMARKS

I conclude this essay with two general observations. The architectures of many popular learning and information processing models are often inadequate because they have not been constrained by the use of design principles whereby they could stably self-organize. Many models are actually incompatible with such constraints and some models utilize physically unrealizable formal mechanisms. Learning models which cannot adaptively cope with unpredictable changes in a complex environment have an unpromising future as models of mind and brain, and provide little hope of solving the outstanding cognitive problems which are not already well-handled by traditional methods of artificial intelligence and engineering.

Models which do embody self-organization constraints in a fundamental way have frequently been shown to have a broader explanatory and predictive range than models which do not. Thus an analysis of learning, in particular of the mechanisms capable of self-stabilizing competitive learning, can lead psychology from metaphorical models to integrative theories which functionally and mechanistically express both a psychological and a neural reality.

## APPENDIX 1

### Competitive Learning Models

Rumelhart and Zipser (1985, pp. 86–87) summarize competitive learning models as follows.

"1.  The units in a given layer are broken into a set of nonoverlapping clusters. Each unit within a cluster inhibits every other unit within a cluster. The clusters are winner-take-all, such that the unit receiving the largest input achieves its maximum value while all other units in the cluster are pushed to the minimum value. We have arbitrarily set the maximum value to 1 and the minimum value to 0.
2.  Every element in every cluster receives inputs from the same lines.
3.  A unit learns if and only if it wins the competition with other units in its cluster.
4.  A stimulus pattern $S_j$ consists of a binary pattern in which each element of the pattern is either *active* or *inactive*. An active element is assigned the value 1 and an inactive element is assigned the value 0.
5.  Each unit has a fixed amount of weight (all weights are positive) which is distributed among its input lines. The weight on the line connecting unit $i$ on the lower (or input) layer of unit $j$ on the upper layer, is designated $\omega_{ij}$. The fixed total amount of weight for unit $j$ is designated $\Sigma_i \omega_{ij} = 1$. A unit learns by shifting weight from its inactive to its active

input lines. If a unit does not respond to a particular pattern no learning takes place in that unit. If a unit wins the competition, then each of its input lines gives up some proportion g of its weight and that weight is then distributed equally among the active input lines. More formally, the learning rule we have studied is:

$$\Delta\omega_{ij} = \begin{cases} 0 & \text{if unit } j \text{ loses on stimulus } k \\ g\dfrac{c_{ik}}{n_k} - g\omega_{ij} & \text{if unit } j \text{ wins on stimulus } k \end{cases} \quad (A1)$$

where $c_{ik}$ is equal to 1 if in stimulus pattern $S_k$ unit $i$ on the lower layer is active and zero otherwise, and $n_k$ is the number of active units in pattern $S_k$ (thus

$$n_k = \sum_i c_{ik}).\text{''} \quad (A2)$$

Rumelhart and Zipser (1985, p. 87) go on to say that "This learning rule was proposed by Von der Malsburg (1973). As Grossberg (1976) points out, renormalization of the weights is not necessary." Actually, this learning rule was proposed by Grossberg (1976a, 1976b), and is not the one used in the important article of Malsburg (1973), as I will show below.

A simple change of notation shows that the Rumelhart and Zipser (1985) model is identical with the Grossberg (1976a, 1976b) model. Equation (6) in Grossberg (1976b) is the learning equation

$$\frac{d}{dt}z_{ij} = (-z_{ij} + \theta_i)x_{2j}, \quad (A3)$$

for the long term memory (*LTM*) trace $z_{ij}$. In (A3), $x_{2j}$ is the activity of the $j$th unit. Activity $x_{2j} = 1$ if unit $j$ wins the competition and $x_{2j} = 0$ if unit $j$ loses the competition, as in equation (A1). In (A3),

$$\theta_i = \frac{I_i}{\sum_m I_m} \quad (A4)$$

where $I_i$ is the $i$th element of the input pattern. Function $\theta_i$ in (A4) is the same as function $c_{ik}n_k^{-1}$ in (A1) and (A2). Function $\theta_i$ is just the normalized input weight. The *LTM* trace $z_{ij}$ in (A3) is identical with the weight $\omega_{ij}$ in (A1). The factor g in (A1) just rescales the time variable, and thus adds no generality to the model.

By contrast, the learning rule used by Malsburg (1973) is (in my notation)

$$\frac{d}{dt}z_{ij} = I_i x_{2j} \quad (A5)$$

subject to the constraint

$$\sum_m z_{mj} = \text{constant}. \quad (A6)$$

Thus Malsburg (1973) normalized the *LTM* traces $z_{mj}$ which abut each unit $j$, not the input weights. The normalization constraint (A6) is, in fact, inconsistent with (A5) unless

$$\sum_m I_m = 0. \tag{A7}$$

Thus a rigorous application of Malsburg's 1973 learning rule forces the choice of both positive and negative inputs, unlike the Grossberg (1976b) model that was used by Rumelhart and Zipser (1985). In his computer simulations, Malsburg implemented equations (A5) and (A6) in alternating time slices. The implication shown in (A7) is then not forced because neither (A5) nor (A6) is true at all times.

Malsburg (1973) needed condition (A6) because he simplified the learning law

$$\frac{d}{dt} z_{ij} = -A z_{ij} + I_i x_{2j} \tag{A8}$$

which was used in the competitive learning model of Grossberg (1972). In fact, the equations used by Malsburg (1973) are identical to the equations used by Grossberg (1972) with this one exception. As Malsburg (1973, p. 88) noted: "To answer these questions we have to write down the equations which govern the evolution of the system. They are summarized in Table 1 (compare Grossberg, 1972)." Term $-Az_{ij}$ in (A8) describes the decay of *LTM*. Malsburg's equation (A5) eliminates *LTM* decay. Since term $I_i x_{2j}$ is non-negative in these applications, the *LTM* trace in Malsburg's equation (A5) can only increase. Without additional constraints, all *LTM* traces could therefore explode to infinity. Malsburg (1973) partially overcame this problem with his constraint (A6). The solution in equation (A3) was to preserve *LTM* decay (term $-Az_{ij}$) while normalizing the inputs $I_i$ to be learned. Then non-negative inputs $I_i$ could freely be used, instead of inputs constrained by (A7).

Rumelhart and Zipser (1985) also mentioned and studied two related models. Equivalent models were introduced in Grossberg (1976b). These alternative models were designed to show that both of them also exhibit temporally unstable learning. Analysis of these variations of the simplest coding model confirmed that its unstable behavior was not an artifact of its simplicity.

Rumelhart and Zipser (1985) attributed one of these modified models to Bienenstock, Cooper, and Munro (1982). They noted that in such a model

> a unit modulates its own sensitivity so that when it is not receiving enough inputs, it becomes increasingly sensitive. When it is receiving too many inputs, it decreases sensitivity. This mechanism can be implemented in the present context by assuming that there is a threshold and that the relevant activation is the degree to which the unit exceeds its threshold. If, whenever a unit fails to win it decreases its threshold, and whenever it does win it increases its threshold,

then this method will also make all of the units eventually respond, thereby engaging the mechanism of competitive learning (Rumelhart and Zipser, 1985, p. 100).

In equations (23)–(24) of Grossberg (1976b), such a variable-threshold model was introduced without changing the basic learning equation

$$\frac{d}{dt}z_{ij} = (-z_{ij} + \theta_i)x_{2j}, \tag{A3}$$

The sensitivity of $x_{2j}$ to its inputs was modified as follows:

$$x_{2j}(t) = \begin{cases} G_j(t) & \text{if } S_j(t)G_j(t) > \max\{S_k(t)G_k(t) : k \neq j\} \\ 0 & \text{if } S_j(t)G_j(t) < \max\{S_k(t)G_k(t) : k \neq j\} \end{cases} \tag{A9}$$

where

$$G_j(t) = g(1 - \int_0^t x_{2j}(v)K(t-v)dv), \tag{A10}$$

$S_j(t)$ is the total input to unit $j$, $g(w)$ is an increasing function such that $g(0) = 0$ and $g(1) = 1$, and $K(w)$ is a decreasing function such that $K(0) = 1$ and $K(\infty) = 0$; for example, $K(w) = e^{-w}$.

The history-dependent threshold is the term

$$\int_0^t x_{2j}(v)K(t-v)dv \tag{A11}$$

in equation (A10). If unit $j$ wins the competition then, by (A9), its activity $x_{2j}$ becomes positive. Consequently, its threshold (A11) increases. If unit $j$ loses the competition then, by (A9), its activity $x_{2j}$ equals zero. Consequently, its threshold (A11) decreases. Thus, "a unit modulates its own sensitivity so that when it is not receiving enough inputs, it becomes increasingly sensitive." By (A9) and (A10), when a unit wins the competition, its activation level $G_j(t)$ "is the degree to which the unit exceeds its threshold." Moreover, by (A3), the learning rate covaries with the activation $G_j(t)$ of unit $j$. Thus if unit $j$ is active for a long time, then its threshold (A11) becomes large, so its learning rate (A10) becomes small. The converse is also true: inactivity increases sensitivity and learning rate.

In the limiting case where the threshold in (A11) equals zero for all time because $K \equiv 0$, this learning model reduces to the simplest competitive learning model. This can be seen as follows. If the threshold is set equal to zero, then $G_j(t) \equiv 1$ in (A10). Hence, by (A9),

$$x_{2j}(t) = \begin{cases} 1 & \text{if } S_j(t) > \max\{S_k : k \neq j\} \\ 0 & \text{if } S_j(t) < \max\{S_k : k \neq j\} \end{cases}. \tag{A12}$$

In other words $x_{2j}(t) = 1$ if unit $j$ wins the competition, and $x_{2j}(t) = 0$ if unit $j$ loses the competition, as in equation (A1).

Thus the model summarized by equations (A3), (A9), and (A10) has all the properties described by Rumelhart and Zipser (1985) for a variable-threshold model and includes the simplest competitive learning model as a special case. In Grossberg (1976b, p. 132), it was noted that "Such a mechanism is inadequate if the training schedule allows $v_{2j}$[unit $j$] to recover its maximal strength." I illustrated this inadequacy by displaying "an ordering of patterns that permits recoding of essentially all populations."

Bienenstock, Cooper, and Munro (1982) studied a formally analogous model with a history-dependent threshold. However, they restricted their analysis to coding by a *single* unit $j$. Grossberg (1982c, p. 332), assumed the viewpoint of competitive learning and considered how that model behaves —in their coding application—when more than one coding unit $j$ exists and the units compete with each other for activation. It was shown that persistent presentation of even a *single* unit pattern could cause temporally unstable coding in this competitive learning situation. This crippling form of instability seems to rule out the use of history-dependent thresholds as a viable learning rule, at least if the thresholds can recover from unit inactivity, which is the main property cited in their favor by Rumelhart and Zipser (1985, p. 100).

Rumelhart and Zipser (1985) call the third model variant that they study the *leaky learning* model. This model is a special case of the *partial contrast* model that was introduced in Grossberg (1976b, p. 132), where it was pointed out that, using such a model, "There can... be a shift in the locus of maximal responsiveness even to a single pattern—that is, recoding." Rumelhart and Zipser (1985, p. 100) consider this a good property, rather than a bad one: "This change has the property that it slowly moves the losing units into the region where the actual stimuli lie, at which point they begin to capture some units and the ordinary dynamics of competitive learning take over." These authors are willing to accept this instability property in order to avoid the even worse problem that "one of the units would have most of its weight on input lines that were never active, whereas another unit may have had most of its weight on lines common to all of the stimulus patterns. Since a unit never learns unless it wins, it is possible that one of the units will never win, and therefore never learn. This, of course, takes the competition out of competitive learning" (pp. 98–99).

This scheme cannot, however, be fully effective without having catastrophic results on code stability. If the recoding is minor, then many nodes may remain unused and too many input patterns may be lumped together. In this case, the scheme cannot solve the problem for which it was introduced. Alternatively, major recoding may be allowed, but this property is just another way to describe a temporally unstable code.

The formal relationship between the leaky learning model and the partial contrast model is now summarized. In the leaky learning model, equation (A1) is replaced by

$$\Delta\omega_{ij} = \begin{cases} g_l \dfrac{C_{ik}}{n_k} - g_\omega \omega_{ij} & \text{if } j \text{ loses on stimulus } k \\ g_\omega \dfrac{C_{ik}}{n_k} - g_\omega \omega_{ij} & \text{if } j \text{ wins on stimulus } k \end{cases} \quad (A13)$$

where

$$g_l \ll g_\omega. \quad (A14)$$

In other words, slower learning occurs at losing units than at winning units. The leaky learning model is a variant of the partial contrast model. The partial contrast model continues to use the basic learning equation

$$\frac{d}{dt} z_{ij} = (-z_{ij} + \theta_i) x_{2j}, \quad (A3)$$

However, $x_{2j}$ is now defined by a partial contrast rule

$$x_{2j} = \begin{cases} \dfrac{f(S_j)}{\sum_{S_m > \epsilon} f(S_m)} & \text{if } S_j > \epsilon \\ 0 & \text{if } S_j < \epsilon \end{cases} \quad (A15)$$

where $f(w)$ is an increasing function of the total input $S_j$ to unit $j$, and $\epsilon$ is a non-negative threshold. By (A15), the learning rate is fastest at the node $x_{2j}$ which receives the largest input $S_j$ and is slower at other nodes, as in the leaky learning model. In summary, all of the types of models described by Rumelhart and Zipser (1985) were shown in Grossberg (1976b) to exhibit a basic problem of learning instability.

## APPENDIX 2

### Stable Code Learning for Sparse Input Patterns

To simplify notation, the simplest competitive learning model is defined again below: Let the input patterns $I_i(t)$ across nodes $v_i$ in $F_1$ be immediately and perfectly normalized; that is, input $I_i(t) = \theta_i I(t)$ generates activity $x_i(t) = \theta_i$ at $v_i$. The signals from a node $v_i$ in $F_1$ to nodes $v_j$ in $F_2$ is chosen to be a linear function of the activity $x_i$. For simplicity, let the signal emitted by $v_i$ equal $\theta_i$. The competition across nodes $v_j$ in $F_2$ normalizes the total activity to the value 1 for definiteness and rapidly chooses that node $v_j$ for *STM* storage which receives the largest input; e.g., design $F_2$ as a cooperative-competitive feedback network with faster-than-linear or (properly chosen) sigmoid signal functions. These properties can be approximated by the simple rule that

$$x_j = \begin{cases} 1 & \text{if } T_j > \max\{\epsilon, T_k : k \neq j\} \\ 0 & \text{if } T_j < \max\{\epsilon, T_k : k \neq j\} \end{cases} \quad (A16)$$

where the total input $T_j$ to $v_j$ is the inner product

$$T_j = \sum_{k \in I} \theta_k z_{kj}. \quad (A17)$$

The *LTM* traces in the $F_1 \to F_2$ pathways sample the pattern $\theta = (\theta_1, \theta_2, \ldots, \theta_n)$ of input signals only when their sampling cell is active. Thus,

$$\frac{d}{dt}z_{ij} = \epsilon x_j(-z_{ij} + \theta_i). \tag{A18}$$

This non-Hebbian associative law was introduced into the neural network literature in Grossberg (1969).

If a single pattern $\theta$ is practiced, it maximizes the input $T_j$ to its coding cell $v_j$. Input $T_j$ increases as the classifying vector $z_j = (z_{ij} : i\in I)$ become parallel to $\theta$ and the length $||z_j||$ of $z_j$ become normalized. Grossberg (1976b) also described circumstances under which a list of input patterns to $F_1$ could generate temporally stable learning capable of parsing these patterns into distinct recognition categories at $F_2$. It was proved that, if not too many input patterns are presented, relative to the number of coding nodes in $F_2$, or if the input patterns are grouped into not too many clusters, then the recognition code stabilizes and the classifying vectors approach the convex hull of the patterns which they code. The latter property shows that the classifying nodes ultimately receive maximal inputs consistent with the fact that the classifying vectors $z_j$ can fluctuate in response to all the input patterns that they code.

To state this theorem, the following notation is convenient. A *partition* $\oplus_{j=1}^J P_j$ of a finite set $P$ is a subdivision of $P$ into nonoverlapping and exhaustive subsets $P_j$. The *convex hull* $H(P)$ of $P$ is the set of all convex combinations of elements of $P$. Given a set $Q \subset P$, let $R = P - Q$ denote the elements of $P$ that are not in $Q$. The distance between a vector $p$ and a set of vectors $Q$, denoted by $||p - Q||$, is defined by $||p - Q|| = \inf(||p - Q|| : q \in Q)$.

Suppose that, at time $t$, the classifying vector $z_j(t) = (z_{ij}(t) : i\in I)$ codes the set of patterns $P_j(t)$; that is, node $v_j$ in $F_2$ would be chosen if any pattern in $P_j(t)$ were presented at that time. Define $P_j^*(t) = P_j(t) \cup z_j(t)$ and $P^*(t) = \cup_{j=1}^J P_j^*(t)$.

**Theorem (Stable Code Learning of Sparse Patterns)**
Let the network practice any finite set $P = (\theta^{(l)} : l = 1, 2, \ldots, L)$ of input patterns. Suppose that at some time $t = T$, the partition $\oplus_{j=1}^J P_j(T)$ of $P$ has the property that

$$\min(u \cdot v : u \in P_j(T), v \in P_j^*(T)) > \max(u \cdot v : u \in P_j(T), v \in P^*(T) - P_j^*(T)) \tag{A19}$$

for all $j = 1, 2, \ldots, J$. Then the network partitions the patterns $P$ into the stable categories $P_j(T)$; that is,

$$P_j(t) = P_j(T) \tag{A20}$$

for all $j = 1, 2, \ldots, J$ and all $t \geq T$. In addition, learning maximizes the input to the classifying nodes; that is, the functions

$$D_j(t) = ||z_j(t) - H(P_j(t))|| \tag{A21}$$

are monotone decreasing for all $j = 1, 2, \ldots, J$ and $t \geq T$. If, moreover, the patterns $P_j(T)$ are practiced in time intervals $[U_{jk}, V_{jk}]$, $k = 1, 2, \ldots$, such that

$$\sum_{k=1}^{\infty} (V_{jk} - U_{jk}) = \infty, \tag{A22}$$

then

$$\lim_{t \to \infty} D_j(t) = 0. \tag{A23}$$

Thus the theorem describes circumstances under which practice of input patterns in $P$ can cause the classifying vectors $z_j$, which may have any initial distribution $z_j(0)$, to be separated well enough, as in (A19), to enable their later tuning to proceed, as in (A23), without disrupting the emergent partition $\oplus_{j=1}^{J} P_j(T)$ of the patterns $P$ into recognition categories.

## REFERENCES

Banquet, J.-P., & Grossberg, S. (1987). Probing cognitive processes through the structure of event-related potentials during learning: An experimental and theoretical analysis. *Applied Optics.*

Bienenstock, E.L., Cooper, L.N., & Munro, P.W. (1982). Theory for the development of neuron selectivity: Orientation specificity and binocular interaction in visual cortex. *Journal of Neuroscience, 2,* 32–48.

Carpenter, G.A., & Grossberg, S. (1986). Associative learning, adaptive pattern recognition, and cooperative-competitive decision making by neural networks. In H. Szu (Ed.) *Hybrid and Optical Computing. SPIE Proceedings.*

Carpenter, G.A., & Grossberg, S. (1987a). Neural dynamics of category learning and recognition: Attention, memory consolidation, and amnesia. In J. Davis, R. Newburgh, & E. Wegman (Eds.), *Brain structure, learning, and memory. AAAS Symposium Series.*

Carpenter, G.A., & Grossberg, S. (1987b). A massively parallel architecture for a self-organizing neural pattern recognition machine. *Computer Vision, Graphics, and Image Processing, 37,* 54–115.

Carpenter, G.A., & Grossberg, S. (1987c). Neural dynamics of category learning and recognition: Structural invariants, reinforcement, and evoked potentials. In M.L. Commons, S.M. Kosslyn, & R.J. Herrnstein (Eds.), *Pattern recognition and concepts in animals, people, and machines.*

Cohen, M.A., & Grossberg, S. (1986a). Neural dynamics of speech and language coding: Developmental programs, perceptual grouping, and competition for short term memory. *Human Neurobiology, 5,* 1–22.

Cohen, M.A., & Grossberg, S. (1987). Masking fields: A massively parallel neural architecture for learning, recognizing, and predicting multiple groupings of patterned data. *Applied Optics.*

Estes, W.K. (1986). Memory storage and retrieval processes in category learning. *Journal of Experimental Psychology: General, 115,* 155–174.

Grossberg, S. (1969). On learning and energy-entropy dependence in recurrent and nonrecurrent signed networks. *Journal of Statistical Physics, 1,* 319–350.

Grossberg, S. (1972). Neural expectation: Cerebellar and retinal analogs of cells fired by learnable or unlearned pattern classes. *Kybernetik, 10,* 49–57.

Grossberg, S. (1975). A neural model of attention, reinforcement, and discrimination learning. *International Review of Neurobiology, 18,* 263–327.

Grossberg, S. (1976a). On the development of feature detectors in the visual cortex with applications to learning and reaction-diffusion systems. *Biological Cybernetics, 21,* 145–159.
Grossberg, S. (1976b). Adaptive pattern classification and universal recoding, I: Parallel development and coding of neural feature detectors. *Biological Cybernetics, 23,* 121–134.
Grossberg, S. (1976c). Adaptive pattern classification and universal recoding. II: Feedback, expectation, olfaction, and illusions. *Biological Cybernetics, 23,* 187–202.
Grossberg, S. (1978). A theory of human memory: Self-organization and performance of sensory-motor codes, maps, and plans. In R. Rosen & F. Snell (Eds.), *Progress in theoretical biology, Vol. 5.* (pp. 233–374). New York: Academic.
Grossberg, S. (1980). How does a brain build a cognitive code? *Psychological Review, 87,* 1–51.
Grossberg, S. (1982a). *Studies of mind and brain: Neural principles of learning, perception, development, cognition, and motor control.* Boston: Reidel Press.
Grossberg, S. (1982b). Processing of expected and unexpected events during conditioning and attention: A psychophysiological theory. *Psychological Review, 89,* 529–572.
Grossberg, S. (1982c). Associative and competitive principles of learning and development: The temporal unfolding and stability of STM and LTM patterns. In S.I. Amari & M. Arbib (Eds.), *Competition and cooperation in neural networks.* New York: Springer-Verlag. Reprinted in Grossberg (1987a).
Grossberg, S. (1984). Unitization, automaticity, temporal order, and word recognition. *Cognition and Brain Theory, 7,* 263–283.
Grossberg, S. (1987c). The adaptive self-organization of serial order in behavior: Speech, language, and motor control. In E.C. Schwab & H.C. Nusbaum (Eds.), *Pattern recognition by humans and machines, Vol. 1: Speech perception.* New York: Academic.
Grossberg, S. (1987a). *The adaptive brain, I: Cognition, learning, reinforcement, and rhythm.* Amsterdam: North-Holland.
Grossberg, S. (1987b). *The adaptive brain, II: Vision, speech, language, and motor control.* Amsterdam: North-Holland.
Grossberg, S., & Gutowski, W. (1987). Neural dynamics of decision making under risk: Affective balance and cognitive-emotional interactions. *Psychological Review.*
Grossberg, S., & Levine, D.S. (1987). Neural dynamics of attentionally modulated Pavlovian conditioning: Blocking, inter-stimulus interval, and secondary reinforcement. Submitted for publication.
Grossberg, S., & Stone, G.O. (1986a). Neural dynamics of word recognition and recall: Attentional priming, learning, and resonance. *Psychological Review, 93,* 46–74.
Grossberg, S., & Stone, G.O. (1986b). Neural dynamics of attention switching and temporal order information in short-term memory. *Memory and Cognition, 14,* 451–468.
Hopfield, J.J. (1982). Neural networks and physical systems with emergent collective computational abilities. *Proceedings of the National Academy of Sciences, 79,* 2554–2558.
Knapp, A.G., & Anderson, J.A. (1984). Theory of categorization based on distributed memory storage. *Journal of Experimental Psychology: Learning, Memory, and Cognition, 10,* 616–637.
Kohonen, T. (1984). *Self-organization and associative memory.* New York: Springer-Verlag.
Malsburg, C. von der (1973). Self-organization of orientation sensitive cells in the striate cortex. *Kybernetik, 14,* 85–100.
McClelland, J.L. (1985). Putting knowledge in its place: A scheme for programming parallel processing structures on the fly. *Cognitive Science, 9,* 113–146.
McClelland, J.L., & Rumelhart, D.E. (1981). An interactive activation model of context effects in letter perception, Part I: An account of basic findings. *Psychological Review, 88,* 375–407.
McClelland, J.L., & Rumelhart, D.E. (1985). Distributed memory and the representation of general and specific information. *Journal of Experimental Psychology: General, 114,* 159–188.

Posner, M.I. (1973). *Cognition: An introduction.* Glenview, IL: Scott, Foresman, and Co.
Posner, M.I., & Keele, S.W. (1968). On the genesis of abstract ideas. *Journal of Experimental Psychology, 77,* 353-363.
Posner, M.I., & Keele, S.W. (1970). Retention of abstract ideas. *Journal of Experimental Psychology, 83,* 304-308.
Rosenblatt, F. (1962). *Principles of neurodynamics.* Washington, DC: Spartan.
Rumelhart, D.E., Hinton, G.E., & Williams, R.J. (1985, September). Learning internal representations by error propagation. *Institute for Cognitive Science Report 8506,* University of California at San Diego.
Rumelhart, D.E., & McClelland, J.L. (1982). An interactive model of context effects in letter perception, Part 2: The contextual enhancement effect and some tests and extensions of the model. *Psychological Review, 89,* 60-94.
Rumelhart, D.E., & Zipser, D. (1985). Feature discovery by competitive learning. *Cognitive Science, 9,* 75-112.
Salasoo, A., Shiffrin, R.M., & Feustal, T.C. (1985). Building permanent memory codes: Codification and repetition effects in word identification. *Journal of Expermental Psychology: General, 114,* 50-77.
Samuel, A.G., van Santen, J.P.H., & Johnston, J.C. (1982). Length effects in word perception: We is better than I but worse than you or them. *Journal of Experimental Psychology: Human Perception and Performance, 8,* 91-105.
Samuel, A.G., van Santen, J.P.H., & Johnston, J.C. (1983). Reply to Matthei: We really is worse than you or them, and so are ma and pa. *Journal of Experimental Psychology: Human Perception and Performance, 9,* 321-322.
Sejnowski, T.J., & Rosenberg, C.R. (1986, January). NETtalk: A parallel-network that learns to read aloud. Johns Hopkins University.
Wheeler, D.D. (1970). Processes in word recognition. *Cognitive Psychology, 1,* 59-85.
Widrow, B. (1962). Generalization and information storage in networks of Adaline neurons. In M.C. Yovits, G.T. Jacobi, & G.D. Goldstein (Eds.), *Self-organizing systems.* Washington, DC: Spartan.

CHAPTER 10

# A Learning Algorithm for Boltzmann Machines*

DAVID H. ACKLEY
GEOFFREY E. HINTON
*Computer Science Department*
*Carnegie-Mellon University*

TERRENCE J. SEJNOWSKI
*Biophysics Department*
*The Johns Hopkins University*

The computational power of massively parallel networks of simple processing elements resides in the communication bandwidth provided by the hardware connections between elements. These connections can allow a significant fraction of the knowledge of the system to be applied to an instance of a problem in a very short time. One kind of computation for which massively parallel networks appear to be well suited is large constraint satisfaction searches, but to use the connections efficiently two conditions must be met: First, a search technique that is suitable for parallel networks must be found. Second, there must be some way of choosing internal representations which allow the preexisting hardware connections to be used efficiently for encoding the constraints in the domain being searched. We describe a general parallel search method, based on statistical mechanics, and we show how it leads to a general learning rule for modifying the connection strengths so as to incorporate knowledge about a task domain in an efficient way. We describe some simple examples in which the learning algorithm creates internal representations that are demonstrably the most efficient way of using the preexisting connectivity structure.

## 1. INTRODUCTION

Evidence about the architecture of the brain and the potential of the new VLSI technology have led to a resurgence of interest in "connectionist" sys-

* The research reported here was supported by grants from the System Development Foundation. We thank Peter Brown, Francis Crick, Mark Derthick, Scott Fahlman, Jerry Feldman, Stuart Geman, Gail Gong, John Hopfield, Jay McClelland, Barak Pearlmutter, Harry Printz, Dave Rumelhart, Tim Shallice, Paul Smolensky, Rick Szeliski, and Venkataraman Venkatasubramanian for helpful discussions.

tems (Feldman & Ballard, 1982; Hinton & Anderson, 1981) that store their long-term knowledge as the strengths of the connections between simple neuron-like processing elements. These networks are clearly suited to tasks like vision that can be performed efficiently in parallel networks which have physical connections in just the places where processes need to communicate. For problems like surface interpolation from sparse depth data (Grimson, 1981; Terzopoulos, 1984) where the necessary decision units and communication paths can be determined in advance, it is relatively easy to see how to make good use of massive parallelism. The more difficult problem is to discover parallel organizations that do not require so much problem-dependent information to be built into the architecture of the network. Ideally, such a system would adapt a given structure of processors and communication paths to whatever problem it was faced with.

This paper presents a type of parallel constraint satisfaction network which we call a "Boltzmann Machine" that is capable of learning the underlying constraints that characterize a domain simply by being shown examples from the domain. The network modifies the strengths of its connections so as to construct an internal *generative* model that produces examples with the same probability distribution as the examples it is shown. Then, when shown any particular example, the network can "interpret" it by finding values of the variables in the internal model that would generate the example. When shown a partial example, the network can complete it by finding internal variable values that generate the partial example and using them to generate the remainder. At present, we have an interesting mathematical result that guarantees that a certain learning procedure will build internal representations which allow the connection strengths to capture the underlying constraints that are implicit in a large ensemble of examples taken from a domain. We also have simulations which show that the theory works for some simple cases, but the current version of the learning algorithm is very slow.

The search for general principles that allow parallel networks to learn the structure of their environment has often begun with the assumption that networks are randomly wired. This seems to us to be just as wrong as the view that *all* knowledge is innate. If there are connectivity structures that are good for particular tasks that the network will have to perform, it is much more efficient to build these in at the start. However, not all tasks can be foreseen, and even for ones that can, fine-tuning may still be helpful.

Another common belief is that a general connectionist learning rule would make sequential "rule-based" models unnecessary. We believe that this view stems from a misunderstanding of the need for multiple levels of description of large systems, which can be usefully viewed as either parallel or serial depending on the grain of the analysis. Most of the key issues and questions that have been studied in the context of sequential models do not magically disappear in connectionist models. It is still necessary to perform searches for good solutions to problems or good interpretations of perceptual input, and to create complex internal representations. Ultimately it will be necessary to bridge the gap between hardware-oriented connectionist descriptions and the more abstract symbol manipulation models that have proved to be an extremely powerful and pervasive way of describing human information processing (Newell & Simon, 1972).

## 2. THE BOLTZMANN MACHINE

The Boltzmann Machine is a parallel computational organization that is well suited to constraint satisfaction tasks involving large numbers of "weak" constraints. Constraint-satisfaction searches (e.g., Waltz, 1975; Winston, 1984) normally use "strong" constraints that *must* be satisfied by any solution. In problem domains such as games and puzzles, for example, the goal criteria often have this character, so strong constraints are the rule.[1] In some problem domains, such as finding the most plausible interpretation of an image, many of the criteria are not all-or-none, and frequently even the best possible solution violates some constraints (Hinton, 1977). A variation that is more appropriate for such domains uses weak constraints that incur a cost when violated. The quality of a solution is then determined by the total cost of all the constraints that it violates. In a perceptual interpretation task, for example, this total cost should reflect the implausibility of the interpretation.

The machine is composed of primitive computing elements called *units* that are connected to each other by bidirectional *links*. A unit is always in one of two states, *on* or *off*, and it adopts these states as a probabilistic function of the states of its neighboring units and the *weights* on its links to them. The weights can take on real values of either sign. A unit being on or off is taken to mean that the system currently accepts or rejects some elemental hypothesis about the domain. The weight on a link represents a weak pairwise constraint between two hypotheses. A positive weight indicates that the two hypotheses tend to support one another; if one is currently accepted, accepting the other should be more likely. Conversely, a negative weight suggests, other things being equal, that the two hypotheses should not both be accepted. Link weights are *symmetric,* having the same strength in both directions (Hinton & Sejnowski, 1983).[2]

---

[1] But, see (Berliner & Ackley, 1982) for argument that, even in such domains, strong constraints must be used only where absolutely necessary for legal play, and in particular must not propagate into the determination of *good* play.

[2] Requiring the weights to be symmetric may seem to restrict the constraints that can be represented. Although a constraint on boolean variables $A$ and $B$ such as "$A \equiv B$ with a penalty of 2 points for violation" is obviously symmetric in $A$ and $B$, "$A => B$ with a penalty of 2 points for violation" appears to be fundamentally asymmetric. Nevertheless, this constraint can be represented by the combination of a constraint on $A$ alone and a symmetric pairwise constraint as follows: "Lose 2 points if $A$ is true" and "Win 2 points if both $A$ and $B$ are true."

The resulting structure is related to a system described by Hopfield (1982), and as in his system, each global state of the network can be assigned a single number called the "energy" of that state. With the right assumptions, the individual units can be made to act so as to *minimize the global energy*. If *some* of the units are externally forced or "clamped" into particular states to represent a particular input, the system will then find the minimum energy configuration that is compatible with that input. The energy of a configuration can be interpreted as the extent to which that combination of hypotheses violates the constraints implicit in the problem domain, so in minimizing energy the system evolves towards "interpretations" of that input that increasingly satisfy the constraints of the problem domain.

The energy of a global configuration is defined as

$$E = -\sum_{i<j} w_{ij} s_i s_j + \sum_i \theta_i s_i \qquad (1)$$

where $w_{ij}$ is the strength of connection between units $i$ and $j$, $s_i$ is 1 if unit $i$ is on and 0 otherwise, and $\theta_i$ is a threshold.

## 2.1 Minimizing Energy

A simple algorithm for finding a combination of truth values that is a *local* minimum is to switch each hypothesis into whichever of its two states yields the lower total energy given the current states of the other hypotheses. If hardware units make their decisions asynchronously, and if transmission times are negligible, then the system always settles into a local energy minimum (Hopfield, 1982). Because the connections are symmetric, the difference between the energy of the whole system with the $k^{th}$ hypothesis rejected and its energy with the $k^{th}$ hypothesis accepted can be determined locally by the $k^{th}$ unit, and this "energy gap" is just

$$\Delta E_k = \sum_i w_{ki} s_i - \theta_k \qquad (2)$$

Therefore, the rule for minimizing the energy contributed by a unit is to adopt the *on* state if its total input from the other units and from outside the system exceeds its threshold. This is the familiar rule for binary threshold units.

The threshold terms can be eliminated from Eqs. (1) and (2) by making the following observation: the effect of $\theta_i$ on the global energy or on the energy gap of an individual unit is identical to the effect of a link with strength $-\theta_i$ between unit $i$ and a special unit that is by definition always held in the *on* state. This "true unit" need have no physical reality, but it simplifies the computations by allowing the threshold of a unit to be treated in the same manner as the links. The value $-\theta_i$ is called the *bias* of unit $i$. If a permanently active "true unit" is assumed to be part of every network, then Eqs. (1) and (2) can be written as:

$$E = - \sum_{i<j} w_{ij} s_i s_j \tag{3}$$

$$\Delta E_k = \sum_i w_{ki} s_i \tag{4}$$

## 2.2 Using Noise to Escape from Local Minima

The simple, deterministic algorithm suffers from the standard weakness of gradient descent methods: It gets stuck in *local* minima that are not globally optimal. This is not a problem in Hopfield's system because the local energy minima of his network are used to store "items": If the system is started near some local minimum, the desired behavior is to fall into that minimum, not to find the global minimum. For constraint satisfaction tasks, however, the system must try to escape from local minima in order to find the configuration that is the global minimum given the current input.

A simple way to get out of local minima is to occasionally allow jumps to configurations of higher energy. An algorithm with this property was introduced by Metropolis, Rosenbluth, Rosenbluth, Teller, & Teller (1953) to study average properties of thermodynamic systems (Binder, 1978) and has recently been applied to problems of constraint satisfaction (Kirkpatrick, Gelatt, & Vecchi, 1983). We adopt a form of the Metroplis algorithm that is suitable for parallel computation: If the energy gap between the *on* and *off* states of the $k^{th}$ unit is $\Delta E_k$ then regardless of the previous state set $s_k = 1$ with probability

$$p_k = \frac{1}{(1 + e^{-\Delta E_k/T})} \tag{5}$$

where $T$ is a parameter that acts like temperature (see Figure 1).

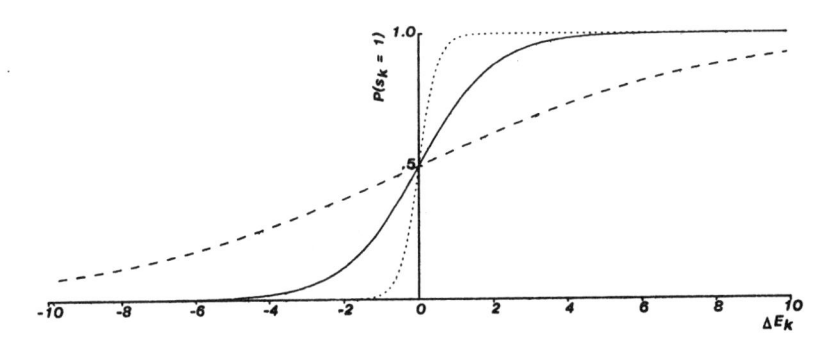

Figure 1. Eq. (5) at $T=1.0$ (solid), $T=4.0$ (dashed), and $T=0.25$ (dotted).

The decision rule in Eq. (5) is the same as that for a particle which has two energy states. A system of such particles in contact with a heat bath at a given temperature will eventually reach thermal equilibrium and the probability of finding the system in any global state will then obey a Boltzmann distribution. Similarly, a network of units obeying this decision rule will eventually reach "thermal equilibrium" and the relative probability of two global states will follow the Boltzman distribution:

$$\frac{P_\alpha}{P_\beta} = e^{-(E_\alpha - E_\beta)/T} \quad (6)$$

where $P_\alpha$ is the probability of being in the $\alpha^{th}$ global state, and $E_\alpha$ is the energy of that state.

The Boltzmann distribution has some beautiful mathematical properties and it is intimately related to information theory. In particular, the difference in the log probabilities of two global states is just their energy difference (at a temperature of 1). The simplicity of this relationship and the fact that the equilibrium distribution is independent of the path followed in reaching equilibrium are what make Boltzmann machines interesting.

At low temperatures there is a strong bias in favor of states with low energy, but the time required to reach equilibrium may be long. At higher temperatures the bias is not so favorable but equilibrium is reached faster. A good way to beat this trade-off is to start at a high temperature and gradually reduce it. This corresponds to annealing a physical system (Kirkpatrick, Gelatt, & Vecchi, 1983). At high temperatures, the network will ignore small energy differences and will rapidly approach equilibrium. In doing so, it will perform a search of the coarse overall structure of the space of global states, and will find a good minimum at that coarse level. As the temperature is lowered, it will begin to respond to smaller energy differences and will find one of the better minima within the coarse-scale minimum it discovered at high temperature. Kirkpatrick et al. have shown that this way of searching the coarse structure before the fine is very effective for combinatorial problems like graph partitioning, and we believe it will also prove useful when trying to satisfy multiple weak constraints, even though it will clearly fail in cases where the best solution corresponds to a minimum that is deep, narrow, and isolated.

## 3. A LEARNING ALGORITHM

Perhaps the most interesting aspect of the Boltzmann Machine formulation is that it leads to a domain-independent learning algorithm that modifies the connection strengths between units in such a way that the whole network develops an internal model which captures the underlying structure of its environment. There has been a long history of failure in the search for such algorithms (Newell, 1982), and many people (particularly in Artificial Intelligence) now believe that no such algorithms exist. The major technical stumbling block which prevented the generalization of simple learning algorithms to more complex networks was this: To be capable of interesting computations, a network must contain nonlinear elements that are not directly constrained by the input, and when such a network does the wrong thing it appears to be impossible to decide which of the many connection strengths is at fault. This "credit-assignment" problem was what led to the demise of perceptrons (Minsky & Papert, 1968; Rosenblatt, 1961). The perceptron convergence theorem guarantees that the weights of a single layer of decision units can be trained, but it could not be generalized to networks of such units when the task did not directly specify how to use all the units in the network.

This version of the credit-assignment problem can be solved within the Boltzmann Machine formulation. By using the right stochastic decision rule, and by running the network until it reaches "thermal equilibrium" at some finite temperature, we achieve a mathematically simple relationship between the probability of a global state and its energy. For a network that is running freely without any input from the environment, this relationship is given by Eq. (6). Because the energy is a *linear* function of the weights (Eq. 1) this leads to a remarkably simple relationship between the log probabilities of global states and the individual connection strengths:

$$\frac{\partial \ln P_\alpha}{\partial w_{ij}} = \frac{1}{T}[s_i^\alpha s_j^\alpha - p_{ij}'] \qquad (7)$$

where $s_i^\alpha$ is the state of the $i^{th}$ unit in the $\alpha^{th}$ global state (so $s_i^\alpha s_j^\alpha$ is 1 only if units $i$ and $j$ are both on in state $\alpha$), and $p_{ij}'$ is just the probability of finding the two units $i$ and $j$ on at the same time when the system is at equilibrium.

Given Eq. (7), it is possible to manipulate the log probabilities of global states. If the environment directly specifies the required probabilities $P_\alpha$ for each global state $\alpha$, there is a straightforward way of converging on a set of weights that achieve those probabilities, provided any such set exists (for details, see Hinton & Sejnowski, 1983a). However, this is not a particularly interesting kind of learning because the system has to be given the required probabilities of *complete* global states. This means that the central question of what internal representation should be used has already been decided by the environment. The interesting problem arises when the environment implicitly contains high-order constraints and the network must choose internal representations that allow these constraints to be expressed efficiently.

## 3.1 Modeling the Underlying Structure of an Environment

The units of a Boltzmann Machine partition into two functional groups, a nonempty set of *visible* units and a possibly empty set of *hidden* units. The visible units are the interface between the network and the environment; during training all the visible units are clamped into specific states by the environment; when testing for completion ability, any subset of the visible units may be clamped. The hidden units, if any, are never clamped by the environment and can be used to "explain" underlying constraints in the ensemble of input vectors that cannot be represented by pairwise constraints among the visible units. A hidden unit would be needed, for example, if the environment demanded that the states of three visible units should have even parity—a regularity that cannot be enforced by pairwise interactions alone. Using hidden units to represent more complex hypotheses about the states of the visible units, such higher-order constraints among the visible units can be reduced to first and second-order constraints among the whole set of units.

We assume that each of the environmental input vectors persists for long enough to allow the network to approach thermal equilibrium, and we ignore any structure that may exist in the *sequence* of environmental vectors. The structure of an environment can then be specified by giving the probability distribution over all $2^v$ states of the $v$ visible units. The network will be said to have a perfect model of the environment if it achieves exactly the same probability distribution over these $2^v$ states when it is running freely at thermal equilibrium with all units unclamped so there is no environmental input.

Unless the number of hidden units is exponentially large compared to the number of visible units, it will be impossible to achieve a *perfect* model because even if the network is totally connected the $(v+h-1)(v+h)/2$ weights and $(v+h)$ biases among the $v$ visible and $h$ hidden units will be insufficient to model the $2^v$ probabilities of the states of the visible units specified by the environment. However, if there are regularities in the environment, and if the network uses its hidden units to capture these regularities, it may achieve a good match to the environmental probabilities.

An information-theoretic measure of the discrepancy between the network's internal model and the environment is

$$G = \sum_\alpha P(V_\alpha) \ln \frac{P(V_\alpha)}{P'(V_\alpha)} \qquad (8)$$

where $P(V_\alpha)$ is the probability of the $\alpha^{th}$ state of the visible units when their states are determined by the environment, and $P'(V_\alpha)$ is the corresponding probability when the network is running freely with no environmental input. The $G$ metric, sometimes called the asymmetric divergence or information gain (Kullback, 1959; Renyi, 1962), is a measure of the distance from the distribution given by the $P'(V_\alpha)$ to the distribution given by the $P(V_\alpha)$. $G$ is zero if and only if the distributions are identical; otherwise it is positive.

The term $P'(V_\alpha)$ depends on the weights, and so $G$ can be altered by changing them. To perform gradient descent in $G$, it is necessary to know the partial derivative of $G$ with respect to each individual weight. In most cross-coupled nonlinear networks it is very hard to derive this quantity, but because of the simple relationships that hold at thermal equilibrium, the partial derivative of $G$ is straightforward to derive for our networks. The probabilities of global states are determined by their energies (Eq. 6) and the energies are determined by the weights (Eq. 1). Using these equations the partial derivative of $G$ (see the appendix) is:

$$\frac{\partial G}{\partial w_{ij}} = -\frac{1}{T}(p_{ij} - p'_{ij}) \qquad (9)$$

where $p_{ij}$ is the average probability of two units both being in the *on* state when the environment is clamping the states of the visible units, and $p'_{ij}$, as in Eq. (7), is the corresponding probability when the environmental input is not present and the network is running freely. (Both these probabilities must be measured at equilibrium.) Note the similarity between this equation and Eq. (7), which shows how changing a weight affects the log probability of a single state.

To minimize $G$, it is therefore sufficient to observe $p_{ij}$ and $p'_{ij}$ when the network is at thermal equilibrium, and to change each weight by an amount proportional to the difference between these two probabilities:

$$\Delta w_{ij} = \epsilon(p_{ij} - p'_{ij}) \qquad (10)$$

where $\epsilon$ scales the size of each weight change.

A surprising feature of this rule is that it uses only *locally available* information. The change in a weight depends only on the behavior of the two units it connects, even though the change optimizes a global measure, and the best value for each weight depends on the values of all the other weights. If there are no hidden units, it can be shown that $G$-space is concave (when viewed from above) so that simple gradient descent will not get trapped at poor local minima. With hidden units, however, there can be local minima that correspond to different ways of using the hidden units to represent the higher-order constraints that are implicit in the probability distribution of environmental vectors. Some techniques for handling these more complex $G$-spaces are discussed in the next section.

Once $G$ has been minimized the network will have captured as well as possible the regularities in the environment, and these regularities will be enforced when performing completion. An alternative view is that the network, in minimizing $G$, is finding the set of weights that is most likely to have generated the set of environmental vectors. It can be shown that maximizing this likelihood is mathematically equivalent to minimizing $G$ (Peter Brown, personal communication, 1983).

## 3.2 Controlling the Learning

There are a number of free parameters and possible variations in the learning algorithm presented above. As well as the size of $\epsilon$, which determines the size of each step taken for gradient descent, the lengths of time over which $p_{ij}$ and $p_{ij}'$ are estimated have a significant impact on the learning process. The values employed for the simulations presented here were selected primarily on the basis of empirical observations.

A practical system which estimates $p_{ij}$ and $p_{ij}'$ will necessarily have some noise in the estimates, leading to occasional "uphill steps" in the value of $G$. Since hidden units in a network can create local minima in $G$, this is not necessarily a liability. The effect of the noise in the estimates can be reduced, if desired, by using a small value for $\epsilon$ or by collecting statistics for a longer time, and so it is relatively easy to implement an annealing search for the minimum of $G$.

The objective function $G$ is a metric that specifies how well two probability distributions match. Problems arise if an environment specifies that only a small subset of the possible patterns over the visible units ever occur. By default, the unmentioned patterns must occur with probability zero, and the only way a Boltzmann Machine running at a non-zero temperature can guarantee that certain configurations *never* occur is to give those configurations infinitely high energy, which requires infinitely large weights.

One way to avoid this implicit demand for infinite weights is to occasionally provide "noisy" input vectors. This can be done by filtering the "correct" input vectors through a process that has a small probability of reversing each of the bits. These noisy vectors are then clamped on the visible units. If the noise is small, the correct vectors will dominate the statistics, but every vector will have some chance of occurring and so infinite energies will not be needed. This "noisy clamping" technique was used for all the examples presented here. It works quite well, but we are not entirely satisfied with it and have been investigating other methods of preventing the weights from growing too large when only a few of the possible input vectors ever occur.

The simulations presented in the next section employed a modification of the obvious steepest descent method implied by Eq. (10). Instead of changing $w_{ij}$ by an amount proportional to $p_{ij} - p_{ij}'$, it is simply incremented by a fixed "weight-step" if $p_{ij} > p_{ij}'$ and decremented by the same amount if $p_{ij} < p_{ij}'$. The advantage of this method over steepest descent is that it can cope with wide variations in the first and second derivatives of $G$. It can make significant progress on dimensions where $G$ changes gently without taking very large divergent steps on dimensions where $G$ falls rapidly and then rises rapidly again. There is no suitable value for the $\epsilon$ in Eq. (10) in such cases. Any value large enough to allow progress along the gently sloping floor of a ravine will cause divergent oscillations up and down the steep sides of the ravine.[3]

## 4. THE ENCODER PROBLEM

The "encoder problem" (suggested to us by Sanjaya Addanki) is a simple abstraction of the recurring task of communicating information among various components of a parallel network. We have used this problem to test out the learning algorithm because it is clear what the optimal solution is like and it is nontrivial to discover it. Two groups of visible units, designated $V_1$ and $V_2$, represent two systems that wish to communicate their states. Each group has $v$ units. In the simple formulation we consider here, each group has only one unit on at a time, so there are only $v$ different states of each group. $V_1$ and $V_2$ are not connected directly but both are connected to a group of $h$ hidden units $H$, with $h < v$ so $H$ may act as a limited capacity bottleneck through which information about the states of $V_1$ and $V_2$ must be squeezed. Since all simulations began with all weights set to zero, finding a solution to such a problem requires that the two visible groups come to agree upon the meanings of a set of codes without any *a priori* conventions for communication through $H$.

To permit perfect communication between the visible groups, it must be the case that $h \geq log_2 v$. We investigated minimal cases in which $h = log_2 v$, and cases when $h$ was somewhat larger than $log_2 v$. In all cases, the environment for the network consisted of $v$ equiprobable vectors of length $2v$ which specified that one unit in $V_1$ and the corresponding unit in $V_2$ should be on together with all other units off. Each visible group is completely connected internally and each is completely connected to $H$, but the units in $H$ are not connected to each other.

Because of the severe speed limitation of simulation on a sequential machine, and because the learning requires many annealings, we have primarily experimented with small versions of the encoder problem. For example, Figure 2 shows a good solution to a "4-2-4" encoder problem in

---

[3] The problem of finding a suitable value for $\epsilon$ disappears if one performs a line search for the lowest value of $G$ along the current direction of steepest descent, but line searches are inapplicable in this case. *Only* the local gradient is available. There are bounds on the second derivative that can be used to pick conservative values of $\epsilon$ (Mark Derthick, personal communication, 1984), and methods of this kind are currently under investigation.

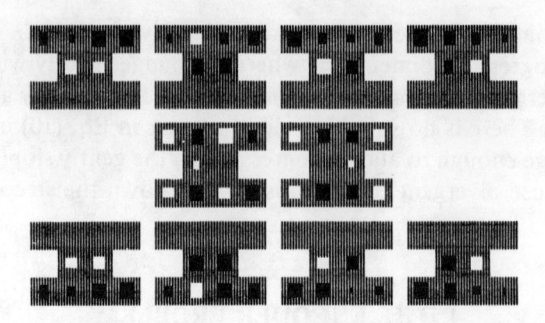

Figure 2. A solution to an encoder problem. The link weights are displayed using a recursive notation. Each unit is represented by a shaded I-shaped box; from top to bottom the rows of boxes represent groups $V_1$, $H$, and $V_2$. Each shaded box is a map of the entire network, showing the strengths of that unit's connections to other units. At each position in a box, the size of the white (positive) or black (negative) rectangle indicates the magnitude of the weight. In the position that would correspond to a unit connecting to itself (the second position in the top row of the second unit in the top row, for example), the bias is displayed. All connections between units appear twice in the diagram, once in the box for each of the two units being connected. For example, the black square in the top right corner of the leftmost unit of $V_1$ represents the same connection as the black square in the top left corner of the rightmost unit of $V_1$. This connection has a weight of $-30$.

which $v=4$ and $h=2$. The interconnections between the visible groups and $H$ have developed a binary coding—each visible unit causes a different pattern of *on* and *off* states in the units of $H$, and corresponding units in $V_1$ and $V_2$ support identical patterns in $H$. Note how the bias of the second unit of $V_1$ and $V_2$ is positive to compensate for the fact that the code which represents that unit has all the $H$ units turned off.

### 4.1. The 4-2-4 Encoder

The experiments on networks with $v=4$ and $h=2$ were performed using the following learning cycle:

1. *Estimation of $p_{ij}$*: Each environmental vector in turn was clamped over the visible units. For each environmental vector, the network was allowed to reach equilibrium twice. Statistics about how often pairs of units were both on together were gathered at equilibrium. To prevent the weights from growing too large we used the "noisy" clamping technique described in Section 3.2. Each *on* bit of a clamped vector was set to *off* with a probability of 0.15 and each *off* bit was set to *on* with a probability of 0.05.
2. *Estimation of $p'_{ij}$*: The network was completely unclamped and allowed to reach equilibrium at a temperature of 10. Statistics about

co-occurrences were then gathered for as many annealings as were used to estimate $p_{ij}$.
3. *Updating the weights:* All weights in the network were incremented or decremented by a fixed weight-step of 2, with the sign of the increment being determined by the sign of $p_{ij} - p'_{ij}$.

When a settling to equilibrium was required, all the unclamped units were randomized with equal probability on or off (corresponding to raising the temperature to infinity), and then the network was allowed to run for the following times at the following temperatures: [2@20, 2@15, 2@12, 4@10].[4] After this annealing schedule it was assumed that the network had reached equilibrium, and statistics were collected at a temperature of 10 for 10 units of time.

We observed three main phases in the search for the global minimum of $G$, and found that the occurrence of these phases was relatively insensitive to the precise parameters used. The first phase begins with all the weights set to zero, and is characterized by the development of negative weights throughout most of the network, implementing two winner-take-all networks that model the simplest aspect of the environmental structure—only one unit in each visible group is normally active at a time. In a *4-2-4* encoder, for example, the number of possible patterns over the visible units is $2^8$. By implementing a winner-take-all network among each group of four this can be reduced to $4 \times 4$ low energy patterns. Only the final reduction from $2^4$ to $2^2$ low energy patterns requires the hidden units to be used for communicating between the two visible groups. Figure 3a shows a *4-2-4* encoder network after four learning cycles.

Although the hidden units are exploited for inhibition in the first phase, the lateral inhibition task can be handled by the connections within the visible groups alone. In the second phase, the hidden units begin to develop positive weights to some of the units in the visible groups, and they tend to maintain symmetry between the sign and approximate magnitude of a connection to a unit in $V_1$ and the corresponding unit in $V_2$. The second phase finishes when every hidden unit has significant connection weights to each unit in $V_1$ and analogous weights to each unit in $V_2$, and most of the different codes are being used, but there are some codes that are used more than once and some not at all. Figure 3b shows the same network after 60 learning cycles.

Occasionally, all the codes are being used at the end of the second phase in which case the problem is solved. Usually, however, there is a third and longest phase during which the learning algorithm sorts out the remaining conflicts and finds a global minimum. There are two basic mechanisms

[4] One unit of time is defined as the time required for each unit to be given, on average, one chance to change its state. This means that if there are $n$ unclamped units, a time period of 1 involves $n$ random probes in which some unit is given a chance to change its state.

involved in the sorting out process. Consider the conflict between the first and fourth units in Figure 3b, which are both employing the code $<-,+>$. When the system is running without environmental input, the two units will be on together quite frequently. Consequently, $p'_{1,4}$ will be higher than $p_{1,4}$ because the environmental input tends to prevent the two units from being on together. Hence, the learning algorithm keeps decreasing the weight of the connection between the first and fourth units in each group, and they come to inhibit each other strongly. (This effect explains the variations in inhibitory weights in Figure 2. Visible units with similar codes are the ones that inhibit each other strongly.) Visible units thus compete for "territory" in the space of possible codes, and this repulsion effect causes codes to migrate away from similar neighbors. In addition to the repulsion effect, we observed another process that tends to eventually bring the unused codes adjacent (in terms of hamming distance) to codes that are involved in a conflict. The mechanics of this process are somewhat subtle and we do not take the time to expand on them here.

The third phase finishes when all the codes are being used, and the weights then tend to increase so that the solution locks in and remains stable against the fluctuations caused by random variations in the co-occurrence statistics. (Figure 2 is the same network shown in Figure 3, after 120 learning cycles.)

In 250 different tests of the *4-2-4* encoder, it always found one of the global minima, and once there it remained there. The median time required to discover four different codes was 110 learning cycles. The longest time was 1810 learning cycles.

### 4.2. The 4-3-4 Encoder

A variation on the binary encoder problem is to give $H$ more units than are absolutely necessary for encoding the patterns in $V_1$ and $V_2$. A simple example is the *4-3-4* encoder which was run with the same parameters as the *4-2-4* encoder. In this case the learning algorithm quickly finds four different codes. Then it always goes on to modify the codes so that they are optimally spaced out and no pair differ by only a single bit, as shown in Figure 4. The median time to find four well-spaced codes was 270 learning cycles and the maximum time in 200 trials was 1090.

### 4.3. The 8-3-8 Encoder

With $v=8$ and $h=3$ it took many more learning cycles to find all 8 three-bit codes. We did 20 simulations, running each for 4000 learning cycles using the same parameters as for the *4-2-4* case (but with a probability of 0.02 of reversing each *off* unit during noisy clamping). The algorithm found all 8

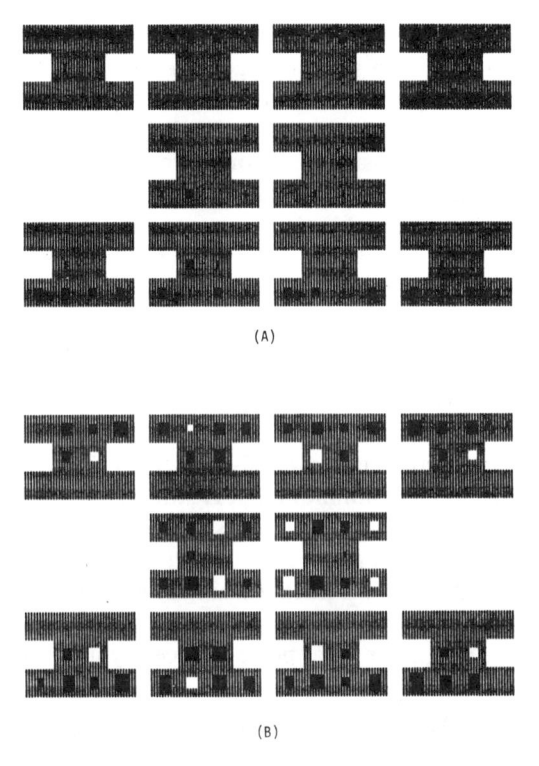

Figure 3. Two phases in the development of the perfect binary encoding shown in Figure 2. The weights are shown (A) after 4 learning trials and (B) after 60 learning trials.

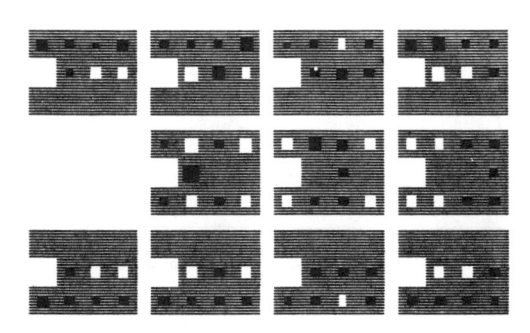

Figure 4. A 4-3-4 encoder that has developed optimally spaced codes.

codes in 16 out of 20 simulations and found 7 codes in the rest. The median time to find 7 codes was 210 learning cycles and the median time to find all 8 was 1570 cycles.

The difficulty of finding all 8 codes is not surprising since the fraction of the weight space that counts as a solution is much smaller than in the *4-2-4* case. Sets of weights that use 7 of the 8 different codes are found fairly rapidly and they constitute local minima which are far more numerous than the global minima and have almost as good a value of $G$. In this type of $G$-space, the learning algorithm must be carefully tuned to achieve a global minimum, and even then it is very slow. We believe that the $G$-spaces for which the algorithm is well-suited are ones where there are a great many possible solutions and it is not essential to get the very best one. For large networks to learn in a reasonable time, it may be necessary to have enough units and weights and a liberal enough specification of the task so that no single unit or weight is essential. The next example illustrates the advantages of having some spare capacity.

### 4.4. The 40-10-40 Encoder

A somewhat larger example is the *40-10-40* encoder. The 10 units in $H$ are almost twice the theoretical minimum, but $H$ still acts as a limited bandwidth bottleneck. The learning algorithm works well on this problem. Figure 5 shows its performance when given a pattern in $V_1$ and required to settle to the corresponding pattern in $V_2$. Each learning cycle involved annealing once with each of the 40 environmental vectors clamped, and the same number of times without clamping. The final performance asymptotes at 98.6% correct.

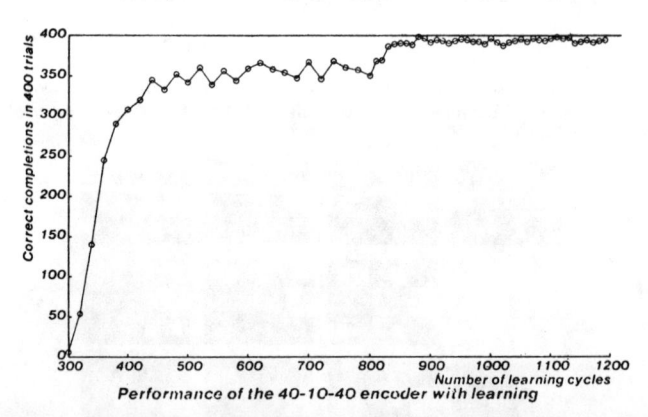

Figure 5. Completion accuracy of a 40-10-40 encoder during learning. The network was tested by clamping the states of the units in $V_1$ and letting the remainder of the network reach equilibrium. If just the correct unit was on in $V_2$, the test was successful. This was repeated 10 times for each of the 40 units in $V_1$. For the first 300 learning cycles the network was run without connecting up the hidden units. This ensured that each group of 40 visible units developed enough lateral inhibition to implement an effective winner-take-all network. The hidden units were then connected up and for the next 500 learning cycles we used "noisy" clamping, switching *on* bits to *off* with a probability of 0.1 and *off* bits to *on* with a probability of 0.0025. After this we removed the noise and this explains the sharp rise in performance after 800 cycles. The final performance asymptotes at 98.6% correct.

The codes that the network selected to represent the patterns in $V_1$ and $V_2$ were all separated by a hamming distance of at least 2, which is very unlikely to happen by chance. As a test, we compared the weights of the connections between visible and hidden units. Each visible unit has 10 weights connecting it to the hidden units, and to avoid errors, the 10 dimensional weight vectors for two different visible units should not be too similar. The cosine of the angle between two vectors was used as a measure of similarity, and no two codes had a similarity greater than 0.73, whereas many pairs had similarities of 0.8 or higher when the same weights were randomly rearranged to provide a control group for comparison.

To achieve good performance on the completion tests, it was necessary to use a very gentle annealing schedule during testing. The schedule spent twice as long at each temperature and went down to half the final temperature of the schedule used during learning. As the annealing was made faster, the error rate increased, thus giving a very natural speed/accuracy trade-off. We have not pursued this issue any further, but it may prove fruitful because some of the better current models of the speed/accuracy trade-off in human reaction time experiments involve the idea of a biased random walk (Ratcliff, 1978), and the annealing search gives rise to similar underlying mathematics.

## 5. REPRESENTATION IN PARALLEL NETWORKS

So far, we have avoided the issue of how complex concepts would be represented in a Boltzmann machine. The individual units stand for "hypotheses," but what is the relationship between these hypotheses and the kinds of concepts for which we have words? Some workers suggest that a concept should be represented in an essentially "local" fashion: The activation of one or a few computing units is the representation for a concept (Feldman & Ballard, 1982); while others view concepts as "distributed" entities: A particular pattern of activity over a large group of units represents a concept, and different concepts corresponds to *alternative* patterns of activity over the same group of units (Hinton, 1981).

One of the better arguments in favor of local representations is their inherent modularity. Knowledge about relationships between concepts is localized in specific connections and is therefore easy to add, remove, and modify, if some reasonable scheme for forming hardware connections can be found (Fahlman, 1980; Feldman, 1982). With distributed representations, however, the knowledge is diffuse. This is good for tolerance to local hardware damage, but it appears to make the design of modules to perform specific functions much harder. It is particularly difficult to see how new distributed representations of concepts could originate spontaneously.

In a Boltzmann machine, a distributed representation corresponds to an energy minimum, and so the problem of creating a good collection of distributed representations is equivalent to the problem of creating a good "energy landscape." The learning algorithm we have presented is capable of solving this problem, and it therefore makes distributed representations considerably more plausible. The diffuseness of any one piece of knowledge is no longer a serious objection, because the mathematical simplicity of the Boltzmann distribution makes it possible to manipulate all the diffuse local weights in a coherent way on the basis of purely local information. The formation of a simple set of distributed representations is illustrated by the encoder problems.

### 5.1. Communicating Information between Modules

The encoder problem examples also suggest a method for communicating symbols between various components of a parallel computational network. Feldman and Ballard (1982) present sketches of two implementations for this task; using the example of the transmission of the concept "wormy apple" from where it is recognized in the perceptual system to where the phrase "wormy apple" can be generated by the speech system. They argue that there appears to be only two ways that this could be accomplished. In the first method, the perceptual information is encoded into a set of symbols that are then transmitted as messages to the speech system, where they are decoded into a form suitable for utterance. In this case, there would be a set of general-purpose communciation lines, analogous to a bus in a conventional computer, that would be used as the medium for all such messages from the visual system to the speech system. Feldman and Ballard describe the problems with such a system as:

- Complex messages would presumably have to be transmitted sequentially over the communication lines.
- Both sender and receiver would have to learn the common code for each new concept.
- The method seems biologically implausible as a mechanism for the brain.

The alternative implementation they suggest requires an individual, dedicated hardware pathway for each concept that is communicated from the perceptual system to the speech system. The idea is that the simultaneous activation of "apple" and "worm" in the perceptual system can be transmitted over private links to their counterparts in the speech system. The critical issues for such an implementation are having the necessary connections available between concepts, and being able to establish new connection pathways as new concepts are learned in the two systems. The main point of this approach is that the links between the computing units carry simple, nonsymbolic information such as a single activation level.

The behavior of the Boltzmann machine when presented with an encoder problem demonstrates a way of communicating concepts that largely combines the best of the two implementations mentioned. Like the second approach, the computing units are small, the links carry a simple numeric value, and the computational and connection requirements are within the range of biological plausibility. Like the first approach, the architecture is such that many different concepts can be transmitted over the same communication lines, allowing for effective use of limited connections. The learning of new codes to represent new concepts emerges automatically as a cooperative process from the $G$-minimization learning algorithm.

## 6. CONCLUSION

The application of statistical mechanics to constraint satisfaction searches in parallel networks is a promising new area that has been discovered independently by several other groups (Geman & Geman, 1983; Smolensky, 1983). There are many interesting issues that we have only mentioned in passing. Some of these issues are discussed in greater detail elsewhere: Hinton and Sejnowski (1983b) and Geman and Geman (1983) describe the relation to Bayesian inference and to more conventional relaxation techniques; Fahlman, Hinton, and Sejnowski (1983) compare Boltzmann machines with some alternative parallel schemes, and discuss some knowledge representation issues. An expanded version of this paper (Hinton, Sejnowski, & Ackley, 1984) presents this material in greater depth and discusses a number of related issues such as the relationship to the brain and the problem of sequential behavior. It also shows how the probabilistic decision function could be realized using gaussian noise, how the assumptions of symmetry in the physical connections and of no time delay in transmission can be relaxed, and describes results of simulations on some other tasks.

Systems with symmetric weights form an interesting class of computational device because their dynamics is governed by an energy function.[5] This is what makes it possible to analyze their behavior and to use them for iterative constraint satisfaction. In their influential exploration of perceptrons, Minsky and Papert (1968, p. 231) concluded that: "Multilayer machines with loops clearly open up all the questions of the general theory of automata." Although this statement is very plausible, recent developments

[5] One can easily write down a similar energy function for asymmetric networks, but this energy function does not govern the behavior of the network when the links are given their normal causal interpretation.

suggest that it may be misleading because it ignores the symmetric case, and it seems to have led to the general belief that it would be impossible to find powerful learning algorithms for networks of perceptron-like elements.

We believe that the Boltzmann Machine is a simple example of a class of interesting stochastic models that exploit the close relationship between Boltzmann distributions and information theory.

> All of this will lead to theories [of computation] which are much less rigidly of an all-or-none nature than past and present formal logic. They will be of a much less combinatorial, and much more analytical, character. In fact, there are numerous indications to make us believe that this new system of formal logic will move closer to another discipline which has been little linked in the past with logic. This is thermodynamics, primarily in the form it was received from Boltzmann, and is that part of theoretical physics which comes nearest in some of its aspects to manipulating and measuring information.
>
> (John Von Neumann, *Collected Works* Vol. 5, p. 304)

## APPENDIX: DERIVATION OF THE LEARNING ALGORITHM

When a network is free-running at equilibrium the probability distribution over the visible units is given by

$$P'(V_\alpha) = \sum_\beta P'(V_\alpha \wedge H_\beta) = \frac{\sum_\beta e^{-E_{\alpha\beta}/T}}{\sum_{\lambda\mu} e^{-E_{\lambda\mu}/T}} \qquad (11)$$

where $V_\alpha$ is a vector of states of the visible units, $H_\beta$ is a vector of states of the hidden units, and $E_{\alpha\beta}$ is the energy of the system in state $V_\alpha \wedge H_\beta$

$$E_{\alpha\beta} = -\sum_{i<j} w_{ij} s_i^{\alpha\beta} s_j^{\alpha\beta}.$$

Hence,

$$\frac{\partial e^{-E_{\alpha\beta}/T}}{\partial w_{ij}} = \frac{1}{T} s_i^{\alpha\beta} s_j^{\alpha\beta} e^{-E_{\alpha\beta}/T}.$$

Differentiating (11) then yields

$$\frac{\partial P'(V_\alpha)}{\partial w_{ij}} = \frac{\frac{1}{T}\sum_\beta e^{-E_{\alpha\beta}/T} s_i^{\alpha\beta} s_j^{\alpha\beta}}{\sum_{\alpha\beta} e^{-E_{\alpha\beta}/T}} - \frac{\sum_\beta e^{-E_{\alpha\beta}/T} \frac{1}{T}\sum_{\lambda\mu} e^{-E_{\lambda\mu}/T} s_i^{\lambda\mu} s_j^{\lambda\mu}}{\left(\sum_{\lambda\mu} e^{-E_{\lambda\mu}/T}\right)^2}$$

$$= \frac{1}{T} \left[ \sum_\beta P'(V_\alpha \wedge H_\beta) s_i^{\alpha\beta} s_j^{\alpha\beta} - P'(V_\alpha) \sum_{\lambda\mu} P'(V_\lambda \wedge H_\mu) s_i^{\lambda\mu} s_j^{\lambda\mu} \right].$$

This derivative is used to compute the gradient of the $G$-measure

$$G = \sum_\alpha P(V_\alpha) \ln \frac{P(V_\alpha)}{P'(V_\alpha)}$$

where $P(V_\alpha)$ is the clamped probability distribution over the visible units and is independent of $w_{ij}$. So

$$\frac{\partial G}{\partial w_{ij}} = -\sum_\alpha \frac{P(V_\alpha)}{P'(V_\alpha)} \frac{\partial P'(V_\alpha)}{\partial w_{ij}}$$

$$= -\frac{1}{T} \sum_\alpha \frac{P(V_\alpha)}{P'(V_\alpha)} \left[ \sum_\beta P'(V_\alpha \wedge H_\beta) s_i^{\alpha\beta} s_j^{\alpha\beta} - P'(V_\alpha) \sum_{\lambda\mu} P'(V_\lambda \wedge H_\mu) s_i^{\lambda\mu} s_j^{\lambda\mu} \right].$$

Now,

$$P(V_\alpha \wedge H_\beta) = P(H_\beta | V_\alpha) P(V_\alpha),$$
$$P'(V_\alpha \wedge H_\beta) = P'(H_\beta | V_\alpha) P'(V_\alpha),$$

and

$$P'(H_\beta | V_\alpha) = P(H_\beta | V_\alpha). \qquad (12)$$

Equation (12) holds because the probability of a hidden state given some visible state must be the same in equilibrium whether the visible units were clamped in that state or arrived there by free-running. Hence,

$$P'(V_\alpha \wedge H_\beta) \frac{P(V_\alpha)}{P'(V_\alpha)} = P(V_\alpha \wedge H_\beta).$$

Also,

$$\sum_\alpha P(V_\alpha) = 1.$$

Therefore,

$$\frac{\partial G}{\partial w_{ij}} = -\frac{1}{T} [p_{ij} - p'_{ij}]$$

where

$$p_{ij} \overset{def}{=} \sum_{\alpha\beta} P(V_\alpha \wedge H_\beta) s_i^{\alpha\beta} s_j^{\alpha\beta}$$

and

$$p'_{ij} \overset{def}{=} \sum_{\lambda\mu} P'(V_\lambda \wedge H_\mu) s_i^{\lambda\mu} s_j^{\lambda\mu}$$

as given in (9).

The Boltzmann Machine learning algorithm can also be formulated as an input-output model. The visible units are divided into an input set $I$ and an output set O, and an environment specifies a set of conditional probabilities of the form $P(O_\beta|I_\alpha)$. During the "training" phase the environment clamps both the input and output units, and $p_{ij}$s are estimated. During the "testing" phase the input units are clamped and the output units and hidden units free-run, and $p'_{ij}$s are estimated. The appropriate $G$ measure in this case is

$$G = \sum_{\alpha\beta} P(I_\alpha \wedge O_\beta) \ln \frac{P(O_\beta|I_\alpha)}{P'(O_\beta|I_\alpha)}$$

Similar mathematics apply in this formulation and $\partial G/\partial w_{ij}$ is the same as before.

## REFERENCES

Berliner, H. J., & Ackley, D. H. (1982, August). The QBKG system: Generating explanations from a non-discrete knowledge representation. *Proceedings of the National Conference on Artificial Intelligence AAAI-82,* Pittsburgh, PA, 213–216.

Binder, K. (Ed.) (1978). *The Monte-Carlo method in statistical physics.* New York: Springer-Verlag.

Fahlman, S. E. (1980, June). The Hashnet Interconnection Scheme. (Tech. Rep. No. CMU-CS-80-125), Carnegie-Mellon University, Pittsburgh, PA.

Fahlman, S. E., Hinton, G. E., & Sejnowski, T. J. (1983, August). Massively parallel architectures for AI: NETL, Thistle, and Boltzmann Machines. *Proceedings of the National Conference on Artificial Intelligence AAAI-83,* Washington, DC, 109–113.

Feldman, J. A. (1982). Dynamic connections in neural networks. *Biological Cybernetics, 46,* 27–39.

Feldman, J. A., & Ballard, D. H. (1982). Connectionist models and their properties. *Cognitive Science, 6,* 205–254.

Geman, S., & Geman, D. (1983). Stochastic relaxation, Gibbs distributions, and the Bayesian restoration of images. Unpublished manuscript.

Grimson, W. E. L. (1981). *From images to surfaces.* Cambridge, MA: MIT Press.

Hinton, G. E. (1977). *Relaxation and its role in vision.* Unpublished doctoral dissertation, University of Edinburgh. Described in D. H. Ballard & C. M. Brown (Eds.), *Computer Vision.* Englewood Cliffs, NJ: Prentice-Hall, 408–430.

Hinton, G. E. (1981). Implementing semantic networks in parallel hardware. In G. E. Hinton & J. A. Anderson (Eds.), *Parallel Models of Associative Memory*. Hillsdale, NJ: Erlbaum.
Hinton, G. E., & Anderson, J. A. (1981). *Parallel models of associative memory*. Hillsdale, NJ: Erlbaum.
Hinton, G. E., & Sejnowski, T. J. (1983a, May). Analyzing cooperative computation. *Proceedings of the Fifth Annual Conference of the Cognitive Science Society*. Rochester, NY.
Hinton, G. E., & Sejnowski, T. J. (1983b, June). Optimal perceptual inference. *Proceedings of the IEEE Computer Society Conference on Computer Vision and Pattern Recognition*. Washington, DC, pp. 448-453.
Hinton, G. E., Sejnowski, T. J., & Ackley, D. H. (1984, May). *Boltzmann Machines: Constraint satisfaction networks that learn*. (Tech. Rep. No. CMU-CS-84-119). Pittsburgh, PA: Carnegie-Mellon University.
Hopfield, J. J. (1982). Neural networks and physical systems with emergent collective computational abilities. *Proceedings of the National Academy of Sciences USA, 79*, 2554-2558.
Kirkpatrick, S., Gelatt, C. D., & Vecchi, M. P. (1983). Optimization by simulated annealing. *Science, 220*, 671-680.
Kullback, S. (1959). *Information theory and statistics*. New York: Wiley.
Metropolis, N., Rosenbluth, A., Rosenbluth, M., Teller, A., & Teller, E. (1953). Equation of state calculations for fast computing machines. *Journal of Chemical Physics, 6*, 1087.
Minsky, M., & Papert, S. (1968). *Perceptrons*. Cambridge, MA: MIT Press.
Newell, A. (1982). *Intellectual issues in the history of artificial intelligence*. (Tech. Rep. No. CMU-CS-82-142). Pittsburgh, PA: Carnegie-Mellon University.
Newell, A., & Simon, H. A. (1972). *Human problem solving*. Englewood Cliffs, NJ: Prentice-Hall, 1972.
Ratcliff, R. (1978). A theory of memory retrieval. *Psychological Review, 85*, 59-108.
Renyi, A. (1962). *Probability theory*. Amsterdam: North-Holland.
Rosenblatt, F. (1961). *Principles of neurodynamics: Perceptrons and the theory of brain mechanisms*. Washington, DC: Spartan.
Smolensky, P. (1983, August). Schema selection and stochastic inference in modular environments. *Proceedings of the National Conference on Artificial Intelligence AAAI-83*, Washington, DC. 109-113.
Terzopoulos, D. (1984). *Multiresolution computation of visible-surface representations*. Unpublished doctoral dissertation, MIT, Cambridge, MA.
Waltz, D. L. (1975). Understanding line drawings of scenes with shadows. In P. Winston (Ed.), *The Psychology of Computer Vision*. New York: McGraw-Hill.
Winston, P. H. (1984). *Artificial Intelligence*. (2nd ed.) Reading, MA: Addison-Wesley.

CHAPTER 11

# Emergence of Grandmother Memory in Feed Forward Networks: Learning With Noise and Forgetfulness*

R. SCALETTAR

*Department of Physics*
*University of California at Santa Barbara*

A. ZEE

*Department of Physics*
*& Institute for Theoretical Physics*
*University of California at Santa Barbara*

We raise the question of whether a feed forward network can learn to be a noise-tolerant memory device. Using the back propagation algorithm, we find that a grandmother-type memory emerges under certain circumstances. Imposing a decay mechanism (forgetfulness) in which a weaker connection decays faster than a stronger connection, the network calls up the grandmother cells one at a time as new patterns to be remembered are presented. (The necessity of introducing forgetfulness may be suggestive for understanding real memories.) The memory capacity exceeds that of the fully connected symmetric network discussed by Hopfield (1982, 1984).

The construction of content-addressable memories (Hebb, 1949; Kohonen, 1984; Little & Shaw, 1978; Rosenblatt, 1962) has attracted a great deal of interest recently. In particular, Hopfield (1982, 1984), building upon earlier work, has invented a strikingly simple network, which can serve as a memory device.

It occurred to us to ask whether a network of a different type, namely a feed forward network, with the architecture shown in Figure 1, can also function as a memory. A layer consisting of $N_{hid}$ "hidden" cells or units (i.e., inter-neurons) processes the input information and sets the output cells

---

R. Scalettar thanks D. Rumelhart for a discussion on back-propagation. A. Zee thanks G. Toulouse for a discussion on network memories and D. Ballard, J. Hopfield, and T. Sejnowski for encouraging comments. We would also like to thank E. Baum, C. von der Malsburg, J. Moody, and F. Wilczek for emphasizing to us some of the attractive features of the grandmother cell solution. This research was supported in part by the National Science Foundation under Grant No. PHY82-17853, supplemented by funds from the National Aeronautics and Space Administration, at the University of California at Santa Barbara.

* Manuscript received October 10, 1986 and revised May 18, 1987.

**Output Patterns**

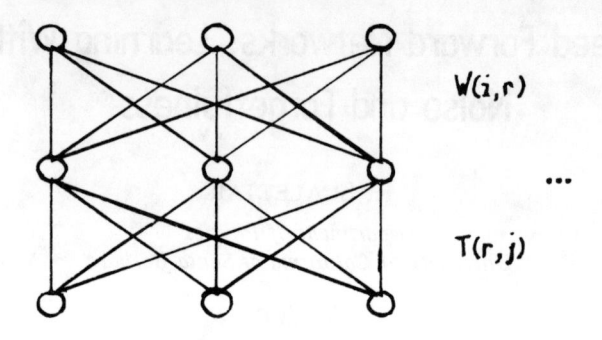

*[Figure: feed-forward network diagram with input, hidden, and output layers. Labels: W(i,r) and T(r,j)]*

**Input Patterns**

**Figure 1.** The geometry of our feed-forward net. The weights $T(r,j)$ connect input and hidden cells. The weights $W(i,r)$ connect hidden and output cells. The hidden cells fire only when their thresholds $\theta(r)$ are exceeded.

to the appropriate values. This is in contrast to a fully-connected iterative network as discussed by Hopfield. When an $N$-bit pattern is fed into the input cells, we want the output pattern to be the one correctly associated with the input pattern. The memory may be hetero-associative: For example, the input may be a name, the output, a phone number. The memory may also be auto-associative so that the input is some error-corrupted pattern while the output is the pattern in question: For example, the input may be a fragment of a name, the output, the complete name.

Our original motivation was to see whether, if we are willing to use more cells than Hopfield, we might be able to construct a memory device with certain other advantages. To remember $N$-bit patterns, the Hopfield network calls for only $N$ cells. In contrast, our feed-forward network requires $2N + N_{hid}$ cells. However, as we will see later, the capacity of our memory is considerably larger.

Let us begin with a summary of our main results.

(1) By requiring tolerance to error in the input, as is the case in the real world, we find that a feed-forward network evolves into a grandmother type memory (to be defined below) when the number of memories $N_{mem}$ is equal to $N_{hid}$.

(2) When $N_{mem} < N_{hid}$, we introduce a mechanism in which the weights decay (i.e., "forgetfulness"). We propose that weaker connections decay faster than stronger connections. As the collection of memories to be

stored grows incrementally, the additional memories become associated with previously unused hidden units in an orderly one-to-one fashion.
(3) The memory capacity exceeds that of a Hopfield type memory and the error tolerance can be comparable.

We hope that these results may be of interest to researchers constructing large capacity content addressable memory devices for practical use and possibly even to brain theorists interested in issues of concept representation and learning in biological networks. (The role of forgetfulness may be particularly intriguing.)

Recently, Rumelhart, Hinton, and Williams (1986) and a number of other workers (Parker, 1985; Le Cun, 1985) have produced a "back-propagation" algorithm whereby one can determine the connection strengths, or weights, and the thresholds appropriate to a feed-forward network. The idea is simple. One forms the energy function

$$E = \frac{1}{2} \sum_{\substack{patterns \\ p}} \sum_n (T_p(n) - O_p(n))^2 \tag{1}$$

which measures the discrepancy between the actual output $O_p$ and the desired output or "target output" $T_p$. Since there may be more than one output unit, $T_p$ and $O_p$ are, in general, vectors (in the sense of being columns of numbers). Hence, we have indexed the different components of $T_p$ and $O_p$ by $n$. Then the square in Equation (1) denotes the scalar product. Denote the weights and thresholds generically by $\omega_i$. One then seeks to minimize $E$ by adjusting the $\omega_i$'s according to the gradient-descent algorithm

$$\delta\omega_i = -\eta \frac{\partial E}{\partial \omega_i} = -\eta \sum_p (T_p - O_p) \frac{\partial O_p}{\partial \omega_i} \tag{2}$$

with $\eta$ a learning-rate parameter. (The index $n$ has been suppressed.) After each presentation, the $\omega_i$'s are changed according to $\omega_i \rightarrow \omega_i + \delta\omega_i$.

Clearly then, the back-propagation algorithm of Rumelhart et al. is capable of turning a feed-forward network into a memory device. However, thus far, the memory is not error resistant: If the input pattern is off by a few bits from the correct input pattern, there is no guarantee that the output will be the correct one.

The solution to this problem lies in the back-propagation algorithm itself. We require that patterns corrupted from a given input pattern give us the same output. The manner in which this is done is detailed below.

The algorithm of Rumelhart et al. insures that the network is capable of associating output and input, at a proficiency level corresponding to the extent to which the global minimum of the energy function is located. What is not clear a priori is how the input information is processed or represented. The state of the hidden cells may be thought of as providing a representation of the input. We found that as a result of our procedure each of the hidden units becomes specialized in remembering a particular input pattern, in a manner to be described below. In other words, the hidden cells assume the role of "grandmother cells."

The existence of grandmother cells has long been a matter of debate in neurobiology. The notion is that a particular neuron in the brain fires when one's grandmother comes into view. Many authors believe that for each of the objects and symbols commonly known to us, a particular neuron specializes in recognizing that object or symbol (Barlow, 1972). The emergence of grandmother cells in this context is perhaps suggestive.

Our net consists of a layer of input cells with values $I(j), j=1,\ldots,N_{in}$, a layer of hidden cells with values $H(r), r=1,\ldots,N_{hid}$, and a layer of output cells with values $O(n), n=1,\ldots,N_{out}$. The values $I(j), H(r)$, and $O(i)$ are equal to either $+1$ or $-1$. Denote the weights between the input and hidden cells by $T(r,j)$ and between hidden and output cells by $W(n,r)$. Then $H(r)$ is determined from $I(j)$ by

$$H(r) = \text{sgn}\left(\sum_j T(r,j)I(j) - \theta(r)\right) \tag{3}$$

where $\theta(r)$ is the threshold for the cell $H(r)$ to fire. Meanwhile $O(n)$ is determined from $H(r)$ via

$$O(n) = \text{sgn}\left(\sum_r W(n,r)\ (H(r)+1)\right) \tag{4}$$

In other words, $H(r)$ is inactive if it is equal to $(-1)$.

In simulation, we replace the sign function $sgn$ by $(2f-1)$ with $f$ some suitably rounded function. We have some freedom[1] in the choice of $f$. Here we follow Rumelhart et al. and let $f$ be the Fermi function

$$f(x) = \frac{1}{e^{-\beta x}+1} \tag{5}$$

where typically we choose the inverse temperature $\beta = 2$. (The output values are not exactly equal to, but close to $\pm 1$.) We found that the rapidity of convergence was quite sensitive to $\beta$, and $\beta = 2$ was roughly optimal. Application of Equations (1) and (2) is straightforward and determines how the weights $W(n,m)$ and $T(n,m)$ are to be changed after presentation of the pattern $p$.

---

[1] Considerable freedom exists in the choice of $f$. For example, nonbinary sequences can be represented by choosing

$$f(x) = \frac{1}{e^{-\beta x}+1} + \frac{1}{e^{-\beta(x-a)}+1} \tag{10}$$

which for large $\beta$ takes on the three variables 0,1,2 according to whether $x<0$, $0<x<a$, or $a<x$. We have used this choice to teach a network to convert numbers from base 2 to base 3.

$$\delta_p W(n,m) = -2\eta \left( O_p(n) - T_p(n) \right) f' \left( \sum_l W(n,l)(H(l)+1) \right) (H(m)+1)$$

$$\delta_p T(n,m) = -2\eta \Delta_p(n) f' \left( \sum_r T(n,r)I(r) - \theta(n) \right) I(m) \qquad (6A)$$

$$\Delta_p(n) = 2\sum_k \left( O_p(k) - T_p(k) \right) f' \left( \sum_l W(k,l)(H(l)+1) \right) W(k,n)$$

A similar equation holds for updating the thresholds.

$$\delta_p \theta(n) = +2\eta \Delta_p(n) f' \left( \sum_r T(n,r)I(r) - \theta(n) \right) \qquad (6B)$$

More precisely, the sum in Equation (1) can be written as $E = \Sigma_p E_p$ with $E_p = 1/2(T_p - O_p)^2$. After a pattern $p$ is presented, $\delta_p \omega_i = -\eta \, \partial E_p / \partial \omega_i$, the change in $\omega_i$ mandated by the pattern $p$, is computed and stored. After all the patterns have been presented, the total change $\delta \omega_i = \Sigma_p \delta_p \omega_i$ is determined and finally the weights in the network are updated. However, following Rumelhart et al. we adopt a slightly different procedure which is perhaps more appropriate for biological systems: The weights are updated after each presentation of a pattern $p$. This procedure is particularly reasonable for learning perceptual tasks where the number of possible patterns may be exponentially large. Out of this very large set, the network is presented with randomly generated patterns one after another. Strictly speaking, this procedure deviates from steepest descent. Rather, each pattern defines its own energy landscape $E_p$ and the network descends in an ever changing landscape. The difference between the two procedures is expected to be slight since at each step the weights are modified only by a small amount. Essentially this interchanges the order of doing the sum over patterns and updating the weights, which is permissible since the derivative of the sum over patterns sum over patterns is the sum over patterns of the derivative.[2]

Actually, in this problem, the energy landscape, that is, the energy $E$ as a function of the weights and thresholds $\omega_i$, consists of a series of almost flat plateaus connected via sheer cliffs. (By using a nonzero temperature, we have rounded the landscape somewhat.) In this type of landscape, the gradient-descent method is clearly not optimal. One alternative is to use Newton's method and change the weights and thresholds according to Newton's method,

$$\delta \omega_i = -\eta \, \frac{\partial E / \partial \omega_i}{|\partial^2 E / \partial \omega_i^2|}, \qquad (7)$$

(Strictly, the inverse of the full second derivative matrix $\partial^2 E / \partial \omega_i \partial \omega_j$ is called for; the form used above was suggested by Sutton (1986) as a compromise.)

---

[2] We have chosen $\eta = 1/4$ and have made on the order of 1000 sweeps through the set of weights and thresholds to reach a point where actual output matches target. Improved descent schemes can reduce this number somewhat. We also choose a maximum bond value $|T(i,j)| < 2$.

Our emphasis in this paper will be on the interesting manner in which the net chooses to represent the task internally. We relegate to an appendix such details as optimzing converging via modified descent methods.

For simplicity, we consider the auto-associative case in which the target pattern is identical to the input pattern so that $N_{out} = N_{in} = N$. Our conclusions carry over trivially into the hetero-associative case. To explore the existence of the grandmother cell solution, let us first take the number of hidden layer cells $N_{hid}$ to equal the number of memories $N_{mem}$ to be stored. Later we will relax this restriction. If no errors are made at input, that is, the pattern is completely uncorrupted, the weights exhibit little special structure. The hidden cells all tend to be active. In Figure 2 we show the weights generated by applying the updating scheme described above to reproduce at output six randomly selected 10-bit patterns. We begin the simulation by setting up the weights at random on $(-1,1)$. One indication that the algorithm is just randomly finding a solution that happens to work, rather than some special pattern, is that the weights are completely changed if the random number seed is different. This will not be the case below.

It is worth mentioning that if $N_{mem}$ becomes large, and $N_{hid}$ equals the number of input and output neurons, then a special weight pattern is generated, namely, the identity map in which nonzero weights extend "vertically" upward and all other weights are zero. However, this pattern is obviously completely error intolerant.

We now ask that the net reproduce the target even if some of the input pattern is missing. In Figure 3 we compare the weights obtained for a completely uncorrupted input (Figure 3a) with that found for the case when three of the ten input bits are missing (Figure 3b). In the latter case when we apply the updating scheme of Equation (6) the target pattern was always chosen to be one of a set of several definite ten-bit memories, but the input to the net was these patterns with $N_{err} = 3$ of the $N_{in}$ bits $I(i)$, randomly chosen, entered as zero instead of their correct values $\pm 1$. As before, we proceed sequentially through the set of patterns, considering only one at a time in updating the weights. Since we also randomly introduce errors into each pattern, we are actually minimizing an energy $E$ of Equation (1) which sums not only over the patterns but over their possible corruptions as well. The exact nature of the faults is unimportant. We could just as well reverse the bit in error as setting it to zero. As in Figure 2, Figure 3a again illustrates a random weight pattern. Figure 3b meanwhile is clearly the grandmother cell solution.[3]

In the grandmother cell solution, one of the hidden cells, say the $p^{th}$, is specialized to recognize the $\nu^{th}$ input pattern. The weights $T(p,j)$ are approximately equal to $I^{\nu}(j)$. In other words, when presented with an input pattern

---

[3] The problem is not particularly CPU intensive, runs even for 20-bit patterns taking on the order of a fraction of a minute on a VAX-750.

# EMERGENCE OF GRANDMOTHER MEMORY

**Figure 2.** The weights evolved by the net to reproduce six randomly selected 10-bit patterns. There is little discernable relation between the memories and the weights.

MEMORIES

| -1 | -1 | -1 | 1 | 1 | -1 | -1 | -1 | 1 | -1 |
|---|---|---|---|---|---|---|---|---|---|
| 1 | 1 | 1 | -1 | -1 | -1 | 1 | 1 | -1 | -1 |
| -1 | 1 | 1 | 1 | 1 | -1 | 1 | -1 | -1 | -1 |
| 1 | 1 | 1 | 1 | -1 | -1 | -1 | 1 | -1 | 1 |
| -1 | 1 | -1 | 1 | -1 | 1 | -1 | 1 | -1 | 1 |
| 1 | -1 | 1 | 1 | -1 | -1 | 1 | -1 | 1 | -1 |

W(i,r)

| 2.0 | -1.1 | 0.4 | -2.0 | 0.4 | 0.9 | 2.0 | -0.5 | 1.0 | -1.3 |
|---|---|---|---|---|---|---|---|---|---|
| -0.2 | -1.5 | 0.8 | 1.1 | 0.1 | 0.5 | -0.4 | 0.5 | 1.9 | -1.2 |
| -1.2 | -0.5 | 0.2 | 1.8 | -2.0 | -1.7 | -0.5 | 2.0 | 0.2 | 2.0 |
| -0.8 | 1.0 | -1.1 | 0.0 | -0.7 | 0.3 | 0.6 | -0.6 | -1.0 | 0.0 |
| 0.7 | -0.2 | 1.3 | -2.0 | 0.5 | -2.0 | -0.1 | -0.8 | -0.1 | 1.3 |
| -2.0 | 1.0 | -2.0 | 0.3 | -1.0 | 0.8 | 1.3 | -0.1 | -0.9 | -0.8 |

T(r,j)

| -1.6 | -0.4 | 0.6 | 1.8 | 2.0 | -1.5 | 1.1 | -2.0 | 0.3 | -2.0 |
|---|---|---|---|---|---|---|---|---|---|
| 2.0 | 0.1 | -0.5 | 1.8 | -2.0 | -0.4 | -2.0 | 2.0 | -0.4 | 2.0 |
| -0.9 | -0.4 | -0.4 | 0.8 | 1.1 | -0.8 | 0.4 | -1.4 | 1.1 | -1.6 |
| 2.0 | 0.6 | 2.0 | -0.8 | -2.0 | -1.9 | 1.9 | 0.1 | -1.2 | -1.7 |
| 0.0 | 2.0 | 2.0 | 0.9 | 0.5 | -1.3 | 0.2 | 0.4 | -2.0 | 0.6 |
| -1.9 | 1.8 | 0.6 | 1.8 | 2.0 | -1.0 | -1.8 | 0.4 | -1.8 | 2.0 |

$I^\mu(j)$, the grandmother cell simply computes the scalar product $\Sigma_j I^\mu(j) I^\nu(j)$ between the two binary vectors $I^\mu$ and $I^\nu$. If the patterns are statistically uncorrelated, then the scalar product above is essentially zero for $\mu \neq \nu$ while equal to $N$ for $\mu = \nu$. This scalar product match represents a well-known elementary memory scheme. Further discussion may be found in the book by Kohonen (1984). The back-propagation algorithm finds an optimal threshold between zero and $N$. Seen in this light, the grandmother cell solution is evidently the best solution to the problem posed. With $N_{mem}$ binary vectors stored, one asks if there is a vector within a certain angle of the input vector.

**Figure 3.** We compare the weights obtained in the case when no errors are made at input (Figure 3a) with the case when we randomly zero three of the ten input bits but still demand convergence to the full original pattern (Figure 3b). As in Figure 2, Figure 3a exhibits a fairly random pattern. In the error resistant case, however, the weights of the individual cells have become specialized to separate memories in the grandmother cell solution, for example $T(1,j)$ remembers pattern 6 while $T(2,j)$ is almost specialized to pattern 5. The threshold $\theta$ evolved to the value 10.

MEMORIES

| -1 | 1 | 1 | 1 | 1 | 1 | -1 | 1 | -1 | -1 |
| -1 | -1 | -1 | -1 | 1 | 1 | -1 | 1 | -1 | -1 |
| 1 | -1 | -1 | 1 | -1 | -1 | -1 | 1 | -1 | 1 |
| -1 | -1 | -1 | -1 | 1 | -1 | 1 | 1 | -1 | -1 |
| 1 | 1 | 1 | -1 | -1 | -1 | 1 | -1 | -1 | -1 |
| 1 | -1 | -1 | 1 | 1 | 1 | 1 | -1 | 1 | -1 |

$W(i,r)$

| 1.0 | 0.4 | 0.0 | -0.5 | -1.7 | -2.0 | 1.0 | 0.4 | -1.4 | 0.6 |
| 0.0 | 1.9 | 2.0 | 0.6 | -0.7 | -0.2 | 0.2 | -0.5 | -1.0 | -0.6 |
| 0.5 | -0.9 | -0.8 | -1.1 | 1.0 | 0.7 | 2.0 | -1.6 | 1.1 | -1.8 |
| 0.8 | -0.4 | -0.5 | 2.0 | 0.4 | 1.0 | -1.1 | 0.4 | 0.0 | 0.0 |
| -2.0 | 0.0 | -0.2 | -1.5 | 2.0 | 0.9 | -1.0 | 1.6 | -1.2 | -1.0 |
| -0.6 | -1.5 | -1.0 | -0.5 | 0.0 | -1.1 | -1.0 | 1.2 | -1.2 | 0.7 |

$T(r,j)$

| -0.7 | -1.0 | -1.7 | -0.9 | 0.1 | -1.3 | -0.2 | 2.0 | -1.8 | 0.5 |
| -2.0 | 0.5 | 0.5 | -2.0 | 0.9 | 0.5 | -0.5 | 2.0 | -1.4 | -1.2 |
| 0.3 | -0.5 | -0.8 | 2.0 | 0.4 | 1.6 | -1.7 | 0.8 | -0.3 | -0.4 |
| -0.1 | -1.3 | 1.2 | -1.7 | 1.8 | 0.0 | 2.0 | -1.4 | 0.3 | -2.0 |
| 0.7 | 2.0 | 2.0 | -0.3 | -0.3 | -0.2 | 0.3 | -0.5 | -1.6 | -0.6 |
| 0.9 | -0.5 | -0.3 | -1.2 | -1.6 | -2.0 | 1.2 | 0.6 | -1.8 | 0.0 |

**Figure 3a.**

Indeed, after we began our investigation, Baum, Moody, and Wilczek (1987) suggested to us that the grandmother cell solution ought to emerge.

The weights $W(i,r)$ are such that when the $r^{th}$ grandmother cell fires, it sets the output cells to have the appropriate values. A particularly nice feature of the grandmother cell solution is that if the input does not correspond

MEMORIES

| -1 | 1  | 1  | 1  | 1  | 1  | -1 | 1  | -1 | -1 |
|----|----|----|----|----|----|----|----|----|----|
| -1 | -1 | -1 | -1 | 1  | 1  | -1 | 1  | -1 | -1 |
| 1  | -1 | -1 | 1  | -1 | -1 | -1 | 1  | -1 | 1  |
| -1 | -1 | -1 | -1 | 1  | -1 | 1  | 1  | -1 | -1 |
| 1  | 1  | 1  | -1 | -1 | -1 | 1  | -1 | -1 | -1 |
| 1  | -1 | -1 | 1  | 1  | 1  | 1  | -1 | 1  | -1 |

W(i,r)

| 1.8  | -1.8 | -1.8 | 1.9  | 1.4  | 1.4  | 1.3  | -1.3 | 1.6  | -1.3 |
|------|------|------|------|------|------|------|------|------|------|
| 1.8  | 1.5  | 1.4  | -1.4 | -1.4 | -1.4 | 1.9  | -1.8 | -1.4 | -1.7 |
| -1.7 | -1.7 | -1.7 | -1.7 | 2.0  | 1.7  | -1.8 | 1.4  | -1.4 | -1.8 |
| -2.0 | 1.6  | 1.6  | 1.5  | 2.0  | 1.6  | -1.7 | 2.0  | -1.8 | -1.9 |
| -1.5 | -2.0 | -2.0 | -2.0 | 1.5  | -0.7 | 0.7  | 1.5  | -2.0 | -1.6 |
| 1.3  | -1.6 | -1.6 | 1.4  | -1.4 | -1.3 | -1.4 | 1.6  | -2.0 | 1.4  |

T(r,j)

| 1.9  | -2.0 | -2.0 | 1.7  | 0.8  | 1.9  | 1.7  | -1.9 | 1.9  | -0.6 |
|------|------|------|------|------|------|------|------|------|------|
| 1.9  | 1.9  | 1.9  | 0.4  | -1.8 | -2.0 | 2.0  | -1.9 | -0.2 | -1.4 |
| -1.3 | -1.3 | -1.7 | -2.0 | 0.5  | 1.0  | -2.0 | 1.8  | -1.4 | -1.5 |
| -1.5 | 1.3  | 1.8  | 1.3  | 1.0  | 1.9  | -2.0 | 1.8  | -1.4 | -1.5 |
| -1.4 | -1.6 | -1.8 | -1.9 | 1.7  | -1.2 | 0.7  | 1.6  | -1.2 | -1.7 |
| 1.3  | -2.0 | -2.0 | 0.6  | -1.9 | -1.6 | -2.0 | 1.9  | -1.4 | 1.6  |

**Figure 3b.**

to one of the patterns already memorized, either more than one grandmother cell or no grandmother cell will fire, thus signaling that the input is novel (Baum, Moody, & Wilczek, 1987).

The grandmother cell solution becomes increasingly likely to emerge as the input pattern contains more errors. If the error rate is only one of the ten bits the solution often appears almost random, while for three of ten bits wrong the grandmother solution is always unmistakeably present, regardless of random number seed, and so forth.[4] On the other hand, if the error rate is too large, the net has difficulty learning.

---

[4] In simulation, 3- or 4-bit errors also seemed sufficient to induce grandmother cell solution on longer memories of 20 bits.

In the results presented above, the threshold for firing the neurons of the output layer was fixed at zero, and the hidden layer thresholds were allowed to vary with the restriction that they all be equal. The grandmother cell solution also emerges if the hidden layer thresholds are allowed to vary independently. This may be a useful freedom in the case when highly correlated memories are to be stored. We will discuss this further below. The grandmother cell solution also emerges if the output neurons are allowed nonzero thresholds, that is if Equation (4) is modified to $O(i) = sgn\ (\Sigma_r W_{ir}(H_r + 1) - \phi_i)$. However, we found $\phi_i$ tended to be small and that the net did not require this freedom to minimize the energy $E$ effectively.

It is important to compare the capacity of this feed-forward memory with the capacity of the fully connected Hopfield network. Here, $(2N + N_{mem})$ cells are needed to store $N_{mem}$ $N$-bit memories. In the fully connected network, $N$ cells can store only about $\sim 0.1N$ $N$-bit memories (Amit, Gutfreund, & Sompolinsky, 1985). It does not appear the feed-forward net is significantly less error tolerant (see Appendix).

There are a variety of interesting issues one can further explore. The first is how grandmother cells behave when a new memory is introduced. To address this question we evolve a net to store $N_{mem}$ memories with errors until the grandmother solution is reached. Then we add an $(N_{mem} + 1)st$ memory and hidden layer cell. What happens to the weights laid down initially for the first $N_{mem}$ memories? Do they remain unchanged, or are they modified by the additional memory? In Figure 4 we show the weight patterns obtained by adding a seventh memory to the six memories of Figure 3. We see that the weights for the first six hidden cells are essentially unchanged. The main difference is a slight strengthening of the weights. Indeed, it is reasonable that the weights should be relatively unaffected by the additional memory. Examining Equation (6) we see that if the new memory is largely uncorrelated with previous ones and the grandmother solution has been found, the first $N_{mem}$ hidden cells will not fire when the added memory is presented. If $H(r) = -1$ then $\delta W(i,r) = \delta T(r,j) = 0$. When the original memories are presented, however, they now have to cope with some additional "noise" from the weights of the added memory, and they respond by strengthening their weights. We made 500 additional presentations (i.e., roughly 70 per memory) of the seven memories in generating Figure 4 from Figure 3, which itself involved 1000 presentations. This indicates why the weights for the seventh memory are weaker than those for the first six.

It is clear that if $N_{hid} < N_{mem}$ the grandmother cell solution cannot emerge. However, we can ask what happens if $N_{hid} > N_{mem}$. Biologically, for the grandmother cell solution to be at all relevant, we imagine that there is a storehouse of cells from which grandmother cells are called up one at a time when needed. This would reflect itself in our net by having the weights of $N_{mem}$ of the $N_{hid}$ hidden cells large, while the remaining hidden cells remain essentially inactive until a new memory is presented. We found that such

**Figure 4.** The modification of the weights of Figure 3b when an additional pattern and hidden cell are evolved. The weights associated with the original six cells remain roughly unchanged, while the added seventh cell develops its weights to recall the new memory.

MEMORIES

| -1 | 1 | 1 | 1 | 1 | 1 | -1 | 1 | -1 | -1 |
|---|---|---|---|---|---|---|---|---|---|
| -1 | -1 | -1 | -1 | 1 | 1 | -1 | 1 | -1 | -1 |
| 1 | -1 | -1 | 1 | -1 | -1 | -1 | 1 | -1 | 1 |
| -1 | -1 | -1 | -1 | 1 | -1 | 1 | 1 | -1 | -1 |
| 1 | 1 | 1 | -1 | -1 | -1 | 1 | -1 | -1 | -1 |
| 1 | -1 | -1 | 1 | 1 | 1 | 1 | -1 | 1 | -1 |
| -1 | -1 | -1 | -1 | -1 | -1 | 1 | -1 | -1 | 1 |

W(i,r)

| 1.9 | -1.8 | -1.8 | 2.0 | 1.7 | 2.0 | 1.3 | -1.4 | 1.9 | -1.7 |
|---|---|---|---|---|---|---|---|---|---|
| 2.0 | 2.0 | 2.0 | -1.4 | -1.4 | -1.4 | 1.9 | -1.9 | -1.4 | -1.9 |
| -1.6 | -0.9 | -0.9 | -0.8 | 2.0 | 1.8 | -1.6 | 2.0 | -1.4 | -1.9 |
| -2.0 | 1.8 | 1.8 | 1.7 | 2.0 | 1.8 | -1.8 | 2.0 | -1.8 | -2.0 |
| -1.5 | -2.0 | -2.0 | -2.0 | 2.0 | -1.8 | 1.5 | 1.9 | -2.0 | -1.9 |
| 1.5 | -1.7 | -1.7 | 1.5 | -1.6 | -1.4 | -1.7 | 1.7 | -2.0 | 1.6 |
| -0.8 | -0.9 | -0.9 | -0.8 | -0.6 | -0.7 | 0.7 | -0.5 | -1.0 | 0.6 |

T(r,j)

| 1.9 | -1.7 | -1.9 | 1.8 | 1.8 | 1.9 | 1.7 | -1.2 | 1.8 | -1.6 |
|---|---|---|---|---|---|---|---|---|---|
| 1.8 | 1.9 | 1.8 | -1.5 | -1.5 | -1.8 | 1.9 | -1.9 | -1.3 | -1.5 |
| -1.7 | -1.7 | -1.5 | -1.5 | 1.9 | 1.9 | -0.6 | 2.0 | -1.4 | -1.9 |
| -1.1 | 1.6 | 1.8 | 1.8 | 2.0 | 1.6 | -1.3 | 2.0 | -2.0 | -1.9 |
| -1.2 | -1.2 | -1.4 | -0.9 | 1.3 | -1.9 | 1.9 | 1.9 | -0.3 | -1.8 |
| 1.7 | -1.5 | -1.1 | 1.9 | -1.1 | -0.9 | -1.8 | 2.0 | -1.1 | 0.9 |
| -1.6 | -1.0 | -0.2 | -1.7 | -0.7 | -1.5 | 1.9 | -2.0 | -1.2 | 1.3 |

localization of the memories was not observed, unless the net was given the ability to "forget", that is, we allow the weights to decay according to the following algorithm. After each updating, we reset the weights to

$$\omega_i \rightarrow \omega_i \left(1 - \frac{1}{A^2 + B^2 \omega_i^2}\right). \tag{8}$$

Thus, weights significantly smaller in magnitude than $A/B$ would decay away while the larger weights persist (In our simulations, we take $A = 5$ and $B = 15$). The introduction of decay naturally tends to allow the cells which are idle to remain idle. In our simulations, we see precisely this effect. With $N_{hid} > N_{mem}$, only $N_{mem}$ of the hidden cells are used to store the memories, while the weights attached to the remaining $N_{hid} - N_{mem}$ stay small. Without the decay mechanism, this was not the case. The net distributed the memories over all available connections.[5] The necessity of introducing "forgetfulness" is perhaps intriguing and possibly suggestive for theories of the brain.

We have seen that the grandmother cell solution is obtained when $N_{hid} = N_{mem}$ if it is induced by corrupted input patterns. We now ask what happens if one of the cells is eliminated. The net then completely forgets the memory attached to that cell, although it still correctly reproduces the others. If we continue to present the lost memory to the truncated net, the weights will gradually re-evolve to store the lost memory again, but the pattern of the weights will no longer be the grandmother cell one.

In some sense, grandmother type memory stands at the opposite extreme from the fully distributed memory discussed by Hopfield and earlier workers. A fully distributed memory is clearly robust in the sense that if some cells are ripped out the memory will continue to function to a large extent, while if a hidden cell in a grandmother type memory is ripped out, the memory which the cell is responsible for is clearly lost. A simple and obvious way of making the grandmother type memory more robust fairly leaps to mind. Suppose that, whenever we speak of *a* hidden cell, that hidden cell actually corresponds to $\gamma$ hidden cells, where $\gamma$ is a number say of order 3. The price for this added robustness is of course a decrease in the memory capacity measured per cell. Hopfield has emphasized to us a second drawback of the grandmother type memory for practical application, namely, that a hidden cell has to "fan out" to a large number of output cells.

Finally, we ask the capability of the net to store correlated memories in the presence of errors. Rather than generating randomly the $N_{mem}$ patterns

---

[5] We were motivated to choose the particular form Equation (8) for the decay of the weights by a desire to suppress small weights while leaving larger ones intact. We have also examined the other forms

$$\omega_i \rightarrow \omega_i(1 - \epsilon) \tag{a}$$

$$\omega_i \rightarrow \omega_i(1 - \epsilon \omega_i^2) \tag{b}$$

in which the decay preferentially reduces larger weights. We found in simulation that both served to localize memories when $N_{hid} > N_{mem}$. Both, however, and especially (b), did this slightly less effectively than (8), i.e., there were larger residual weights in the "unused" hidden cells, although these were still clearly less large than the really active cells. This was as expected since these decay schemes tend to democratize the connection strengths. On the other hand, both forms also appeared to allow the addition of new patterns more rapidly.

**Figure 5.** The weights evolved for $N_{hid} > N_{mem}$ when there is no decay are shown in Figure 5a. The memories are distributed over the entire net. In the presence of decay (Figure 5b) the memories are localized. Only $N_{mem}$ of the hidden cells are active, and they are in the grandmother cell configuration. In Figure 5c we add a new pattern and observe a new hidden cell become utilized.

MEMORIES

| -1 | 1 | -1 | -1 | -1 | 1 | -1 | -1 | 1 | 1 |
|----|----|----|----|----|----|----|----|----|----|
| -1 | -1 | 1 | -1 | -1 | 1 | -1 | 1 | 1 | -1 |
| 1 | 1 | 1 | 1 | -1 | -1 | 1 | -1 | 1 | -1 |
| 1 | 1 | 1 | 1 | -1 | -1 | 1 | -1 | -1 | 1 |

W(i,r)

| -0.5 | 0.6 | -0.6 | -0.5 | -0.6 | 0.5 | -0.3 | -0.3 | 2.0 | -2.0 |
|------|-----|------|------|------|-----|------|------|-----|------|
| -0.9 | 0.3 | -0.8 | -0.5 | -0.5 | 0.5 | -0.6 | -1.0 | -1.3 | 1.8 |
| 0.0 | -1.2 | 1.9 | 0.3 | -0.4 | 0.0 | 0.0 | 1.3 | 1.1 | -1.4 |
| -1.0 | -1.4 | 0.4 | -1.3 | -0.7 | 1.0 | -1.2 | 1.3 | 1.5 | 0.2 |
| 1.3 | 1.5 | 2.0 | 1.2 | -0.4 | -1.3 | 1.2 | -1.5 | -0.9 | 0.8 |
| 0.8 | 1.9 | -0.4 | 0.9 | -0.8 | -0.6 | 0.7 | -1.9 | -0.6 | 1.5 |

T(r,j)

| 0.5 | 2.0 | -0.9 | 1.3 | -1.8 | -0.2 | 0.2 | -1.9 | -0.2 | 2.0 |
|-----|-----|------|-----|------|------|-----|------|------|-----|
| 1.1 | 1.8 | 1.2 | 1.7 | -1.3 | -2.0 | 2.0 | -0.9 | -1.7 | -0.5 |
| -1.4 | -0.6 | -0.6 | -1.6 | -0.8 | 1.2 | -1.4 | 0.6 | 1.7 | 0.0 |
| 0.5 | -1.3 | 0.9 | 0.1 | -1.2 | -0.7 | 0.7 | 1.3 | 2.0 | -2.0 |
| 0.1 | 1.4 | -1.1 | 0.0 | -1.0 | 0.2 | -0.2 | -0.9 | -1.9 | 2.0 |
| -0.4 | 1.0 | -0.6 | -1.2 | -1.0 | -0.2 | -1.0 | -1.2 | 2.0 | -2.0 |

**Figure 5a.**

to be stored, we instead fix the first $N_{cor}$ of the bits to be identical, and allow only the remaining $N - N_{cor}$ to differ. There are three questions. How large can $N_{cor}$ be such that the net still correctly discriminates between memories? Provided the net can discriminate, how does it adapt itself to the increased degree of correlation? What happens when errors are introduced?

Without the presence of errors, we found that the net can distinguish between memories even with $N_{cor}$ as high as $N-1$, although as might be ex-

**MEMORIES**

| | | | | | | | | | |
|---|---|---|---|---|---|---|---|---|---|
| -1 | 1 | 1 | -1 | -1 | -1 | -1 | -1 | -1 | 1 |
| 1 | 1 | 1 | -1 | 1 | -1 | -1 | 1 | 1 | 1 |
| 1 | 1 | 1 | -1 | -1 | 1 | -1 | -1 | 1 | -1 |
| 1 | 1 | 1 | 1 | 1 | 1 | 1 | 1 | -1 | 1 |

$W(i,r)$

| | | | | | | | | | |
|---|---|---|---|---|---|---|---|---|---|
| -0.7 | 0.7 | 0.7 | -0.7 | -0.7 | -0.7 | -0.7 | -0.7 | -0.7 | 0.7 |
| 0.0 | 0.0 | 0.0 | 0.0 | 0.0 | 0.0 | 0.0 | 0.0 | 0.0 | 0.0 |
| 0.7 | 0.7 | 0.7 | 0.7 | 0.7 | 0.7 | 0.7 | 0.7 | -0.7 | 0.7 |
| 0.0 | 0.0 | 0.0 | 0.0 | 0.0 | 0.0 | 0.0 | 0.0 | 0.0 | 0.0 |
| 0.7 | 0.7 | 0.7 | -0.7 | 0.7 | -0.7 | -0.7 | 0.7 | 0.7 | 0.7 |
| 0.7 | 0.7 | 0.7 | -0.7 | -0.7 | 0.7 | -0.7 | -0.7 | 0.7 | -0.7 |

$T(r,j)$

| | | | | | | | | | |
|---|---|---|---|---|---|---|---|---|---|
| -1.6 | 0.9 | 1.6 | -1.4 | -1.7 | -1.5 | -1.4 | -0.4 | -1.2 | 0.6 |
| 0.0 | 0.1 | 0.0 | 0.0 | -0.1 | 0.0 | 0.0 | 0.0 | -0.1 | 0.0 |
| 2.0 | 0.3 | 1.3 | 1.9 | 2.0 | 1.7 | 2.0 | 2.0 | -1.4 | 1.9 |
| 0.0 | 0.0 | 0.0 | 0.0 | -0.3 | 0.0 | 0.0 | 0.0 | 0.2 | 0.0 |
| 1.0 | 1.7 | 1.4 | -1.5 | 0.8 | -0.4 | -1.2 | 1.6 | 1.8 | 1.0 |
| 1.7 | 0.7 | 1.6 | -1.7 | -1.5 | 0.6 | -1.9 | -1.9 | 1.9 | -1.8 |

**Figure 5b.**

pected the probability of falling into a metastable state or a local minimum[6] and thus being unable to discriminate increases with $N_{cor}$.

The net adapts to $N_{cor}$ large by increasing its threshold $\theta$. Naively one expects that

$$T_{av}(N_{cor} + \sqrt{N - N_{cor}}) < \theta < T_{av}N \tag{9}$$

where $T_{av}$ is the average bond weight entering the hidden cells. Indeed, the

---

[6] These metastable states can arise even in small 2-bit problems like XOR (Rumelhart, Hinton & Williams, 1986; Rumelhart, McClelland, & PDP Research Group, 1986). Interestingly, we find that the improved descent scheme Equation (7), tends to alleviate somewhat the metastability problem, possibly by moving the weights around somewhat more effectively. We have not investigated this question fully, but believe this may be useful in cases where metastability difficulties are pronounced.

MEMORIES

| -1 | 1 | 1 | -1 | -1 | -1 | -1 | -1 | -1 | 1 |
|---|---|---|---|---|---|---|---|---|---|
| 1 | 1 | 1 | -1 | 1 | -1 | -1 | 1 | 1 | 1 |
| 1 | 1 | 1 | -1 | -1 | 1 | -1 | -1 | 1 | -1 |
| 1 | 1 | 1 | 1 | 1 | 1 | 1 | 1 | -1 | 1 |
| 1 | 1 | -1 | 1 | -1 | 1 | -1 | 1 | -1 | -1 |

$W(i,r)$

| -0.9 | 0.9 | 0.9 | -0.9 | -0.9 | -0.9 | -0.9 | -0.9 | -0.9 | 0.9 |
|---|---|---|---|---|---|---|---|---|---|
| 0.0 | 0.0 | 0.0 | 0.0 | 0.0 | 0.0 | 0.0 | 0.0 | 0.0 | 0.0 |
| 0.9 | 0.9 | 0.9 | 0.9 | 0.9 | 0.9 | 0.9 | 0.9 | -0.9 | 0.9 |
| 0.6 | 0.6 | -0.6 | 0.6 | -0.6 | 0.6 | -0.6 | 0.6 | -0.6 | -0.6 |
| 0.9 | 0.9 | 0.9 | -0.9 | 0.9 | -0.9 | -0.9 | 0.9 | 0.9 | 0.9 |
| 0.9 | 0.9 | 0.9 | -0.9 | -0.9 | 0.9 | -0.9 | -0.9 | 0.9 | -0.9 |

$T(r,j)$

| -1.7 | 1.2 | 1.4 | -1.7 | -1.9 | -2.0 | -1.3 | -1.2 | -1.2 | 0.9 |
|---|---|---|---|---|---|---|---|---|---|
| 0.1 | 0.0 | 0.3 | 0.0 | 0.0 | 0.0 | -0.1 | 0.0 | 0.2 | 0.0 |
| 2.0 | 0.3 | 1.3 | 1.9 | 2.0 | 1.7 | 2.0 | 2.0 | -1.4 | 1.9 |
| 0.9 | 0.9 | -0.8 | 1.1 | -1.3 | 0.6 | -0.8 | 1.4 | -1.2 | -0.3 |
| 1.0 | 1.7 | 1.4 | -1.5 | 0.8 | -0.4 | -1.2 | 1.6 | 1.8 | 1.0 |
| 1.7 | 0.7 | 1.6 | -1.7 | -1.5 | 0.6 | -1.9 | -1.9 | 1.9 | -1.8 |

**Figure 5c.**

thresholds evolved in the data of Figure 3b, where $N_{cor}=0$, are consistent with this. We find that $\theta$ increases with $N_{cor}$ in roughly this fashion, although we note that if we allow the evolution of the weights to continue substantially beyond the point at which satisfactory convergence has been attained, $\theta$ will occasionally continue to evolve to high levels even for $N_{cor}$ small.

The inclusion of error resistance produces in the net a behavior more or less as expected. The net can perform poorly for highly correlated memories if errors are present. This is to be expected since if the bits distinguishing two memories are absent, there *really* is no way to discriminate the memories. From another perspective we note that the net responds to the demand of error tolerance by decreasing its thresholds, but if the threshold is lowered

**# Presentations per Pattern**

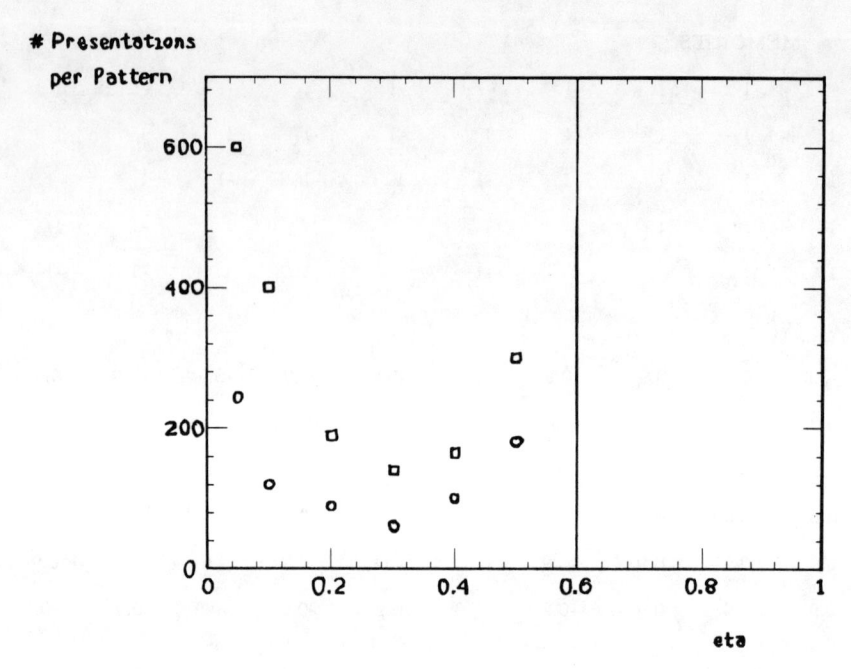

**Figure 6.** The number of presentations required per memory as a function of $\eta$. The data shown is for six 10-bit memories, but the results are typical for other choices. The simple gradient-descent (Equation (2)) is shown by ($\square$). The modified descent scheme (Equation (7)) is shown by (O). The latter algorithm is, however, more time consuming to implement. The vertical line at $\eta=0.6$ indicates where the algorithm became unstable.

excessively one will tend to activate more than a single hidden cell, especially if the memories are highly correlated.

Finally, we note that if the thresholds are allowed to evolve independently, the hidden cells specializing in memories that happen to possess highly correlated partners tend to develop larger thresholds. In this sense the net might do "better" than a naive uniform threshold grandmother cell solution by doing some optimization of independent thresholds.

While any connections between this work and neurobiology are remote at best, we are nevertheless tempted to say that the results may be vaguely suggestive. If we see a pattern repeatedly, a cell may become specialized to that pattern, but only if the pattern is presented with noise. We are aware that the notion of grandmother cells is the subject of considerable controversy among neurobiologists. Nevertheless, it appears to us not unreasonable to imagine that there are cells specialized in recognizing common symbols, such as numbers and letters. We have shown that the grandmother cell arrangement is "optimal" in the context of back-propagation. Perhaps more interestingly, we have shown that grandmother cells can be called up in an orderly fashion when new patterns are presented, provided a decay mechanism (= forgetfulness?) operates.

**Figure 7.** The number of presentations required per memory as a function of number of bits. Results are shown for nets learning 10 memories at $\eta=0.3$ with $N_{err}=\frac{1}{2}N_{in}$. The number of presentations is roughly independent of memory length.

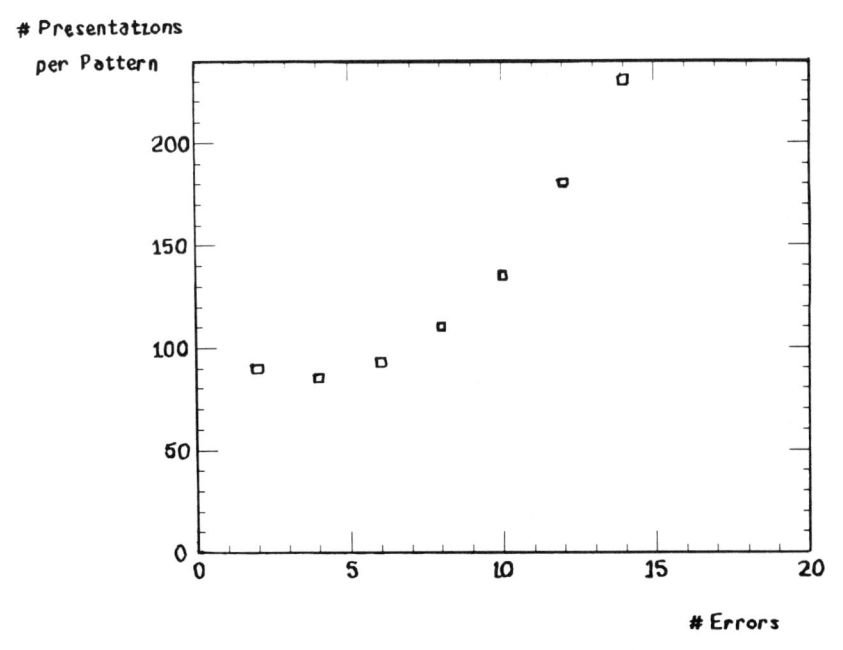

**Figure 8.** The number of presentations required per memory as a function of $N_{err}$ for learning twenty 20-bit memories. The memories are uncorrelated. When $N_{err} \geq 16$ the algorithm had convergence problems.

## APPENDIX

Here we present a few comments on the rate of convergence of the algorithm. Even with the simplest gradient-descent, the convergence rate is strongly affected by the choice of $\eta$. As has been noted previously, (Rumelhart, Hinton, & Williams, 1986) a reasonably optimal choice appears to be $\eta = 0.25$. We find that roughly 100 presentations per pattern suffice for convergence, for a typical run with six 10-bit patterns with an error rate of 3 of the 10 bits. Our criterion for satisfactory convergence was that the energy $E$ give by Equation (1) should be less than .01, when normalized by the number of bits and the number of patterns. Roughly speaking this means that if the target is $T$ for a given bit of a given pattern, the net's output $O$ obeys $|T - O| < 0.1$ which represents a rather stringent requirement.

How does this rate scale with the number of patterns and the pattern size? We found that for six 40-bit patterns, again roughly 100 presentations were required, that is, the convergence rate was not significantly affected by the number of bits. Meanwhile, if we ask the net to recall 24 ten-bit patterns, rather than six, we find that more presentations are required since for such a large number of patterns we begin to see some correlated ones. However, if we scale $N$ and $N_{mem}$ simultaneously this doesn't occur. Figures 6–8 summarize our results for convergence rates as a function of $\eta$, memory length, and fault tolerance, for a variety of typical nets we have simulated.

The above comments refer to the naive gradient-descent, Equation (2). We also found that applying the slightly more complicated Equation (7) could accelerate convergence by up to a factor of four. However, it should be noted that the additional computations nearly triple the CPU time per iteration required. Since the tasks reported here were not particularly computer intensive, the results have generally been obtained with the simpler gradient-descent form. One should however keep the possibility of applying these other schemes in mind, particularly if one is interested in minimizing the number of presentations required, for example if one is interested in comparing the number of times a net must "see" something to that required by a real biological system.

## REFERENCES

Amit, D.J., Gutfreund, H.S., & Sompolinsky, H. (1985). Spin glass models of neural networks. *Physical Review, A32*, 1007–1014.

Barlow, H.B. (1972). Single units and sensation: A neuron doctrine for perceptual psychology. *Perception*, 1, 371–394.

Baum, E., Moody, J., & Wilczek, F. (1987). Paper in preparation about grandmother cell at the University of California at Santa Barbara.

Block, H.D. (1962). The Perceptron: A model for brain functioning: I. *Reviews of Modern Physics*, 34, 123–135.

Block, H.D., Knight, B.W., & Rosenblatt, F. (1962). Analysis of a four-layer series-coupled Perceptron II. *Reviews of Modern Physics*, 34, 135–142.

Hebb, D.O. (1949). *The organization of behavior*. New York: Wiley.
Hopfield, J.J. (1982). Neural networks physical systems with emergent collective computational abilities. *Proceedings of the National Academy of Sciences, 79*, 2554-2558.
Hopfield, J.J. (1984). Neurons with graded response have collective computational properties like those of two-state neurons. *Proceedings of the National Academy of Sciences, 81*, 3088-3092.
Hopfield, J.J. (1986). Physics, biological computation and complementarity. In J. de Boer, E. Dal, & O. Ulsbeck (Eds.), *The lesson of quantum theory* (pp. 295-314). New York: Elsevier.
Kohonen, T. (1984). *Self organization and associative memory*. New York: Springer-Verlag.
Le Cun, Y. (1985). Unpublished reports as cited by Rumelhart, D.E., Hinton, G.E., & Williams, R.J. (1986).
Little, W., & Shaw, G. (1978). Analytic study of the memory storage capacity of a neural network. *Mathematical Bioscience, 39*, 281-290.
Mezard, M., Nadal, J.P., & Toulouse, G. (in press). Article to appear in *Europhysics letters*.
Parker, D.P. (1985). Unpublished reports as cited by Rumelhart, D.E., Hinton, G.E., & Williams, R.J. (1986).
Rosenblatt, F. (1962). *Principles of neurodynamics: Perceptrons and the theory of brain mechanisms*. Washington, DC: Spartan Books.
Rumelhart, D.E., Hinton, G.E., & Williams, R.J. (1986). Learning internal representation by error propagations. In D.E. Rumelhart, J.L. McClelland, & the PDP Research Group (Eds.), *Parallel distributed processings: Explorations in the microstructure of cognition*. Cambridge, MA: Bradford Books.
Sutton, R.S. (1986). Two problems with back-propagation and other steepest-descent learning procedures for networks. GTE Laboratory Report.
Von der Malsburg, C. (1986). Principles of neurodynamics: Perceptrons and the theory of brain mechanisms. In G. Palm & A. Aerten (Eds.), *Brain theory*. (pp. 245-261) New York: Springer-Verlag.

CHAPTER 12

# An Implementation of Network Learning on the Connection Machine*

CHARLES R. ROSENBERG

Princeton University
Princeton, New Jersey 08542

GUY BLELLOCH

M.I.T. Artificial Intelligence Laboratory

Connectionist networks are powerful models of cognition which promise to transform many of our traditional notions of knowledge representation and learning. Inspired by the parallel architecture of the brain, however, they are not well suited for implementation on the serial architecture of most conventional computers. In this paper, we discuss the first implementation of a connectionist learning algorithm, error back-propagation, on a fine-grained *parallel* computer, the Connection Machine. As an example of how the system can be used, we present a parallel implementation of NETtalk, a connectionist network that learns to the mapping from English text to the pronunciation of that text. Currently, networks containing more than 65,000 links can be simulated on the Connection Machine at speeds nearly twice that of the Cray-2. Networks of up to 16 million links can be simulated using virtual processors with some sacrifice in speed. The major impediment to further speed-up was found to be communications between processors, and not processor speed per se. We believe that this advantage for parallel computers will become even clearer as developments in parallel computing continue.

## 1 INTRODUCTION

Massively parallel, connectionist networks have undergone a rediscovery in artificial intelligence and cognitive science, and may constitute a major paradigm shift for these fields (Schneider, 1987). Connectionist networks have already led to important theoretical insights and broad application in many areas, including knowledge representation in semantic networks (Hinton, 1986; Rumelhart, 1986), bandwidth compression by dimensionality reduction (Saund, 1986; Zipser, 1986), speech recognition (Ellman & Zipser, Watrous, Shastri, & Waibel, 1986), and backgammon (Tesauro & Sejnowski, in prep.). The resulting learned encodings can often be analyzed to discover new facts about the modelled domain (Rosenberg, 1987).

---

* This research was supported by Thinking Machines Corporation, Cambridge, MA. The authors wish to thank Terry Sejnowski, David Waltz, and Craig Stanfill for their assistance.

Connectionist network models are dynamic systems composed of a large number of simple processing units arranged into a network structure, where the state of any unit in the network depends on the states of the units to which this unit connects. The values of the links or connections determine how the units affect each other: A positive weight indicates that if one unit turns on, the second unit will also be influenced to turn on. A negative weight between two units will tend to influence the units to be in different states. The overall behavior of the system is modified by adjusting the values of these connections, or weights, through the repeated application of a learning rule. These rules allow these systems to learn by experience, or by example.

One of the most successful network models to date is NETtalk (Sejnowski & Rosenberg, 1987a, b), a network which learns the mapping from English text to the pronunciation of that text. The units composing NETtalk were arranged into three completely-connected layers. The first, or input, layer of units receives information regarding the input text. The pronunciation of that text is computed by the network, and is encoded as the pattern of activation of the units in the third, or output, layer of the network.

Unfortunately, connectionist networks are computationally intensive; days or weeks are often required to train networks using the fastest serial computers. Practical exploration of networks consisting of more than a million links or connections is currently not possible using standard architectures. Further progress in connectionist modelling depends on making the exploration of large networks computationally feasible.

One option that has begun to be explored is the development of special-purpose hardware (Alspector & Allen, 1987; Graf et al., 1986; Silviotti, Emerling, & Mead, 1985), where links between units are realized as physical connections using VLSI technology. Initial estimates of these so-called neuromorphic systems indicate that tremendous speed-ups may be achievable, perhaps up to five to six orders of magnitude over a VAX780 implementation. However, one problem with special purpose hardware is precisely that it is special purpose: Gone is the programmability that allows one to explore different patterns of connectivity and learning algorithms.

A more flexible alternative is seen in the development of general purpose, fine-grained, parallel computers. The Connection Machine (CM) is a massively parallel computer consisting of up to 65,536 ($2^{16}$) one-bit processors arranged in a hypercube architecture. In this paper we discuss the implementation of a connectionist network learning algorithm, back-propagation (Rumelhart, Hinton, & Williams, 1986), on the Connection Machine. We then present a Connection Machine version of NETtalk, which we call CM-NETtalk, that uses this implementation. Currently, the Connection Machine offers a factor of 500 speedup over a previous implementation on a VAX780 and a factor of two speed-up over an implementation of a similar network on a Cray-2. Considering that parallel computing is only in its infancy, we expect that the advantages of parallel over serial computers for the implementation of connectionist networks will become even clearer as progress continues.

## 2 ERROR BACK-PROPAGATION

The back-propagation learning algorithm is an error-correcting learning procedure that generalizes the Wildrow-Hoff algorithm (Widrow & Hoff, 1960) to multi-layered networks. This learning algorithm is intended for networks whose units are arranged into layers, such that two units in the same layer cannot be directly connected. However, there may be an arbitrary number of layers in the network. Information flows forward from the first or input layer, through the hidden layer(s), to the last or output layer. If there are $J$ units in the input layer, the value of unit $i$ is determined by first computing the weighted sum of the values of the units connected to this unit

$$E_i = \sum_{j=1}^{J} w_{ij} s_j \tag{1}$$

and then applying the logistic function to the result:

$$s_i = P(E_i) = \frac{1}{1+e^{-E_i}} \tag{2}$$

The logistic, or sigmoid, is a continuous function with asymptotically approaches zero for large negative values and one for large positive values. This forward propagation rule is recursively applied to successively determine the unit values for each layer.

The goal of the learning procedure is to iteratively minimize the average squared error between the values of the output units and the correct pattern, $s_i^*$, provided by a teacher. This is accomplished by first computing the error gradient for each unit on the output layer, which is proportional to the difference between the target value and the current output value:

$$\delta_i = (s_i^* - s_i)P'(E_i) \tag{3}$$

The error gradient is then recursively determined for layers from the output layer to the input by computing the weighted sum of the errors at the previous layer:

$$\delta_i = \sum_{j=1}^{J} \delta_j w_{ji} P'(E_i) \tag{4}$$

where $J$ is the number of units at the previous layer and $P'(E_i)$ is the first derivative of the function $P(E_i)$. The error gradients are then used to update the weights:

$$w_{ij}(t+1) = w_{ij}(t) + \epsilon \delta_i s_j \tag{5}$$

where $\epsilon$ is the learning rate (typically between 1.0 and 3.0).[1]

The forward and backward propagation steps are very similar from a computational standpoint. Forward propagation consists of four basic steps: distributing the activation values of the units to their respective fan-out weights, multiplying the activations by the weight values, summing these values from the weights into the next layer of units, and applying the logistic function to this value. The backward propagation of error consists of four similar steps: distributing the error values of the units to their respective fan-in weights, multiplying the error by the weight values, summing these values from the weights into the previous layer of units, and evaluating the derivative of the logistic function. In addition to forward and backward propagation, the inputs and outputs must be clamped to the appropriate values. In the next section, we will show how each of these steps is executed on the Connection Machine.

## 3 THE CONNECTION MACHINE

The Connection Machine is a highly parallel computer configurable with between 16,384 and 65,536 processors. Each processor has two single-bit arithmetic logic units (ALUs), and some local memory—currently 64K bits. In addition, every 32 processors shares a floating point unit. All the processors are controlled by a single instruction stream (SIMD) broadcast from a microcontroller. Figure 1 shows a block diagram of the Connection Machine. Processors can communicate using a few different techniques—the only two of concern in this paper are the *router* and the *scan* operations. The *router* operations allow any processor to write into the memory or read from the memory of any other processor. The *Scan* operations allow a quick[2] summation of many values from different processors into a single processor, or the copying of a single value to many processors (Blelloch, 1987).

In our implementation of back-propagation, we allocated one processor for each unit and two processors for each weight.[3] The processor for each unit is immediately followed by all the processors for it's outgoing, or fan-out, weights, and immediately preceded by all of it's incoming, or fan-in, weights (see Figure 2). The beginning and ending of these contiguous segments of processors are marked by flags. This layout enables us to use the fast *segmented scan* operations to distribute the unit activation values from

---

[1] Actually, the moving average of the weight gradient is often used rather than the gradient itself. See the original sources for details.
[2] Usually faster than a router cycle.
[3] Later in this paper we will show how the processors can be shared between units and weights, requiring only one processor per weight.

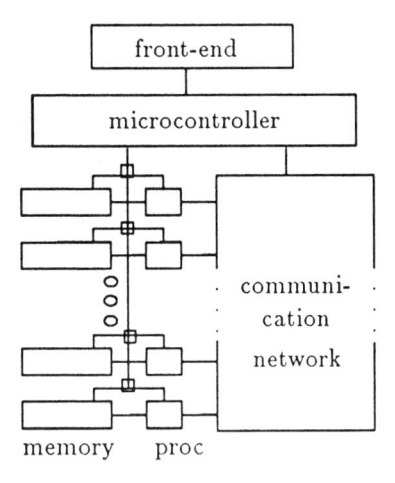

[Figure 1: diagram of The Connection Machine, showing front-end connected to microcontroller, which connects to memory and proc units alongside a communication network.]

**Figure 1.** The Connection Machine.

the units to their fan-out weights (and vice versa) and to sum values along contiguous segments of fan-in or fan-out weights. A scan operation called *segmented copy-scan* is used to quickly copy a value from the beginning or end of a segment of contiguous processors to all the other processors in the segment and an operation called *segmental plus-scan* is used to quickly sum all the values in a segment of processors and leave the result in either the first or last processor of the segment (Blelloch, 1987).

The forward propagation step proceeds as follows. First, the activations of the seven groups of input units are clamped according to the input text. These activations are then distributed to their fan-out weights using a *copy-scan* operation (step A in Figure 3), and multiplied by the values of these weights in parallel. The result is sent from the fan-out weights to the fan-in weights of the units in the next layer (step B) using the router. A *plus-scan* then sums these input values into the next layer of units (step C), and the logistic function $P$ (equation 2) is applied to the result at all the unit processors to determine the unit activations. This forward propagation step is performed once for each hidden layer of units in the network (if any), and then once again for the output layer.

The back-propagation step is similar to running forward-propagation in the reverse direction. First, the error gradient at each output unit, $\delta_i$, is computed by taking the difference of the unit values (computed by the forward-propagation step) and the correct target values (provided by the "teacher") and multiplying the result by $P'(E_i)$ (equation 3). Using the *segmented copy scan* operation (running now in the reverse direction), the deltas are copied from the output units into their respective fan-in weights. The deltas are then sent, using the router, to the fan-out weights of the units at the previous

Simple Network

(A)

Processors

0  1  2 | 3  4  5 | 6  7  8  9 | 10 11 12 13 | 14 15 16
○ ▥ ▥   ○ ▥ ▥   ▤ ▤ ○ ▥       ▤ ▤ ○ ▥       ▤ ▤ ○
a         b         c             d             e

Units

▥  fan-out weight

▤  fan-in weight

(B)

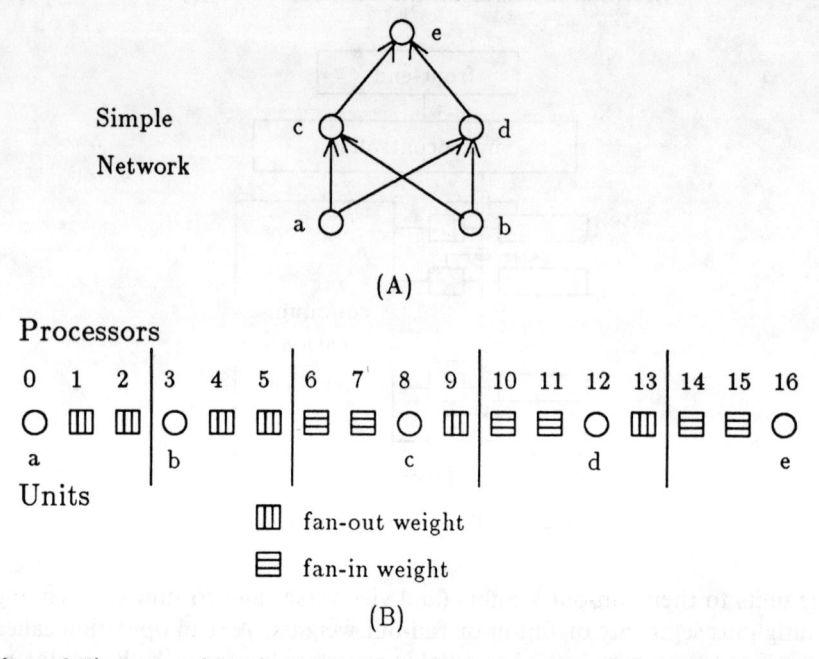

**Figure 2.** The Layout of Weights and Units of a Simple Network on the Connection Machine. (A) A simple two layer network. (B) The layout of the network on the processors of the Connection Machine.

Processors

0  1  2  3  4  5  6  7  8  9  10 11 12 13 14 15 16

○ ▥ ▥  ○ ▥ ▥  ▤ ▤ ○ ▥  ▤ ▤ ○ ▥  ▤ ▤ ○
a        b        c        d        e

A) copy-scan

○ ▥ ▥ ○ ▥ ▥ ▤ ▤ ○ ▥ ▤ ▤ ○ ▥ ▤ ▤ ○

B) send

○ ▥ ▥  ○ ▥ ▥  ▤ ▤ ○ ▥  ▤ ▤ ○ ▥  ▤ ▤ ○

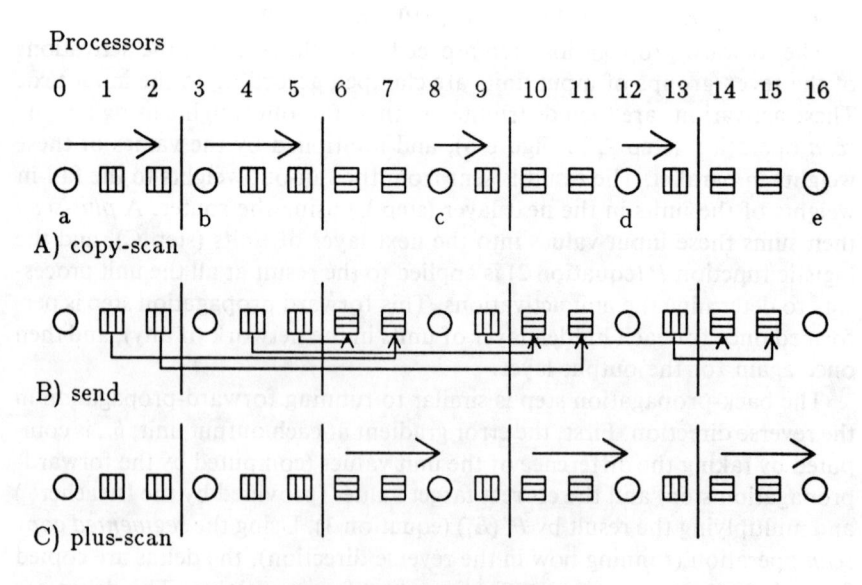

C) plus-scan

**Figure 3.** Forward Propagation.

layer and multiplied in parallel by the weight values. These values, now residing at the fan-out weights, are then summed using *segmented plus-scan* in the reverse direction into the units at the previous layer. The derivative of the logistic function, $P'(E_j)$, is then evaluated at the units to determine the deltas of units at the previous layer. As in forward propagation, the back-propagation step must be applied once for each layer of units ($-1$).

The algorithm described uses the processors inefficiently for two reasons. First, we use two processors for each weight when only one of them is busy at a time, and second, when we are running one layer, the processors for all the other layers are idle. To overcome the first inefficiency and use one processor per weight, we overlap the input and output weights. We also overlap the units with the weights. To overcome the second problem and keep the processors for each layer busy we pipeline the layers as follows. Given a set of $n$ input vectors $v_i$ ($0 \leq i < n$), and $m$ layers $l_j$, pipelining consists of propagating the $i^{th}$ input vector across the first layer, while propagating the previous input vector ($v_{i-1}$) across the second layer, $v_{i-2}$ across the third layer, ..., and $v_{i-m}$ across the last layer. We also interleave the back-propagation with the forward-propagation so that immediately after presenting $v_i$ to the input, we start back-propagation $v_{i-m}$ backward from the ouput. The depth of the whole pipe for $m$ layers is $2m$.

This implementation has some important advantages over other possible implementations.

1. The method we described with pipelining unwraps all the potential concurrency.[4]
2. The method works well with sparse connectivity. Methods based on dense matrix multiplies, such as some of the serial implementations, although faster for dense connectivity, are extremely inefficient with sparse connectivity.
3. The time taken by the algorithm is independent of the largest fan-in and the user does not have to worry about building fan-in trees.
4. All processors are kept active even if different units have different fan-ins. This would not be true if we, for example, serially looped over the fan-ins of each unit.
5. We could have used the *send-with-add* instruction of the Connection Machine instead of the *plus-scan* instruction, but since the time taken by the *send-with-add* instruction is very dependent on the message pattern, in some cases it would be much slower than the *plus-scan*.

Networks with more links than physical processors can be simulated in the Connection Machine using an abstraction called the *virtual processor*

---

[4] This is not strictly true since we could get another factor of 2 by running the forward and backward propagation concurrently.

(VP). A *virtual processor* is a slice of memory within a physical processor. A large number of VPs can therefore exist in the memory of each physical processor, and when an operation is sent to the Connection Machine at run time, the physical processors loop over each of the VPs executing the operation, thus simulating many more processors than actually exist. This looping is invisible to the user—the user must only specify how many VPs are needed. For operations on local memory (within a processor), looping over the VPs will cause a slow-down of execution time proportional to the number of VPs. For operations which require communications, this slow-down is generally non-linear (scans do better with VPs while routing does worse).

Similar layouts of static networks have been used to implement a rule-based system (Blelloch, 1986), a SPICE circuit simulator and a maximum-flow algorithm.

## 4  CM-NETtalk

NETtalk is a connectionist network that uses back-propagation to learn the pronunciations of English words. We have implemented NETtalk on the Connection Machine, and present the results of our implementation here.[5]

NETtalk is composed of three layers of units, an input layer, and output layer, and an intermediate or hidden layer. Each unit in each layer is connected to each unit in the layer immediately above and/or below it. The representations at the input and output layers are fixed to be representations of letters and phonemes, respectively. The representation at the hidden layer, on the other hand, is constructed automatically using back-propagation.

NETtalk uses a fixed-size input window of seven letters to allow the textual context of three letters on either side of the current letter to be taken account in the determination of that letter's pronunciation (see Figure 4). This window is progressively stepped through the text. At each step, the output of the network represents the guess or estimate for the pronunciation of the middle, or fourth, letter of the sequence of letters currently within the input window. This guess is compared to the correct pronunciation, and the values of the weights are iteratively adjusted using back-propagation to minimize this difference. Good pronunciations (over 95% of the phonemes correct) of a thousand-word corpus of high-frequency words are typically achieved after ten passes through the corpus.

We have experimented with a network consisting of 203 input units (7 letters with 29 units each), 60 hidden units and 26 output units. This required a total of 13826 links (processors)—12180 in the first layer, 1560 in the sec-

---

[5] Since we present here only those aspects of NETtalk relevant to the present discussion, interested readers should consult the original sources for details.

Teacher:
| t | r | @ | n | z | l | e | S | x | - | n |

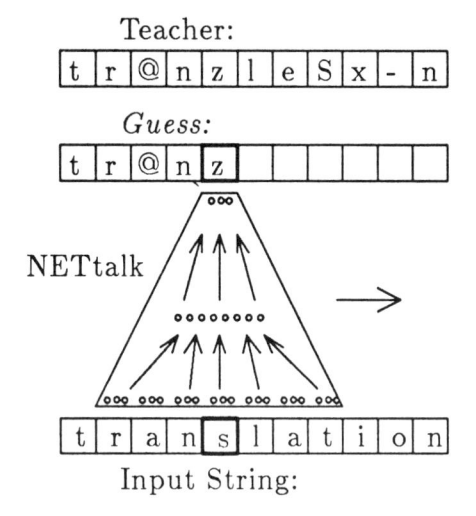

Guess:
| t | r | @ | n | z |   |   |   |   |   |   |

NETtalk

| t | r | a | n | s | l | a | t | i | o | n |

Input String:

**Figure 4.** The Seven Letter Window Used by NETtalk.

ond layer and 86 to the true units.[6] The learning rate was approximately the same as that achieved by a C implementation on various serial machines using floating-point computations.

In the current implementation, using a 16,384 processor machine, the time required for each letter during the learning stage was 5 milliseconds. This includes the forward propagation, the backward propagation, the time necessary to clamp the input and output, and the time required to calculate the error. The time is broken up by the type of operation as follows:

- Scanning 30%—This includes two segmented *plus-scans* and two segmented *copy-scans*.
- Routing 40%—This includes two routing cycles.
- Arithmetic 20%—This includes seven multiplies and several additions, subtractions and comparisons.
- Clamping 5%—This is the time needed to get the characters to the input units and the expected phonemes to the output units.
- Other 5%—Mostly for moving values around.

With improvements in the coding of the implementation and in microcode, we expect that this time could be improved by a factor of three or more.

Table 5 shows comparative running times of the error back-propagation algorithm for several machines. On an existing implementation of back-propagation on a VAX 780, the same network required 650 microseconds per

---

[6] True units are units that are always kept in the active state. Their function is to allow the thresholds of the other units in the network to be modified in a simple way.

|  | MLPS | Relative Times |
|---|---|---|
| MicroVax | .008 | 1 |
| Sun 3/75 | .01 | 1.3 |
| Vax 780 | .027 | 3.4 |
| Sun 160 with FPA | .034 | 4.2 |
| Dec 8600 | .06 | 7.5 |
| Convex | .80 | 98 |
| Cray-2 | 7 | 860 |
| 65,536 Processor CM | 13 | 1600 |

**Figure 5.** Comparison of Running Times for Various Machines. MLPS stands for Millions of Links Per Second. Some of these times are from (Personal communications, Kukich, 1986; McClelland & Lang, 1987; Sejnowski, 1987).

letter. This represents a 130 to 1 improvement in speed. On a 64K machine and larger networks, we could get a 500 to 1 improvement. This is about twice the speed of an implementation of the algorithm on a Cray-2, yet a Connection Machine costs a quarter of the price. Fanty (Fanty, 1986) has implemented a connectionist network using the BBN Butterfly, a coarse grained parallel computer with up to 128 processors, but because the type of networks he used were considerably different, we cannot compare the performances.

Using virtual processors, on the Connection Machine it is possible to simulate up to 16 million links in physical memory. With software currently being developed to use the Connection Machine's disks, it will be able to simulate many more than this.

## 5 CONCLUSIONS

We have discussed the first implementation of a connectionist learning network on a fine-grained parallel machine. Our experiments indicate that the Connection Machine can currently simulate networks of over 65,000 connections at speed over twice as fast as the most powerful serial machines such as the Cray-2. The method outlined here should generalize to any highly concurrent computer with a routing network and, with small modifications, can be used with other kinds of units and transfer functions, such as sigma-pi units (Rumelhart & McClelland, 1986). The basic techniques used here for back-propagation are also compatible with other learning algorithms, such as Boltzmann machine algorithm (Ackley, Hinton, & Sejnowski, 1985) and related architectures such as that proposed by Hopfield (Hopfield, 1982). Unlike neuromorphic, hardware-based systems, our method places no restrictions on the computations performed at the links or the units, nor on the topology of the network.

In our implementation we were able to keep all of the processors busy most of the time using a single instruction stream; multiple instruction streams do not seem to be necessary. Rather, communication was the bottleneck—at least on the current Connection Machine. Effort needs to be spent designing faster routing networks.

Arguably, one of the major reasons for the decline of connectionism following much active research in the 1950s and 1960s, by Rosenblatt and others, was a lack of computational power. Alternative, symbolic techniques were more successful, in part, because they better fit the computational resources available at the time. Today, this situation is beginning to change. It is not yet clear which of many technologies will be most useful for exploring the large-scale connectionist networks of the future. There are many possibilities: programmable fine-grained parallel computers such as the Connection Machine, programmable coarse-grained parallel computers, special-purpose hardware, and optical computers. The future will depend upon the uncertainties of the marketplace and as well as a host of other factors. Nevertheless, one thing is clear: The exploration of truly large-scale connectionist networks is becoming computationally feasible. The scale of parallelism exhibited by the human brain, consisting of more than $10^{10}$ neurons, suggests that this may be a promising direction in which to search for a science of cognition.

## REFERENCES

Ackley, D.H., Hinton, G.E., & Sejnowski, T.J. (1985). A learning algorithm for Boltzmann Machines. *Cognitive Science, 9,* 147-169.

Alspector, J., & Allen, R.B. (1987). A neuromorphic VLSI learning system. In Paul Losleben (Ed.), *Advanced research in VLSI: Proceedings of the 1987 Stanford Conference* (pp. 313-349). Cambridge, MA: MIT Press.

Blelloch, G.E. (1986, November). *AFL-1: A programming language for massively concurrent computers.* (Tech. Rep. No. 918). Cambridge, MA: Massachusetts Institute of Technology.

Blelloch, G.E. (1987, August). The scan model of parallel computation. *Proceedings of the International Conference on Parallel Processing.*

Ellman, J.L., & Zipser, D. (1987). *Learning the hidden structure of speech.* (Tech. Rep. ICS Report No. 8701). La Jolla, CA: University of California at San Diego, Institute for Cognitive Science.

Graf, H.P., Jackel, L.D., Howard, R.E., Straughn, B., Denker, J.S., Hubbard, W., Tennant, D.M., & Schwartz, D. (1986). VLSI implementation of a neural network memory with several hundreds of neurons. In *Proceedings of the Neural Networks for Computing Conference,* Snowbird, UT.

Fanty, M. (1986, January). *A connectionist simulator for the BBN Butterfly Multiprocessor.* (Tech. Rep. Butterfly Project Report 2). Rochester, NY: University of Rochester, Computer Science Dept.

Hinton, G.E. (1986). Learning distributed representations of concepts. In *Proceedings of the Cognitive Science Society,* (pp. 1-12). Hillsdale, NJ: Erlbaum.

Hopfield, J.J. (1982). Neural networks and physical systems with emergent collective computational abilities. *Proceedings of the National Academy of Sciences USA, 79,* 2554-2558.

Rosenberg, C.R. (1987, July). Revealing the structure of NETtalk's internal representations. In *Proceedings of the Cognitive Science Society.* Seattle, WA.

Rumelhart, D.E. (1986, September). Multiple agents, parallelism and learning. Presentation at the Symposium on Connectionism: Geneva, Switzerland.

Rumelhart, D.E., Hinton, G.E., & Williams, R.J. (1986). Learning internal representations by error propagation. In *Parallel distributed processing: Explorations in the microstructure of cognition. Vol. 1: Foundations* (pp. 318-362). Cambridge, MA: MIT Press.

Rumelhart, D.E. & McClelland, J.L. (1986). On learning the past tenses of English verbs. In *Parallel distributed processing: Explorations in the microstructure of cognition. Vol. 2: Psychological and biological models* (pp. 216-217). Cambridge, MA: MIT Press.

Saund, E. (1986). Abstraction and representation of continuous variables in connectionist networks. In *Proceedings of the Fifth National Conference on Artificial Intelligence* (pp. 638-644). Morgan Kaufmann.

Schneider, W. (1987). Connectionism: Is it a paradigm shift of psychology? *Behavior Research Methods, Instruments, & Computers, 19,* 73-83.

Sejnowski, T.J., & Rosenberg, C.R. (1987a). Connectionist models of learning. In M.S. Gazzaniga (Ed.), *Perspectives in memory research and training.* Cambridge, MA: MIT Press.

Sejnowski, T.J., & Rosenberg, C.R. (1987b). Parallel networks that learn to pronounce English text. *Complex Systems, 1,* 145-168.

Silviotti, M., Emerling, M., & Mead, C. (1985, March). A novel associative memory implemented using collective computation. In *Proceedings of the 1985 Chapel Hill Conference on Very Large Scale Integration,* Chapel Hill, NC (pp. 329).

Tesauro, G., & Sejnowski, T.J. (1987, in prep.). A parallel network that learns to play backgammon.

Watrous, R.L., Shastri, L., & Waibel, A.H. *Learned phonetic discrimination using connectionist networks.* (Tech. Rep. No. ). Philadelphia, PA: University of Pennsylvania Department of Electrical Engineering and Computer Science.

Widrow, G., & Hoff, M.E. (1960). Adaptive switching circuits. *Institute of Radio Engineers, Western Electronic Show and Convention, Convention Record, Part 4,* 96-104.

Zipser, D. (1986). *Programming networks to compute spatial functions.* (Tech. Rep. No. ) La Jolla, CA: University of California at San Diego, Institute for Cognitive Science.

CHAPTER 13

# Connectionist Representation of Concepts

JEROME A. FELDMAN

*Computer Science Department*
*University of Rochester*

One of the most attractive features of connectionist systems is their potential to support models that combine constraints from different disciplines. The method of converging constraints is paradigmatic for Cognitive Science and connectionist models which are having some of their greatest impact in this new discipline. The methodology can help constrain speculation about neural representation and computation even without detailed anatomical and physiological knowledge. Some proposed solutions to classical questions can be precluded on first order computational grounds. This seems to be the case for one of the most basic questions about mental activity—how does conceptual memory work?

The nature of the neural substrate of memory is of compelling scientific interest and considerable practical importance. There are obvious applications in neurology, but it goes well beyond that. Our understanding of how human minds incorporate and process information has a marked influence on many aspects of social intercourse including formal and informal education, psychotherapy, public information, and interpersonal communication (Minsky, 1987; Roediger, 1980). Implicit assumptions about neural encoding can have a profound influence on experimental research in many fields.

Recent advances in the behavioral, biological, and computational sciences may yield a major improvement in our understanding of the neural coding of memory. Neurobiology is making remarkable strides in elucidating the structure of the nervous system and the details of its functioning. The behavioral sciences are developing deep structural models of mental operations and employing increasingly sophisticated experimental and simulation techniques to refine them. Computer science has produced powerful devices and principles of representation and computation that are supporting the study of how the complex structural theories of the behavioral scientists could be carried out by the information processing mechanisms revealed by neurobiology. Taken together, these developments are leading to a theoretical neuroscience that can integrate and inform experimental work. The use of connectionist computational models is likely to be central to these efforts. Figure 1 depicts one example of a connectionist model that captures and predicts a large number of regularities in behavioral studies of reading

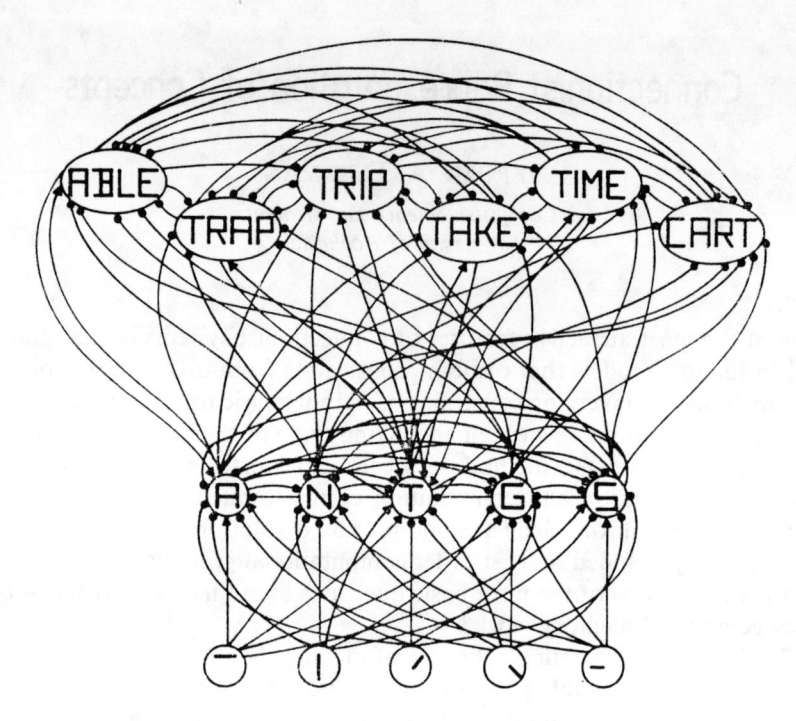

**Figure 1.** Reading model after McClelland and Rumelhart (1981). Units representing letter features (bottom), letters and words form a hierarchical excitation-inhibition network. Feedback links from word nodes to letter nodes facilitate identification of a letter in the context of a word. (See also Figure 1 of Chapter 2, p. 16)

(Rumelhart & McClelland, 1981). It is simply a hierarchy of excitatory and inhibitory (circle-tipped) links among units representing letter-features, letters, and words. Among the successes of this model is an elegant explanation of the word superiority effect previously considered to be paradoxical. It turns out that subjects can more reliably recognize a letter, for example, "A," in the context of a word, for example "TAKE," than in isolation. The model suggests that word superiority is due to feedback links from words to letters that provide additional activation for letters presented in the context of words; the same mechanism accounts for other behavioral data discussed below. We will also return to the fact that the model uses exactly one unit to represent each item of interest. Connectionist systems like Figure 1 allow us to combine directly results from various disciplines and evaluate alternative models of conceptual knowledge.

For concreteness, let us suppose that the problem is to describe how the neurons of a human brain could represent and exploit the information in a standard encyclopedia. The representation of such knowledge must include the concepts (e.g., people, places, events) involved as well as how they relate to one another. Formal models follow our intuition in assuming that the totality of knowledge can be captured as a (large) collection of relations among primitive concepts. The literature contains a wide range of notions on the nature of concepts and their neural representation. The simplest view restricts consideration to very concrete nouns such as "horse" or "chair," ignoring more complex concepts such as "game," "yesterday," "active," or "love." Even in this quite restricted context, the representation of concepts has proven to be a deep problem for philosophers, linguists, and psychologists. The related fields of artificial intelligence and cognitive psychology have provided considerable detail on how people use concepts and the kinds of information processing required (Smith & Medin, 1981). As a minimum, any representation must support the answering of questions about concepts and about their relations. One simple kind of question concerns the structure of the object itself, for example, how many legs has some chair. Theories that treat concepts as unstructured collections of attributes will need to provide additional mechanisms for answering even these simple questions.

The range of possible representations of concepts in the brain is constrained by a combination of computational considerations and neurobiological findings. Neurons with a 5 millisecond firing rate support our ability to solve nontrivial perception and conceptual problems in about 500 milliseconds. This remarkable performance in just 100 sequential time steps contrasts with conventional computers which require millions of time steps. Other pertinent facts include the relatively small number of neurons (about $10^{11}$ or 100 billion), the large number of connections between neurons (about $10^4$ per unit), and the low rate of information transfer. It may seem that $10^{11}$ is not a small number, but when one considers the $10^6$ input fibres from each eye, a computer scientist immediately detects a major constraint. For example, having a separate neuron to test for a possible line between any pair of points in the retina would require more neurons ($(10^6)^2$) than there are in the cortex. The information rate between individual neurons at a firing rate of 100 spikes per second is about five bits or enough to encode one letter of the alphabet. Thus, if complex messages are being conveyed, it is not by individual neurons. Much of the computational power of the brain derives from its great connectivity and a challenge to connectionist theory is to explain how this power is realized.

When one combines the constraints from biology, computation, and psychology, the range of plausible encodings of conceptual knowledge is relatively small. Two simple theories embody the extreme ends of the range of possible answers. The most compact representation possible would be to have a unique unit dedicated to each concept. If we assume that a unit corresponds to one neuron then this punctate model is the "grandmother cell" or "pontifical cell" theory. The other extreme would be to have each concept represented as a pattern of activity in all of the units in the system. This is known as the "holographic" model of memory and is the most highly distributed theory that we will consider. In addition to the pure theory based on optical holograms, we will call holographic any model that has all the units in a system encoding each concept (Golden, 1986; Ackley, Hinton, & Sejnowski, 1985; Willshaw, 1981); most of these are matrix formulations.

We should first dispense with an abstract argument that equates the punctate and hologram models. It is true, in a sense, that an encoding having one active unit per concept is a pattern of activity in the mass. But this identification is too abstract to be meaningful. Another proposed way to identify holographic and punctate representations comes from linear algebra and the idea of alternate coordinate axes for a vector space. If, as in many models, the output of a unit is the (thresholded) linear combination of its inputs, one can view this unit as a (very large) vector, $v$, whose coordinates are the outputs of each predecessor unit. There is, in principle, another set of bases for the vector space for which this vector, $v$, is an axis and can therefore be represented by one non-zero coordinate. This argument fails for three reasons. Even for strictly linear input combinations, the output threshold destroys the applicability of linear algebra and there is no biologically plausible way to eliminate nonlinearity. In addition, no single transform would work unless the set of concepts were independent and therefore small. More importantly, the computational properties of the two representations are radically different.

The extreme oposite models of neural representation lead to radically different views of many aspects of Cognitive Science. Table 1 presents a number of contrasting terms that arise, respectively, from the punctate and fully distributed views of neural coding. While the two extreme models have traditionally been seen as fundamentally incompatible, the method of converging constraints applied to either extreme yields essentially the same answer. From this perspective, the two styles of model can be seen as arising from different research strategies. To greatly oversimplify the discussion of the rest of the chapter: Punctate models have focused on structural issues while diffuse models have concentrated on error tolerance and generalization.

TABLE 1
Contrasting Terms

| | |
|---|---|
| punctate | diffuse |
| local | highly distributed |
| grandmother neuron | hologram, spin-glass |
| disjoint codes | homogeneous code |
| detector | filter |
| labelled line | pattern of activity |
| active memory | passive memory |
| reduction | emergence |
| hierarchy | complete connectivity |
| recruiting | adapting |
| general computation | correlation |

## COMPACT MODELS

The extreme compact representation position is to assume each concept is represented by exactly one neuron. This view received considerable support from single unit recording research, which found that units in sensory areas responded best to a relatively narrow class of stimuli (Hubel & Wiesel, 1979). The punctate encoding is also called "labelled lines" emphasizing the fact that, in this encoding, each axon will be conveying a specific message when it is active. The most influential expression of this position was Horace Barlow's "neuron doctrine" (Barlow, 1972). His five dogmas are:

> The following five brief statements are intended to define which aspect of the brain's activity is important for understanding its main function, to suggest the way that single neurons represent what is going on around us, and to say how this is related to our subjective experience. The statements are dogmatic and incautious because it is important that they should be clear and testable.
>
> *First dogma*
>
> A description of that activity of a single nerve cell which is transmitted to and influences other nerve cells, and of a nerve cell's response to such influences from other cells, is a complete enough description for functional understanding of the nervous system. There is nothing else 'looking at' or controlling this activity, which must therefore provide a basis for understanding how the brain controls behaviour.
>
> *Second dogma*
>
> At progressively higher levels in sensory pathways information about the physical stimulus is carried by progressively fewer active neurons. The sensory system is organized to achieve as complete a representation as possible with the minimum number of active neurons. ([cf. Figure 4])
>
> *Third dogma*
>
> Trigger features of neurons are matched to the redundant features of sensory stimulation in order to achieve greater completeness and economy of representation. This selective responsiveness is determined by the sensory stimulation to which neurons have been exposed, as well as by genetic factors operating during development.
>
> *Fourth dogma*
>
> Just as physical stimuli directly cause receptors to initiate neural activity, so the active high-level neurons directly and simply cause the elements of our perception.
>
> *Fifth dogma*
>
> The frequency of neural impulses codes subjective certainty: A high impulse frequency in a given neuron corresponds to a high degree of confidence that the cause of the percept is present in the external world.

Figure 2 (after Shastri (Shastri & Feldman, 1986)) provides one example of how a punctate system might encode and exploit conceptual knowledge. The memory network is a category-based hierarchy with each concept and property-name represented by a rectangular unit. The triangular nodes stand for intermediate units which become active when two of their three input

**Figure 2.** A connectionist retrieval system after Shastri (Shastri & Feldman, 1986). Figure 2a is a network fragment whose behavior is described in the text, 2b depicts the time course of activation in selected units.

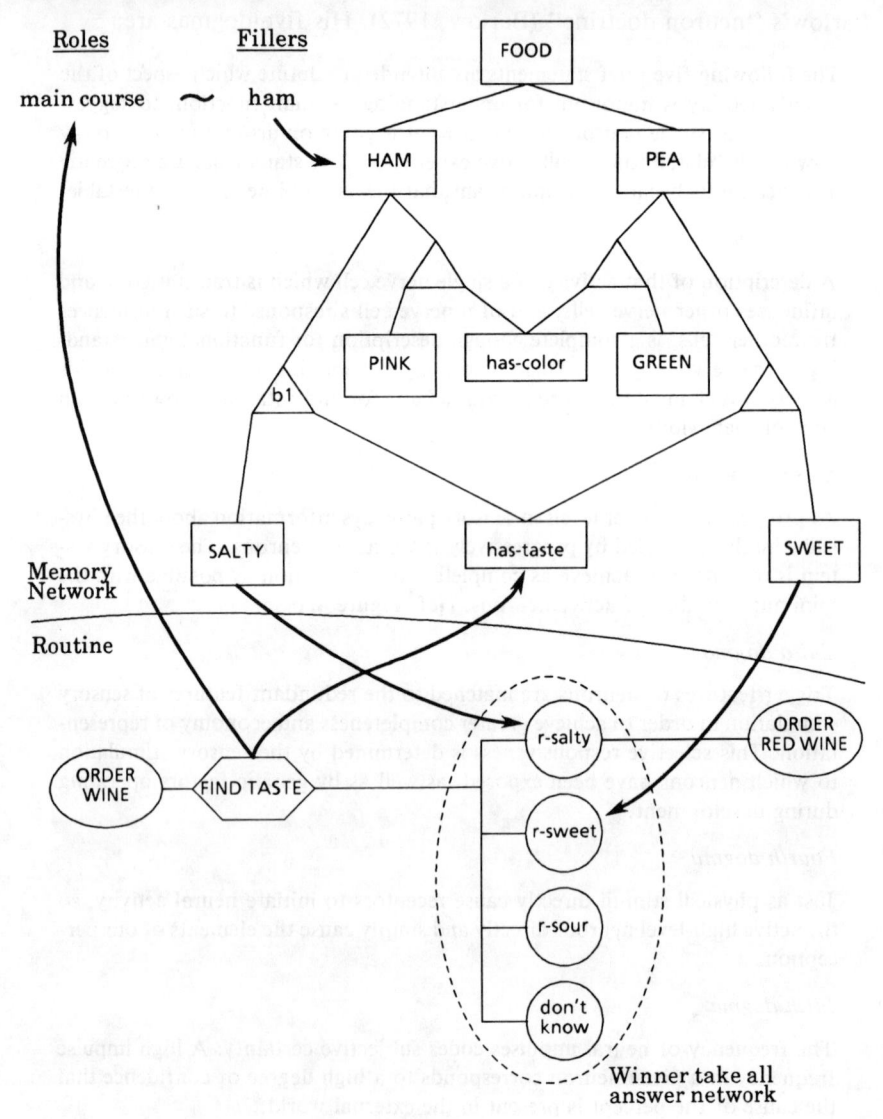

**Figure 2a**

**Figure 2b**

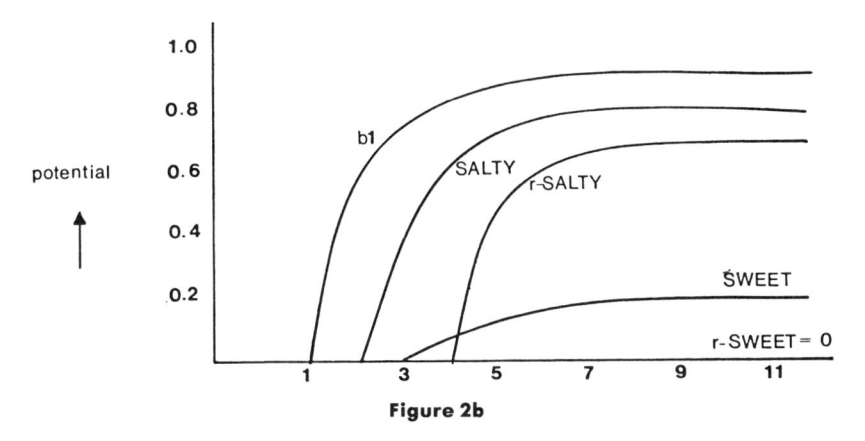

lines are active. Suppose the system has a routine that retrieves its knowledge of food tastes as an aid to ordering wine, such as that cartooned in the lower half of the figure. If activation is spread simultaneously to the "main course" of the meal and to the desired property "has-taste," exactly one triangular evidence node, $b_1$, will receive two active inputs leading to the activation of the concept "salty." This is the required answer, but for technical reasons an intervening clean-up network is needed to actually use the result. Figure 2b shows an activation record of the "potentials" of units in the network responding to the query. One interesting feature of the model is that the same memory network is able to categorize a salty, pink food as ham—the triangular evidence nodes work in both directions.

While oversimplified, Figure 2 does convey much of the flavor of punctate (and other compact) connectionist models and their appeal to some scientists and rejection by others. The main point is that everything is quite explicit; the concepts, properties, and even the rules of operation are simple and direct. This makes it relatively easy to express and test specific models either at the neural level or more abstractly as in our example. No one believes that the brain uses exactly the structure of Figure 2, but any highly compact concept representation could behave in essentially the same way. Notice that even in this extreme case many units (representing features and components) will be active for each concept. A great deal of interesting work is being done with such models, but there are also a variety of arguments why the neural encoding must be more complex.

The first point is that a large number of neurons ($\sim 10^5$) die each day and these are distributed throughout the brain. If each concept were represented by exactly one neuron, one would expect to lose at least some concepts (at random) each day. This argument is often taken to be conclusive evidence against any compact model, but a slight variant of the punctate view is stable under the death of individual units. Suppose that instead of one unit per concept, the system dedicated three or five, distributed somewhat physically. All of the theories and experiments would look the same as in the single unit case, but the redundancy would solve the problem of neuron death. While the number of neurons dying is large, the fraction is quite small ($\sim 10^{-6}$) so the probability of losing two of the representatives of a concept in a lifetime is quite low.

Another argument used against compact models of neural representation arises from bulk activity experiments, which show that single small stimulus can give rise to activity in a significant fraction of the total population. There are several reasons why this fact does not preclude compact representations. For one thing, the "simplicity" of a stimulus looks different from various encoding schemes. A small dot, in a Fourier encoding, activates receptors for all spatial frequencies. In a parameter-space encoding (Ballard, 1986), a single feature provides evidence for many higher-level features and these may be suppressed slower than when more information is present (cf. Figure 1). Finally, there appears to be some non-specific evoked potentials (Barber, 1980) which activate an entire structure on any input.

While the cell-death and bulk activities arguments against punctate representation are easily answered, one can see that there really could not be enough neurons to have one for each concept of interest. One example arises in early vision. It is well known that the visual system is sensitive to at least the following local stimulus properties: orientation, intensity, hue, depth, motion direction, and size. A system that could resolve 10 values for each of the six dimensions would require $10^6$ units to represent all combinations of values. But there are about $10^6$ separate points at the narrowest (retinal ganglion) level that must be represented so the total requirement would be $10^{12}$, which is too many.

Experimentalists have also been reframing their view of the activity of single units. For some time, the idea of a neuron as a "detector" of one event type has been declining but no alternative term has yet evolved. The notion of a "filter" is the diffuse equivalent (all units filter all signals) and equally misleading. Although no new word has been established, experimentalists now view sensory neurons as having responsivity of different coarseness to a variety of stimulus dimensions. Studies in single unit neurophysiology are finding effects of stimuli beyond the classical receptive field (Allman, Miezen, & McGuiness, 1985) as seems inevitable in an interacting system of units. A proposed revision of Barlow's dogmas that takes these developments into account is given in the Appendix.

There is a concise computational account that both helps organize the experimental data and contributes to our goal of constraining the possible models of neural encoding. The basic idea is depicted in Figure 3 for the case of the two stimulus dimensions. Suppose (as appears to be the case in the brain) that units respond nonuniformly to various stimulus dimensions. For example, the vertical rectangle in the lower left depicts the responsiveness of a cell that is five times more sensitive to size than to orientation and the horizontal rectangle the opposite. The nice point is that the joint activity of two such cells can code the stimulus space as finely as the finest dimension of either (cf. the crossed rectangles), while requiring significantly fewer units. This computational mechanism goes by the name "coarse-fine coding" and appears to characterize a good deal of neural computation. In general, given $K$ stimulus dimensions, each with a desired resolution of $N$ values, the punctate encoding requires $N^K$ units. A coarse-fine encoding with the coarse dimensions $D$ requires a total of

$$T = K \cdot N \left(\frac{N}{D}\right)^{K-1}$$

units. This is because there will be $K$ separate covers of the $K$-dimensional space, but each will be covered coarsely and this requires $(N/D)$ units in all dimensions but one, which has $N$ units. This formula still grows exponentially, but is significantly smaller for the cases of interest. For example, our early vision example had $K = 6$ and $N = 10$. With $D = 5$, this yields 1920 units per point instead of the 1,000,000 for the pure punctate encoding. An additional saving arises from having the converging inputs each go to a separate site (dendritic subtree).

The critical point in the construction is the overlap of receptive fields, not their assymmetric shape. Essentially the same arguments can be made for symmetric overlapping fields and this is known as "coarse-coding" (Hinton, McClelland, & Rumelhart, 1986). Computational ideas of this kind have been known for some time to provide a nice account of hyperacuity, the ability of people to resolve details finer than the spacing of their receptors. In both coarse and coarse-fine coding there is a price to be paid for saving all these units. If two stimuli that overlap occur simultaneously, the system will be unable to resolve them. In Figure 3, one stimulus is encoded by the two rectangles containing $X$ and the other by the two rectangles containing $Y$. The intersections labelled $X$ and $Y$ encode the desired information, but the "ghosts" labelled $G$ would be equally active. One of these expresses the conjunction of $Y$-size and $X$-orientation and the other the converse. This is an instance of "cross-talk" in neural encodings. Cross-talk is the fundamental problem of shared encodings and appears to be a critical limiting factor on distributed models.

The coarse-fine coding examples used overlapping encodings, but were based on the minimum possible number of cells to cover some feature space. Suppose instead we allowed for redundancy in the coverage, say three separate tesselations of the space. In terms of Figure 3, a second covering could be similar bars at 45° and 315° to the axes. We could still use separate receiving sites for each desired stimulus, but the computation would be not just a logical $AND$. In fact, a thresholded sum of activity might be quite plausible as the sites' way of computing the likelihood of its combination

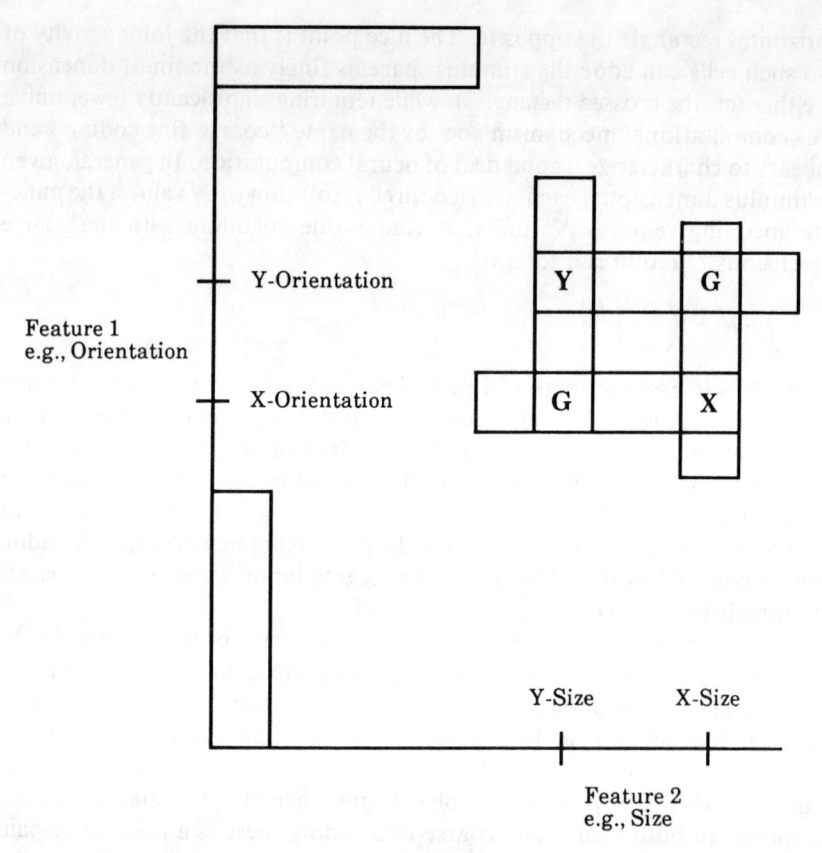

**Figure 3.** Coarse-Fine coding. Both computational and biological considerations suggest overlapping representations. When two stimuli that are too close appear together, confusing (ghost) activation can result.

being present. This would combine the error tolerance and information reduction ideas in a simple and plausible way. Edelman has come to essentially the same kind of model through a very different route, starting from his expertise in immunology (Edelman, 1981).

## HOLOGRAPHIC MODELS

Holographic models have been fervently supported by biologists, psychologists and theoreticians. There is no comparable fervor for compact models. In fact, several researchers advocate fully distributed models while employing punctate ones, (Hopfield, 1982; Hopfield & Tank, 1985) sometimes in the same paper (McClelland, 1985, this volume; Hinton, 1986). One major contributing factor in the popularity of this view was the early work of Karl

Lashley who found that for a variety of tasks, the deficit exhibited by lesioned rats was best explained by the total amount of cortex removed—the "law of mass action." Lashley summarized his view of memory representation in the classic 1950 paper "In Search of the Engram." The following quotation continues to motivate much current work (Lashley, 1950):

> It is difficult to interpret such findings, but I think that they point to the conclusions that the associative connections or memory traces of the conditioned reflex do not extend across the cortex as well-defined arcs or paths. Such arcs are either diffused through all parts of the cortex, pass by relay through lower centres, or do not exist. [Lashley, 1950, p. 461]

Recent work (Thompson, 1986) has established that conditioned reflexes are supported by remarkably compact cerebellar structures, but the general idea of holographic memory continues to attract attention. Since all of the primary sensory and motor areas have been found to have specialized structure, the holographic hypothesis is currently restricted to "higher" brain areas whose functional organization is not yet understood. Another historical source of motivation for diffuse models was the landmark 1949 book, *The Organization of Behavior*, by Donald Hebb (Hebb, 1949). Hebb introduced the notion of cell assemblies, but was (appropriately for the time) vague about how they actually encode knowledge. Hebb definitely envisioned a dynamic pattern of activity, but there are two interpretations of this. The most literal would be to assume concepts are purely dynamic and are not tied to any particular tissue; this idea has been pursued somewhat (von der Malsburg, 1985) but without much success and the experimental data is not encouraging. A more general notion of dynamic activity of cell assemblies is inherent in all current connectionist theories; the compact-diffuse question is about what fraction of a system is involved in representing a concept.

Holographic models are theoretically attractive because of two properties: fault-tolerance and generalization. The brain clearly has these properties and it is easy to see informally why holographic models do also. If all of the units of a large system are involved in coding one concept, the failure of some of them can be tolerated. Furthermore, two concepts that share much of their activity patterns will tend to behave similarly. An excellent compendium of holographic-style models can be found in (Hinton & Anderson, 1981). More recently, there has been a flurry of interest in spin-glasses as holographic memory models in the theoretical physics community (Hopfield, 1982; cf. Chapter 11, this volume).

The purest holographic model is, unsurprisingly, the optical hologram itself. A nice presentation can be found in (Willshaw, 1981) where it is also shown that the purely linear hologram is undesirable on both computational and biological grounds. The models typically studied are large rectangular matrices representing all the possible connections between $m$ input units and

$n$ output units, which need not be distinct. The most common case, and one of the easiest to consider, is where each unit compares the sum of its weighted inputs to its threshold and emits 1 if the sum of inputs is greater and 0 otherwise. A concept is represented as a binary vector over all the input lines. Each element of the vector may be uninterpreted or can be thought of as the presence or absence of microfeature characterizing the concept.

If the input and output are identified, the matrix becomes a pattern completion machine. The basic idea is simple and derives from the mathematical notion of correlation. The correlation of a binary vector, $u$, with another, $v$, is:

$$\sum_{j=1}^{n} u_j v_j$$

and this is obviously maximized when $u = v$. In a well distributed set of vectors, an input distorted by modest noise or omission will correlate best with the appropriate complete vector. This is the basic source of error resistance and generalization in holographic-style models. What makes them interesting is that a connection matrix with the appropriate weights can be learned with a simple, local procedure. For the auto-associative, pattern completion case, the learning procedure should yield weights such that each output unit is most likely to be active when the corresponding input is active. The natural local weight-change rule is to increase the weight between two units when they both are active (output = 1). This is essentially Hebb's rule (Hebb, 1949) and can be interpreted as increasing the weight of units whose firing is correlated.

The resulting matrix can be shown to be optimal in some cases (orthogonal vectors) and is a good estimator in many others (Kohonen, 1984). Thus a simple and biologically plausible local learning rule yields a matrix which reliably finds the best match to an input pattern under some distortion conditions (but not others). The binary, linear auto-associator is the simplest of a wide variety of essentially similar models. Many variations have been tried on all aspects of the model (Hinton & Anderson, 1981). The more refined models use a variant of Hebb's rule which allows feedback from some external result to affect the weight change process (Kohonen, 1984; Barto, Sutton, & Brouwer, 1981). The shortcomings of the methodology do not lie in the learning rules, which demonstrate that essentially anything that can be represented as a linear threshold matrix can be learned by correlation. While something like associative matrices may occur in nature (Lynch, 1986), the representation is much too weak to support human concept memory.

The basic problems with any holographic representational scheme are cross-talk, communication, invariance, and the inability to capture structure. Essentially the same problems have prevented the development of holographic computer memories or recognition systems despite considerable effort. Consider the problem of representing a concept like "grandmother" as a pattern of activity in all the units in some memory network. Notice what would happen if two (or more) concepts were presented at the same time, for example, grandmother at the White House. The encodings for the two concepts would overlap and the system would get garbled. This is a massive instance of the cross-talk problem of Figure 3. We can reduce the probability of cross-talk by having fewer units active for each pattern. If one assumes that the cross-talk is randomly distributed, Willshaw (1981) has shown that the system will be reliable (even for single entries) only if the number of units active for each pattern is proportional to the logarithm of the number of units in the diffuse memory. This means that a network of 1,000,000 units should use an encoding with about 20 active units per concept—this is interestingly close to the number for a redundant, coarse-coded compact approach as derived in the previous section.

A related problem with diffuse representations is that only one concept at a time could be transmitted between subsystems. The sequential nature of diffuse representations is particularly troublesome when we consider how information about a complex scene could be transferred from vision to other systems such as language and motor control. There appears to be no alternative to assuming that, at least for simultaneous communication, representation of concepts must be largely disjoint and thus compact.

The same communication problem arises within the concept memory itself if one tried to build a knowledge structure like Figure 2 with a diffuse representation. If a concept like "salty" were represented only by a large pattern, the links for this entire pattern would have to go to all the places that related to saltiness—and be treated correctly at each of these. If a concept encoded by $N$ units needed to be linked to $M$ other concepts, a total of $M*N$ links would be needed. The more distributed the representation, the more serious this problem becomes. Again, any serious reduction in this wiring requirement constitutes a compact representation. And, as in the intermodal case, unless these representations were largely disjoint, concept processing would have to be sequential to avoid cross-talk. This eliminates the massively parallel processing that is required by the 100-step constraint.

Moreover, no one has suggested how to represent any but the simplest concepts in the holographic style. For example, the problem of all the different historical views of one's grandmother is as difficult for the hologram as for the punctate model. In fact, it is far from obvious how to make the same distributed pattern active for alternative views of a chair. All concepts in any holographic structure that have been proposed are totally without internal structure. There is an idea of associating the components of a representation with microfeatures (Anderson & Mozer, 1981) but these are still unstructured. Nor do any of the holographic proposals provide a way of answering even simple questions like the color of grandmother's hair. The only suggestion on how to encode structured knowledge in a holographic system is to encode each proposition (e.g., Dave likes candy) as a separate "memory" (Touretsky & Hinton, 1985). One can build up arbitrary structures in this way, but at the cost of losing all the advantages that led to connectionist models in the first place. There may be some technical advantages to exploring such models, but they have nothing to do with human concept memory—for example, they violate the 100-step constraint by many orders of magnitude.

The biological evidence against anything vaguely like a holographic model is equally compelling. This is only fair, since the holographic hypothesis denies any relevance to neuroanatomy and physiology. Since intricate specificity and detailed visuotopic, tonotopic, and so forth, maps have been discovered everywhere in the brain in which they have been sought, the only hope for the hologram is for some higher association areas. Even there, the anatomical structure has been found to be much like the sensory areas (Goldman-Rakic & Schwartz, 1982) and nothing like the connection of each input to each output required by matrix models.

Another argument for highly distributed representations derives from the large number of input fibers ($\sim 10^4$) to cortical neurons. If all of these fibers participate actively then, *ipso facto,* the representation is diffuse. While there has been no definitive study on the number of presynaptic events required for neural firing, estimates gleaned from papers and conversations run from one event to a few dozen (Shepard, 1986). No one has suggested that several thousand synapses must fire at once for an action potential. Also, we can see from Figure 4 that many of the connections could represent alternative ways of activating the same concept (e.g., from different points in space). Another way of looking at this is that the thousand-fold connectivity is capturing an "OR" of activation conditions rather than an "AND." Finally, as we discuss below, learning in a connectionist system requires the potential for many more connections than are ever made functional. The most striking physiological demonstration of this principle is in neural reorganization studies (Merzenich *et al.,* 1984).

## THE MIDDLE GROUND AND BEYOND

A variety of findings have all converged on the idea that concepts are represented by overlapping activity among a modest number of units and that the structure within and across these groups can not be uniform or arbitrary. The fact that both the punctate and holographic approaches yield essentially the same result is also reflected in current research on neural representation.

Consider the punctate model of the appearance of horses shown in Figure 4 (Feldman, 1985). At the lowest level are feature pairs, which are assumed to be derived by early vision networks. The remaining structure is

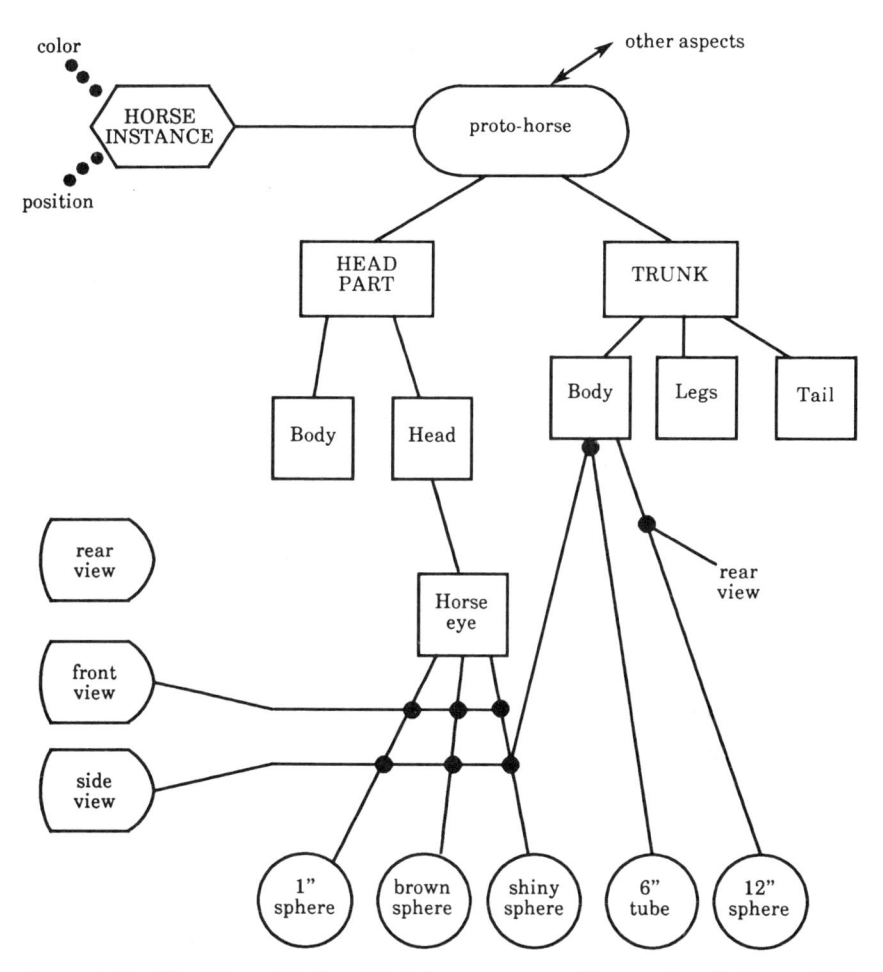

**Figure 4.** Simplified hierarchical model of the appearance of horses after (Feldman, 1985). Spreading activation selects the concept best matching the visual features. Circular junctions depict gated inputs.

organized as a hierarchy (cf. Table 1 and dogma 2) where activation propagates towards the top pontifical cell which is active when a horse is recognized. Even in this unrealistically punctate version, the recognition of a horse is a "pattern of activation" in many of these units. Since the connections are bidirectional, mentioning the name of a horse will cause (some) activation in all the nodes comprising visual and other descriptions of horses. Notice that it makes no sense to talk about the activation of a single concept in such structure—activation automatically spreads to encompass a subnetwork. Notice furthermore that this structure will recognize a horse even when some features are missing or distorted if the other features plus context are sufficiently strong. This captures the error tolerance and some of the generalization ability that were most attractive about correlation matrices. Computer experiments of moderate complexity along these lines have been successfully carried out (Cooper & Hollbach, 1987; Sabbah, 1985). The main current versions of holographic correlation theory use microfeatures essentially identical to the bottom row of Figure 4. The difference is that a holographic model would not have the hierarchical structure but would represent horse exclusively by the pattern of activity of the feature-pair units. The previous section has suggested that feature vectors without the hierarchical structure of Figure 4 will not support the requirements of conceptual memory. In fact, pure feature representations have no way at all to represent particular concepts such as a horse named *Secretariat*. Nevertheless, some very interesting work has been done by suppressing structure and focussing on the properties of feature vectors.

There are two important problems that have been effectively studied using the structureless approach—generalization and learning. Recall that the reading model of Figure 1 helped explain why people can recognize a letter better in context than in isolation. The model actually did much more than this; it turns out the model (and the subjects) also recognize letters better in the context of pronounceable nonwords, but not generally in unpronounceable nonwords. Since there are no units representing nonwords, the network exhibits an important generalization ability. What happens is that pronounceable letter combinations moderately activate units representing similar words and the combined effect of these word-units produces the enhancement. From the perspective of diffuse modelling one can view this as a "distributed representation" of nonwords by the network. But there are many aspects of the nonwords (size, pronunciation, etc.) which are not represented and it is more accurate to say that the system behaves *for this task* as if it had explicit representation of nonwords. System ("emergent") properties of neural networks have fascinated many investigators (Rumelhart & McClelland, 1986) and remain one of the major motivations for studying diffuse models. Correlation-based learning schemes, like those discussed above, can be used to generate systems that capture important generalizations including some of the ones discussed above. Golden (Chapter 5) discusses the relationship between the two styles of model for the reading task.

There is no inherent conflict between structured, compact models such as Figure 4 and generalization but a model that is *strictly* hierarchical will be weak at generalization. In the horse example, one might want to capture the generalization that animals with hair are mammals. We could link horse to mammal, but this is of no direct use in classifying novel hairy animals. The point is that if all uses of a concept were forced to go through some focus, the system would be unable to yield many crucial generalizations. There are systematic relationships among properties that are best captured directly. The relations between syntactic and case roles, between pronunciation and morphology and between hairiness and mammalhood are systematic and most compactly encoded by circumventing particular instances of words or mammals. There is an interesting duality here. If one knew enough hairy mammals, it is likely that activation of "hairy" would lead (indirectly) to the activation of "mammal." One could view this as all the individual mammals constituting a distributed representation of the hair-mammal relation. A great deal of generalization to totally new examples probably takes this form, but our conceptual memory also has the ability to represent much richer evidential relations among concepts (Shastri & Feldman, 1986).

## LEARNING

The preceding discussion has explored the ways in which conceptual knowledge might be represented in neural networks, suppressing the critical problem of learning. It is clear that evolution provides humans with an enormous amount of prewired structure, which develops a great deal more through interaction with the environment, and that this continues throughout life. This presents a severe methodological problem for computational learning research.

The basic difficulty is that we know human learning is based on an elaborate existing structure, but we know very little about the exact nature of this structure, particularly in the area of conceptual knowledge. Any study of learning either assumes some existing structure (at the risk of trivializing the learning aspects) or assumes no existing structure and is restricted to quite simple problems. The correlation matrix models of memory have presented an attractive research vehicle. The total initial connectivity and linear threshold rule are a minimal a priori structure, and the correlation method of updating is an easily analyzed learning rule.

These correlation matrices employ a diffuse encoding and the appeal of the two ideas is strongly interrelated. But the ideas of correlation, feedback, and weight change are not restricted to diffuse representations, linear threshold rules, or complete connectivity matrices. In fact, complete connectivity is not biologically plausible and the other assumptions are similarly problematical.

It has been known for some time that any effective learning rule must include an input from the ultimate result of the computation; Hebb's correlation rule has no way to punish a connection that leads to a disaster. In a network that directly links inputs and outputs, like a matrix model or a 1-layer perception (Fogelman-Soulie, Gallinari, LeCun, & Thiria, in press), it is easy to punish the offending links. For a system with more structure and indirect links, it has not been obvious how to assess credit and blame.

Recent developments in learning theory, while not totally solving this problem, have facilitated the study of learning in much richer structures (Rumelhart, Hinton, & Williams, 1986; Parker, 1985).

The development of back-propagation learning techniques (Rumelhart, Hinton, & Williams, 1986) has had a complex effect on theories of representation. Although no one has suggested that animal learning works the same way, several studies have used the results of back-propagation experiments to pursue the compact-diffuse issue. In this volume, Scalleter and Zee (Chapter 11, this volume) establish conditions under which punctate representation is optimal while Blelloch and Rosenberg's results (Chapter 12) yield a somewhat more distributed representation. The fundamental structure requirements described in this chapter are independent of whether the information is prespecified or learned and suggest that if learning systems ever reach the complexity of Figure 2, they will have compact representations.

Learning in structured connectionist systems has been studied directly. A major problem in this formulation is "recruiting" the compact representation for new concepts (Fanty, 1987; Feldman, 1982; Wickelgren, 1979). It is all very well to show the advantages of representational schemes like Figures 1, 2, and 4, but how could they arise? This question is far from settled, but there are some encouraging preliminary results. The central question is how a system that grows essentially no new connections could recruit compact groups of units to capture new concepts and relations. One relevant result concerns the probability of finding compact groups that link nodes in a random graph (Feldman, 1982). It turns out that, for biologically reasonable parameters, the probability of a compact cluster ($\sim 20$ units) is quite high. The brain is not random, except perhaps very locally, but that increases the probability of recruiting compact concept representations. Current work on learning in more structured networks (Hinton, 1986; Fanty, 1987) is examining this more closely.

In summary, connectionist studies of pure generalization and learning minimize the pre-existing structure and tend to study diffuse models. This has turned out to be very valuable and will continue to be. Like the studies of overall convergence, for example, (Cohen & Grossberg, 1983; Hopfield, 1982; Wilson & Cowan, 1972), these are best done assuming no particular structure of the network. But the structure exists and only compact representations can capture it.

Structured networks of evidence-combining simple units have a number of attractive computational properties. Error resistance, context dependence, and the ability to assimilate conflicting information are natural properties of such systems. A reasonable degree of generalization follows from these properties. Simple weight-change rules can enable these systems to improve their performance significantly. The collective behavior of such systems can produce powerful computations not easily anticipated, as is true of any complex circuit. While certain concepts are represented explicitly, the system is not restricted to dealing only with those. A properly structured network can behave, in certain situations, as if it had concepts and relations implicitly represented. An explicit, compact representation of concepts is required when they have disjunction or internal structure or when they participate in relations or are communicated among subsystems. In addition, the nature of any emergent system properties depends heavily on which concepts are explicitly represented and the detailed structure of the representation.

From this perspective, connectionist concept representation presents no paradigmatic conflict or unapproachable mystery but a deep and fascinating set of scientific questions. The issues discussed in this paper arose from considering only the simplest concepts and properties. Human conceptual memory extends to much richer concepts, to relations among them, and to knowledge of space, time, and intensionality. All of these issues have been studied seriously in several disciplines and an imposing collection of problems has been elucidated (Brachman & Levesque, 1985). The hope of connectionist research is that explicit consideration of the computational mechanisms used by the brain will help clarify how it does its marvelous work.

## APPENDIX

### Interim Neuron Doctrine

The following five brief statements are intended to define which aspect of the brain's activity is important for understanding its main function, to suggest the way that single neurons represent what is going on around us, and to say how this is related to our subjective experience. The statements are dogmatic and incautious because it is important that they should be clear and testable.

*First dogma*

> A description of that activity of a single nerve cell which is transmitted to and influences other nerve cells, and of a nerve cell's response to such influences from other cells, is a complete enough description for functional understanding of the nervous system. There is nothing else 'looking at' or controlling this activity, which must therefore provide a basis for understanding how the brain controls behaviour.

**Since significant behaviors involve many individual nerve cells, functional understanding of the nervous system will require scientific languages for characterizing the behavior of networks of neurons.**

## Second dogma

**Efficient coding of information is a central problem of the sensory system. At progressively higher levels in the sensory pathways information about the physical stimulus is more abstract and is represented by progressively fewer active neurons.**

## Third dogma

Trigger features of neurons are matched to the redundant features of sensory stimulation in order to achieve greater completeness and economy of representation. This selective responsiveness is determined by the sensory stimulation to which neurons have been exposed, as well as by genetic factors operating during development.

## Fourth dogma

Just as physical stimuli directly cause receptors to initiate neural activity, so the active **networks of intermediate and** high-level neurons directly and simply cause the elements of our perception.

## Fifth dogma

**Frequency coding is the primary basis of neural communication. Sensory neurons respond with high frequency discharge to external stimuli which fit into a narrow range of possibilities; the higher the discharge the more narrow the range of possible causes.**

## Zeroth dogma

**Intelligent behavior and its neural realization are incredibly complex. A functional understanding of this will require organizational principles from the behavioral and computational sciences as well as biology.**

## REFERENCES

Ackley, D.H., Hinton, G.E., & Sejnowski, T.J. (1985). A learning algorithm for Boltzmann machines, *Cognitive Science 9*, 147–169; also Chapter 10, this volume.

Allman, J., Miezen, F., & McGuiness, E. (1985). Stimulus specific responses from beyond the classical receptive field: Neurophysiological mechanisms for local-global comparisons in visual neurons, *Annual Review of Neuroscience, 8*, 407–430.

Amari, S., & Arbib, M.A. (1982). Competition and cooperation in neural networks, Vol. 45. In S. Levin (Ed.), *Lecture Notes in Biomathematics*. New York: Springer-Verlag.

Anderson, J.A., & Hinton, G.E. (1981). Models of information processing in the brain. In G.E. Hinton & J.A. Anderson (Eds.), *Parallel models of associative memory*. Hillsdale, NJ: Erlbaum.

Anderson, J.A., & Mozer, M.C. (1981). Categorization and selective neurons, In G.E. Hinton & J.A. Anderson (Eds.), *Parallel models of associative memory.* Hillsdale, NJ: Erlbaum.

Ballard, D.H. (1986). Cortical connections and parallel processing: Structure and function, *The Behavioral and Brain Sciences, 9,* 67-120.

Barber, C. (1980). *Evoked potentials.* Baltimore, University Park Press.

Barlow, H.B. (1972). Single units and sensation: A neuron doctrine for perceptual psychology? *Perception, 1,* 371-392.

Barto, A.G., Sutton, R.S. & Brouwer, P.S. (1981). Associative search network: A reinforcement learning associative memory, *Biological Cybernetics, 40,* 201-211.

Barto, A.G., & Anandan, P. (1984). Pattern recognizing stochastic learning automata, (Tech. Rep. No. 84-30). University of Massachusetts at Amherst, Computer and Information Science.

Brachman, R., & Levesque, H. (1985). *Readings in knowledge representation,* Palo Alto, CA: Morgan-Kaufman Publishers.

Cohen, M.A., & Grossberg, S. (1983). Absolute stability of global pattern formation and parallel memory storage by competitive neural networks. *IEEE Transactions: Systems, man and cybernetics, 13,* 815-825.

Cooper, P.R., & Hollbach, S.C. (1987, February). Parallel recognition of objects comprised of pure structure, *Proceedings,* DARPA Image Understanding Workshop, Los Angeles, CA.

Desimone, R., Schein, S.J., Moran, J., & Ungerleider, L.G. (1985). Contour, color, and shape analysis beyond the striate cortex. *Vision Research, 25,* 441-452.

Edelman, G.M. (1981). Group selection as the basis for higher brain function. In F.O. Schmitt, F.G. Worden, G. Adelman, & S.G. Dennis (Eds.), *The Organization of Cerebral Cortex.* Cambridge, MA: MIT Press.

Fanty, M.A. (1987). Learning in structured connectionist networks, forthcoming doctoral dissertation. Computer Science Dept., University of Rochester, New York.

Feldman, J.A. (1985). Four frames suffice: A provisional model of vision and space, *The Behavioral and Brain Sciences, 8,* 265-289.

Feldman, J.A. (1982). Dynamic connections in neural networks, *Biological Cybernetics, 46,* 27-39.

Feldman, J.A., & Ballard, D.H. (1982). Connectionist models and their properties, *Cognitive Science, 6,* 205-254; also, Chapter 1, this volume.

Fogelman-Soulie, F., Gallinari, P., LeCun, Y., & Thiria, S. (in press). Automata networks and artificial intelligence. In F. Fogelman-Soulie, Y. Robert, & M. Tchuente, (Eds.), *Automata networks in computer science, theory and applications.* Manchester, England: Manchester University Press.

Golden, R.M. (1986). A developmental neural model of visual word perception. *Cognitive Science, 10,* 241-276; also, Chapter 5, this volume.

Goldman-Rakic, P.S., & Schwartz, M.L. (1982). Interdigitation of contralateral and ipsilateral columnar projections to frontal association cortex in primates, *Science, 216,* 755-757.

Hebb, D.O. (1949). *The Organization of Behavior.* New York: John Wiley.

Hinton, G.E. (1986, August). Learning distributed representations of concepts, *Proceedings of the Eighth Annual Conference of Cognitive Science Society,* Amherst, MA.

Hinton, G.E. (1981). Implementing semantic networks in parallel hardware. In G.E. Hinton & J.A. Anderson (Eds.), *Parallel models of associative memory.* Hillsdale, NJ: Erlbaum.

Hinton, G.E., & Anderson, J.A. (1981). *Parallel models of associative memory.* Hillsdale, NJ: Erlbaum.

Hinton, G.E., McClelland, J.L. & Rumelhart, D.E. (1986). Distributed representations. In D.E. Rumelhart & J.L. McClelland (Eds.), *Parallel distributed processing, explorations in the microstructure of cognition, Vol. 1: Foundations.* Cambridge, MA: Bradford Books/MIT Press.

Hopfield, J.J. (1982). Neural networks and physical systems with emergent collective computational abilities. *Proceedings of the National Academy of Sciences of the United States of America 79,* 2554-2558.
Hopfield, J.J., & Tank, D.W. (1985). Neural computation in optimization problems, *Biological Cybernetics.*
Hubel, D.H., & Wiesel, T.N. (1979, March). Brain mechanisms of vision. *Scientific American, 241,* 150-162.
Kohonen, T. (1984). *Self-organization and associative memory, Springer Series in Information Sciences, 8,* New York: Springer Verlag.
Lashley, K. (1950). In search of the engram. In Symposia of the Society for Experimental Biology, No. 4, *Physiological mechanisms in animal behavior,* (pp. 454-483). New York: Academic Press.
Lashley, K. (1929). *Brain mechanisms and intelligence.* Chicago: University of Chicago Press.
Lynch, G. (1986). *The neurobiology of learning and memory.* Cambridge, MA: MIT Press.
McClelland, J.L., & Rumelhart, D.E. (Eds.) (1986). *Parallel distributed processing: Explorations in the microstructure of cognition. Vol. 2: Applications.* Cambridge, MA: Bradford Books/MIT Press.
McClelland, J.L., Rumelhart, D.E. (1981). An interactive activation model of the effect of context on language learning (Part I), *Psychological Review, 88,* 375-401.
Merzenich, M.M., Nelson, R.J., Stryker, M.P., Cynader, M.S., Schoppmann, A., & Zook, J.M. (1984). Somatosensory cortical map changes following digit amputation in adult monkeys, *The Journal of Comparative Neurology, 224,* 591-605.
Minsky, M. (1987). *The society of mind.* New York: Simon and Schuster.
Minsky, M., & Papert, S. (1972). *Perceptrons,* (2nd ed.). Cambridge, MA: MIT Press.
Mishkin, M. (1987, June). The anatomy of memory, *Scientific American.*
Newell, A. (1983). Intellectual issues in the history of artificial intelligence. In F. Machlup & U. Mansfield (Eds.), *The study of information: Interdisciplinary messages.* New York: John Wiley & Sons, Inc.
Palm, G. (1980). On associative memory, *Biological Cybernetics, 36,* 19-31.
Parker, D.B. (1985, April). Learning-logic, TR-47, Center for Computational Research in Economics and Management Science. Massachusetts Institute of Technology.
Phillips, C.G., Zeki, S., & Barlow, H.B. (1984). Localization of function in the cerebral cortex, *Brain, 107,* 328-361.
Pribram, K.H., Nuwer, M., & Baron, R. (1974). The holographic hypothesis of memory structure in brain function and perception. In R.C. Atkinson, D.H. Krantz, R.C. Luce, & P. Suppes (Eds.), *Contemporary developments in mathematical psychology.* San Francisco: W.H. Freeman.
Roediger, H.L. III (1980). Memory metaphors in cognitive psychology. *Memory and Cognition, 8,*(1), 231-246.
Rumelhart, D.E., & McClelland, J.L. (Eds.) (1986). *Parallel distributed processing: Explorations in the microstructure of cognition. Vol. 1: Foundations.* Cambridge, MA: Bradford Books/MIT Press.
Rumelhart, D.E., & McClelland, J.L. (1981). An interactive activation model of the effect of context in language learning (Part 2), *Psychological Review, 89,* 60-94.
Rumelhart, D.E., Hinton, G.E., & Williams, R.J. (1986). Learning internal representations by error propagation. In D.E. Rumelhart & J.L. McClelland (Eds.), *Parallel distributed processing, explorations in the microstructure of cognition, Vol. 1: Foundations.* Cambridge, MA: Bradford Books/MIT Press.
Sabbah, D. (1985). Computing with connections in visual recognition of Origami objects, *Cognitive Science, 9,* 25-50.
Sejnowski, T.J. (1986). Open questions about computation in cerebral cortex. In J.L. McClelland & D.E. Rumelhart (Eds.), *Parallel distributed processing: Explorations in the*

*microstructure of cognition, Volume 2: Applications.* Cambridge, MA: Bradford Books/MIT Press.

Shastri, L. (1985, September). Evidential reasoning in semantic networks: A formal theory and its parallel implementation, Ph.D. thesis and TR166, Computer Science Department, University of Rochester.

Shastri, L., & Feldman, J.A. (1986). Neural nets, routines and semantic networks. In N. Sharkey (Ed.), *Advances in Cognitive Science*. Ellis Horwood Publishers.

Shaw, G.L., Silverman, D.J., & Pearson, J.C. (1985). Model of cortical organization embodying a basis for a theory of information processing and memory recall. *Proceedings of the National Academy of Sciences of the United States of America, 82,* 2364-2368.

Shepard, G. (1986). *The neurobiology of learning and memory,* G. Lynch (Ed.), Cambridge, MA: MIT Press.

Smith, E.E., & Medin, D.L. (1981). *Categories and Concepts.* Cambridge, MA: Harvard University Press.

Thompson, R.F. (1986, August). The neurobiology of learning and memory, *Science, 233,* 941-947.

Touretzky, D.S., & Hinton, G.E. (1985, August). Symbols among neurons: Details of a connectionist inference architecture, *Proceedings, IJCAI,* Los Angeles, CA, 238-243.

van Heerden, P.J. (1963). A new optical method of storing and retreiving information, *Applied Optics, 2,* 387-392.

von der Malsburg, C. (in press). Nervous structures with dynamical links. *Berichte der Bunsen-Gesellschaft fur Physikalische Chemie.*

Wickelgren, W.A. (1979). Chunking and consolidation: A theoretical synthesis of semantic networks, configuring in conditioning, S-R versus cognitive learning, normal forgetting the amnesic syndrome, and the hippocampal arousal system. *Psychological Review, 86,* 44-60.

Willshaw, D. (1981). Holography, associative memory, and inductive generalization. In G.E. Hinton & J.A. Anderson (Eds.), *Parallel models of associative memory.* Hillsdale, NJ: Erlbaum.

Wilson, H.R., & Cowan, J.D. (1972). Excitatory and inhibitory interactions in localized populations of model neurons. *Biophysical Journal, 12,* 1-24.

# Author Index

## A

Ackley, D.H., 5, *10*, 93, *95*, 129, *152*, 250, 287, 303, *306, 307*, 338, *339*, 344, *360*
Adams, M.J., 97, 111, *116*, 119, 122, 134, 136, 139, 144, 147, 148, *152*
Albano, J.E., 28, 54, *62*
Allen, R.B., 330, *339*
Allman, J., 348, *360*
Alspector, J., 330, *339*
Amari, S., *360*
Amit, D.J., 318, *326*
Anandan, P., 6, *10, 361*
Anderson, A., 4, 5, *10*
Anderson, J.A., 15, 17, *60, 61*, 66, *95*, 119, 124, 125, 127, 128, *152, 153*, 201, *201*, 250, *282*, 286, *307*, 351, 352, 353, *360, 361*
Arbib, M.A., 15, 47, 57, *60*, *360*
Asanuma, C., 2, 6, *10*, 71, *95*

## B

Baars, B.J., 104, 109, 112, 113, 114, *116, 117*
Backus, J., 3, *10*, 200, *201*
Ballard, D.H., 2, 7, *10*, 15, 19, 26, 28, 39, 44, 48, 49, 57, *60*, 63, 70, *95*, 99, *116*, 120, *153*, 156, 157, 161, 168, 173, 177, *178, 179, 202*, 286, 301, 302, *306*, 348, *361*
Banquet, J.P., 253, *281*
Barber, C., 348, *361*
Barlow, H.B., 312, *326*, 345, *361, 362*
Baron, J., 122, 136, *153*
Baron, R., *362*
Barrow, H.G., *179*
Barto, A.G., 6, *10, 11*, 352, *361*
Baudet, G.M., 177, *179*
Baum, E., 316, 317, *326*
Berg, T., 109, *116*
Berlin, B., 4, *10*
Berliner, H.J., 287, *306*
Bernstein, J., 207, 212, *242*
Berwick, R.C., 189, *201*
Bien, J., 199, *204*
Bienenstock, E.L., 230, *242*, 276, 278, *281*
Bienkowski, M.A., 110, *117*
Binder, K., 289, *306*

Blelloch, G., 3, 5, *10*, 332, 333, 336, *339*
Block, H.D., *326*
Bock, J.K., 107, *116*
Brachman, R., 359, *361*
Bresnan, J., 191, *202*
Brouwer, P.S., 352, *361*
Byrnes, J., 133, *153*

## C

Caplan, D., 57, *60*
Carpenter, G.A., 201, *203*, 249, 253, 255, 264, 266, 268, *281*
Chabot, R.J., 131, *153*
Charniak, E., 182, 183, 185, *201*
Chomsky, N., 189, *201*
Church, K.W., 191, *201*
Clark, W.A., 208, *242*
Cohen, M.A., 251, 253, 264, 271, *281*, 358, *361*
Collins, A.M., *60*, 185, 186, *201*
Cooper, L.N., 230, *242*, 276, 278, *281*
Cooper, P.R., 356, *361*
Cottrell, G.W., 3, 4, *10*, 184, 194, *202*
Cowan, J.D., 358, *363*
Cowey, A., 124, *153*
Crain, S., 191, *202*
Crick, F., 2, *10*, 71, *95*
Cutler, A., 115, *116*
Cynader, M.S., 354, *362*

## D

Davis, L.S., *179*
Debrunner, C., 200, *202*
DeJong, G.F., 194, *202*
Delcomyn, F., 54, *60*
Dell, G.S., *60*, 67, *95*, 100, 101, 102, 103, 105, 109, 111, 112, 113, 115, *116*
Denker, J.S,. 330, *339*
Desimone, R., *361*
Didday, R.L., 54, *60*
Duda, R.O., 48, *60*
Durr, W.K., 123, *153*

## E

Edelman, G.M., *60*, 350, *361*
Ellman, J.L., 67, *95*, 329, *339*
Emerling, M., 330, *340*
Estes, W.K., 136, *153*, 258, *281*

# AUTHOR INDEX

## F

Fahlman, S.E., 5, *10,* 14, 31, 49, *60,* 200, *202,* 301, 303, *306*
Fanty, M.A., 3, *10,* 338, *339,* 358, *361*
Farley, B.G., 208, *242*
Fay, D., 115, *116*
Feldman, J.A., 3, 5, 7, 8, *10, 11,* 17, 18, 19, 24, 25, 26, 39, 52, 59, *60,* 63, 70, *95,* 99, *116,* 119–120, *153,* 156, 168, 173, 178, *179,* 185, 201, *202,* 286, 301, 302, *306,* 346, 354, 355, 357, 358, *361*
Fennel, R.D., 184, 201, *202*
Feustal, T.C., 252, *283*
Fischler, M.A., 4, *11, 61*
Fodor, J., 7, *10,* 110, *116,* 189, 191, *202*
Fogelman-Soulie, F., 357, *361*
Forbus, K.D., 57, *60*
Ford, M., *202*
Forster, K.I., 110, *116*
Frazier, L., 183, 191, 201, *202*
Freuder, E.C., *60*
Fromkin, V.A., 115, *116*
Fukushima, K., 206, *242*

## G

Gallinari, P., 357, *361*
Garrett, M.F., 115, *116*
Garvey, T.D, 57, *61*
Gazdar, G., 189, *202*
Gelade, G., *62,* 83, *96*
Gelanter, *10*
Gelatt, C.D., 289, 290, *307*
Geman, D., 303, *306*
Geman, S., 303, *306*
Gibson, E.J., 125, 143, 149, *153*
Gigley, H.M., 191, *202*
Goddard, N., 3, *10*
Golden, R.M., 127, *153,* 344, *361*
Goldman, N., 184, 189, *203*
Goldman-Rakic, P.S., 354, *361*
Goodman, B., 183, *202*
Gordon, P.C., 115, *116*
Graf, H.P., 330, *339*
Grimson, W.E.L., 286, *306*
Grossberg, S., 3, 4, 6, *10,* 36, *61,* 66, *95,* 206, 216, 217, 219, *242,* 243, 247, 248, 249, 250, 251, 252, 253, 255, 258, 262, 264, 266, 268, 271, 273, 275, 276, 277, 278, 279, 280, *281, 282,* 358, *361*

Gutfreund, H.S., 318, *326*
Gutowski, W., 253, *282*

## H

Hanson, A.R., 15, *61, 179*
Harley, T.A., 115, *116*
Hart, P.E., 48, *60*
Hebb, D.O., 207, *242,* 309, *327,* 351, 352, *361*
Henderson, L., 143, *153*
Hendler, J., 5, *10*
Hewitt, C., 2, *10,* 184, *202*
Hicks, R.E, 15, *61*
Hillis, W.D., 5, *10,* 14, 49, *61,* 200, *202*
Hinton, G.E., 2, 4, 5, 6, 8, *10, 11,* 17, 30, 41, *61,* 66, 70, 86, 93, *95, 96,* 129, *152,* 158, 178, *179,* 194, 196, 201, *202,* 250, 267, *283,* 286, 287, 291, 301, 303, *306, 307,* 311, 322, 326, *327,* 329, 330, 338, *339, 340,* 344, 349, 350, 351, 352, 354, 358, *360, 361, 362, 363*
Hirst, 4
Hoff, M.E., 331, *340*
Hollbach, S.C., 356, *361*
Hopfield, J.J., 5, *10,* 250, *282,* 288, *307,* 309, *327,* 338, *340,* 350, 351, 358, *362*
Horn, B.K.P., *61*
Hough, P.V.C., 157, *179*
Howard, R.E., 331, *339*
Hubbard, W., 330, *339*
Hubel, D.H., 28, *61,* 156, *179,* 345, *362*
Hummel, R.A., 15, *62,* 158, *179*

## I

Ikeuchi, K., 177, *179*

## J

Jackel, L.D., 330, *339*
Ja'Ja, J., 14, *61*
Johnston, J.C., 122, 134, 136, 138, 139, 145, 147, *153, 154,* 252, *283*
Jones, R.S., 15, *60,* 119, 124, 125, 127, 128, 142, *153*
Juola, J.F., 122, 131, 133, 138, 139, 147, *153, 154*
Jusczyk, P.W., 38, *61*
Just, M.A., 201, *203*

## K

Kanade, T., 155, 156, 169, 176, *179*
Kandel, E.R., 25, 30, *61*

# AUTHOR INDEX

Kaplan, R., 191, *202*
Kawamota, A., 124, *153*
Kay, M., 189, *202*
Kay, P., 4, *10*
Keele, S.W., 258, *283*
Kender, J.R., 159, *179*
Kimball, J., 183, 191, *202*
Kimball, O.A., 44, *60*
Kimme, C., 48, *61*
Kinsbourne, M., 15, *61*
Kirkpatrick, S., 289, 290, *307*
Klatt, D.H., 115, *117*
Klein, R.M., 38, *61*
Knapp, A.G., 250, *282*
Knight, B.S., *326*
Kohonen, T., 124, *153*, 206, *242*, 250, 267, *282*, 309, 315, *327*, 352, *362*
Kornfeld, W., 176, *179*
Kosslyn, S.M., *61*
Kuffler, S.W., 21, *61*
Kullback, S., 293, *307*
Kwasny, S.C., 183, *202*

## L

Lakoff, G., 4, *10*
Lashley, K., 351, *362*
Le Cun, Y., 311, *327*, 357, *361*
Lefton, L.A., 133, 134, *153*
Leiman, J.M., 110, *117*, 186, 201, *203*
Lesser, V.R., 184, 201, *202*
Levesque, H., 359, *361*
Levine, D.S., 253, *282*
Little, W., 309, *327*
Loftus, E.F., *60*, 185, 186, *201*
Lowrance, J.D., *61*
Luce, R.D., 78, *96*
Lynch, G., 352, *362*
Lynne, K., 3, *10*

## M

MacKay, D.G., 100, 103, 104, 109, 112, 114, *116*, 182, *202*
Marcus, M.P., 182, 183, 191, *202, 203*
Marr, D.C., 15, 20, *61*, 156, 162, *179*
Marslen-Wilson, W., 182, *203*
Mason, M., 122, 133, 150, *153*
Massaro, D.W., 122, 142, 143, 144, 147, *153*
McCarthy, J., 211, *242*
McCaughey, M.W., 131, *153*

McClelland, J.L., 5, 8, *10*, 15, 16, *61*, 64, 65, 67, 76, 78, 83, 86, *95, 96*, 97, 98, 100, *116, 117*, 119, 120, 122, 124, 134, 136, 137, 138, 139, 143, 145, 147, 148, *153, 154*, 166, *179*, 185, 186, 201, *203*, 228, 243, 246, 247, 248, 250, 251, 253, *282, 283*, 322, 338, *340*, 342, 349, 350, 356, *361, 362*
McCulloch, W.S., 207, *242*
McDonald, J.E., 143, *154*
McGuiness, E., 348, *360*
Mead, C., 330, *340*
Medin, D., 4, *11*, 343, *363*
Mervis, C., 4, *11*
Merzenich, M.M., 354, *362*
Metropolis, N., 289, *307*
Meyer, D.E., 110, 115, *116, 117*
Mezard, M., *327*
Miezen, F., 348, *360*
Miller, G.A., 1, *10*
Miller, T.J., 122, 133, 138, 139, 147, *154*
Milne, R.W., 183, 191, *203*
Minsky, M.L., 1, 7, *10*, 20, *61*, 194, 196, *203*, 207, 211, 212, 213, *242*, 291, 303, *307*, 341, *362*
Mishkin, M., *362*
Moody, J., 316, 317, *326*
Moran, J., *361*
Motley, M.T., 104, 109, 112, 113, 114, *116, 117*
Mountcastle, B., *60*
Mozer, M.C., 67, 80, 81, 83, 84, *96*, 127, 128, 142, *153*, 353, *361*
Mulloney, B., *61*
Munro, P.W., 230, *242*, 276, 278, *281*

## N

Nadal, J.P., *327*
Nelson, R.J., 354, *362*
Newell, A., 1, *11*, 287, 291, *307, 362*
Newsome, S.L., 143, *154*
Nicholls, J.G., 21, *61*
Nii, H.P., 5, *11*
Nilsson, N.J., 4, *11*
Nooteboom, S.G., 109, *117*
Norman, D.A., 15, *61*, 100, *117*
Nuwer, M., *362*

## O

Ortony, A., 186, *203*

## P

Paap, K.R., 143, *154*
Palm, G., *362*
Palmer, S.E., 172, *179*
Papert, S., 1, *10,* 20, *61,* 212, 303, *242, 362*
Parker, D.B., 358, *362*
Parker, D.R., 311, *327*
PDP Research Group, 5, *11,* 322
Pearson, J.C., *363*
Pentland, A.P., 4, *11*
Perkel, D.H., 23, *61*
Phillips, C.G., *362*
Pikulski, J.J., 123, *153*
Pitts, W., *242*
Poggio, T., 15, 21, *61,* 70, *96*
Pollack, J.B., 3, 4, 5, 8, *11,* 67, *96,* 189, 200, *203*
Pollatsek, A., 133, 138, 147, *154*
Posner, M.I., 2, *11,* 14, 51, *61,* 258, *283*
Prager, J.M., *61*
Pribam, K.H., 1, *10, 362*
Pullum, G., 189, *202*
Pylyshyn, Z.W., 3, *11,* 182, *203*

## Q

Quillian, M.R., 5, *11,* 184, 185, *203*

## R

Radin, D., 186, *203*
Ratcliff, R., 15, *61,* 301, *307*
Reich, P.A., *60,* 100, 103, 109, 115, *116*
Reicher, G.M., 122, 130, 134, 136, *154*
Renyi, A., 293, *307*
Rieger, C., 184, 189, *203*
Riesbeck, C., 184, 189, *203*
Rips, L.J., *62*
Riseman, E.M., 15, *61, 179*
Ritz, S.A., 15, *60,* 119, 124, 125, 127, 128, 142, *153*
Roediger, H.L., III., 341, *362*
Rosch, E., 4, *11*
Rosenberg, C.R., 3, 5, 6, *10, 11,* 250, 267, 273, *283,* 329, 330, *340*
Rosenblatt, F., 1, *11,* 208, 210, 211, 212, *242,* 267, *283,* 291, *307,* 309, *326, 327*
Rosenbluth, A., 289, *307*
Rosenbluth, M., 289, *307*
Rosenfeld, A., 15, *62,* 158, *179*
Rosinski, R.R., 133, *154*

Rumelhart, D.E., 3, 5, 6, 8, 9, *10, 11,* 15, 16, *61,* 64, 65, 66, 76, 78, 83, 86, 93, *96,* 97, 98, 100, *117,* 119, 120, 124, 136, 137, 148, *154,* 166, *179,* 185, 186, *203,* 228, *242,* 243, 244, 247, 248, 249, 250, 251, 253, 267, 274, 275, 276, 278, 279, *282, 283,* 311, 322, 326, *327,* 329, 330, 338, *340,* 342, 349, 356, 358, *361, 362*

## S

Sabbah, D., 3, *11,* 15, 19, 21, 24, 28, 44, 57, *60, 62,* 159, 161, 168, 169, 172, 177, *179, 180,* 356, *362*
Sager, N., 183, *203*
Salasoo, A., 252, *283*
Samuel, A.G, 252, *283*
Saund, E., 329, *340*
Schadler, M., 131, *153*
Schank, R.C, 184, 189, *203*
Schein, S.J., *361*
Schindler, R.M., 133, 138, 147, *154*
Schneider, W., 329, *340*
Schoppman, A., 354, *362*
Schunck, B.G., *61*
Schvaneveldt, R.W., 110, *117,* 143, *154*
Schwartz, D., 330, *339*
Schwartz, M.L., 354, *361*
Seidenberg, M.S., 110, *117,* 186, 201, *203*
Sejnowski, T.J., 2, 5, 6, *10, 11,* 20, *62,* 70, 93, *95, 96,* 129, *152,* 250, 267, 273, *283,* 287, 291, 303, *306, 307,* 329, 330, 338, *339, 340,* 344, *360, 362*
Selman, 4
Shastri, L., 3, 5, 7, *11,* 329, *340,* 346, 357, *363*
Shattuck-Hufnagel, S., 115, *117*
Shaw, G., 309, *327, 363*
Shepard, G., 354, *363*
Shepherd, G.M., 71, *96*
Shieber, S.M., 183, 191, *203*
Shiffrin, R.M., 252, *283*
Shoben, E.J., *62*
Silverman, D.J., *363*
Silverstein, J.W., 15, *60,* 119, 124, 125, 127, 128, 142, *153*
Silviotti, M., 330, *340*
Simon, H.A., 287, *307*
Simon, J., *61*
Sklansky, J., 48, *61*
Small, S.L., 3, 4, *10, 11,* 182, 184, 186, 194, *202, 203*

Smith, E.E., 4, *11, 62,* 343, *363*
Smolensky, P., 7, *11,* 127, *154,* 303, *307*
Sompolinsky, H., 318, *326*
Sondheimer, N.K., 183, *202*
Spragins, A.B., 133, 134, *153*
Stanfill, C., 4, *11*
Stanovich, K.E., 110, *117*
Steedman, M., 191, *202*
Stefik, M., 57, *62*
Stemberger, J.P., 100, 103, 115, *117*
Stent, G.S., *62*
Stone, G.O., 251, 253, 258, 271, *282*
Straughn, B., 330, *339*
Stryker, M.P., 354, *362*
Sunshine, C.A., 24, *62*
Sutton, R.S., 6, *11,* 313, *327,* 352, *361*
Swinney, D.A., 186, 201, *203*

## T

Tanenhaus, M.K., 110, *117,* 186, *203*
Tank, D.W., 350, *362*
Taylor, G.A., 122, 133, 138, 139, 144, 147, *153, 154*
Teller, A., 289, *307*
Teller, E., 289, *307*
Tenenbaum, J.M., *179*
Tennant, D.M., 330, *339*
Terrace, H.S., 233, *242*
Terzopoulos, D., 286, *307*
Tesauro, G., 329, *340*
Thibadeau, R., 201, *203*
Thiria, S., 357, *361*
Thompson, R.E., 351, *363*
Thurston, I., 122, 136, *153*
Torioka, T., *62*
Torre, V., 70, *96*
Toulouse, G., *327*
Touretzky, D.S., 354, *363*
Triesman, A.M., 51, *62,* 83, *96*
Tyler, L.K., 182, *203*

## U

Ullman, S., *62*
Ungerleider, L.G., *361*

## V

van Heerden, P.J., *363*
van Santen, J.P.H., 252, *283*
Vecchi, M.P., 289, 290, *307*
Venezky, R.L., 122, 144, *153*
von der Malsburg, C., 59, *62,* 66, *96,* 206, 216, 217, *242,* 248, 275, 277, *282, 327,* 351, *363*
von Neumann, J., 304

## W

Waibel, A.H., 329, *340*
Waltz, D.L., 3, 4, 5, 8, *11,* 67, *96,* 158, 172, *180,* 183, 189, 194, 195, *202, 203,* 287, *307*
Wanner, E., 191, *203*
Watrous, R.L., 329, *340*
Well, A.D., 133, 138, 147, *154*
West, R.F., 110, *117*
Wheeler, D.D., 136, *154,* 252, *283*
Wheeler, K.E., 133, *154*
Wickelgren, W.A., *62,* 111, *117,* 358, *363*
Widrow, B., 268, *283,* 331, *340*
Wiesel, T.N., 28, *61,* 156, *179,* 345, *362*
Wilczek, F., 316, 317, *326*
Wilks, Y., 199, *204*
Williams, R.J., 6, *11,* 250, 267, *283,* 311, 322, 326, *327,* 330, *340,* 358, *362*
Willshaw, D.J., 59, *62,* 86, *96,* 344, 351, 353, *363*
Wilson, H.R., 358, *363*
Winograd, T., 191, 201, *204*
Winston, P.H., 287, *307*
Woods, W.A., 182, 198, *204*
Wurtz, R.H., 28, 54, *62*

## Z

Zeki, S., 28, *62,* 362
Zipser, D., 3, 9, *11,* 66, 93, *96,* 120, *154,* 243, 244, 247, 248, 249, 250, 251, 274, 275, 276, 278, 279, *283,* 329, *339, 340*
Zook, J.M., 354, *362*
Zucker, S.W., 15, *62,* 158, *179, 180*

# Subject Index

## A

ACT, 201
Accumulation of evidence, 166
Action potential, models of, 21
Actor formalisms, 184
Adaline model, 268
Adam's model of letter-within-word
  perception, 144-145
Adaptability of networks, 249
Adaptive filtering, bottom-up, 259-260
Adaptive resonance theory (ART), 6, 243,
    253-255
  attentional subsystem in, 254-255
  comparison with back propagation,
    267-272
  code learning in, example of, 266-269
  direct access to learned codes in, 257
  models, 248
  properties of, 256-259
  self-adjusting memory search, 257
  speech model, 273-274
  stable code learning in, 264
Algorithm, maximum-flow, 336
Alternating interpretations, 183
Ambiguity in natural language, 182
Ambiguity, effect on processing load, 182
Analog spreading activation, 184
Animal models, 15
Animal vision, 156
Application of connectionist networks,
  controlling physical motion, 54-57
  motor control of the eye, 52
  object recognition, 47-52
  language production, 97-117
ART (*see* adaptive resonance theory)
Artificial Intelligence (AI),
  applications to problems in, 155
  history, with relation to connectionism,
    1-3
  relationship to connectionist models, 210,
    343
  successes and failures of the standard
    approach, 211
Associative learning, 273
Associative maps, distributed, 267
Associative memory (*see also* auto-
    associative memory), 4-5

ATN's (*see* augmented transition networks),
    2, 182, 191
Attention, visual, role of, 177
Attentional,
  and orienting subsystems, interactions
    between, 262
  gain control in ART, 255, 263-264
  priming in ART, 255
  subsystem in ART, 254-255
  vigilance in ART, 255
  vigilance, modulation of, 258-259
Attractors, 5
Augmented transition networks (ATN's), 2,
    182, 191
Auto-associative memory, 310, 314-319
Auto-associative networks, 194
Auto-associator paradigm, 214
Auto-associator, binary, linear, 352
Autonomy of syntax, 184, 189

## B

Back propagation error learning, 6, 243,
    267-273, 329-340
  algorithm, 273, 309-327
  comparison with ART, 267-272
  implementation of, 329-340
  learning techniques, 358
Backgammon, 329
Backtracking,
  prevention in garden path sentences, 191
  elimination of, 176, 182
Bandwidth compression, 329
Barlow's neuron doctrine, 345
Bigram detectors, position-specific, 141
Biological concept representation, 311
Biological learning, 311
Biologically plausible models, 17-19
Blackboard systems, 2, 5, 201
Boltzmann distribution, 290
Boltzmann machines, 6, 285-307, 388
  encoder problems, solution on, 295-301
  learning algorithm, derivation, 304-306
  learning algorithms for, 285-307
  representation of concepts in a, 301-303
Bottom-up,
  adaptive filtering, 259-260
  and top-down connections, 48

Bottom-up, (cont.)
  and top-down processing, 120
  connections in origami world, 163
  processing in connectionist networks, 97
Brain state in a box (BSB) model, 119, 125
  harmony function for the, 127
  information processing in the, 125
  perceptual invariants in the, 142
Breadth-first chart parser, 189
BSB model (*see* Brain state in a box model), 119, 125

## C

C' perceptrons, 210
Cancellation links, 31
Capacity of feed-forward vs. Hopfield networks, 319
Cascaded ATN's, 198
Case frames, 194
Categories, recognition, 248
Categorization, 4
Category-based hierarchy, an example of representing a, 346-347
Causality, 6
Cell assemblies, dynamic activity of, 351
Centroids, use in representing objects, 160-161
Chaotic relaxation, 177
Chart parser, breadth-first, 189
CID (connection information distributor), 68-96
  computer simulation of, 75
  essential features of, 89-92
  information processing capacity of, 85-89
  networks, programmability, 68-69
  programmable modules in, 71
CIP (*see* Context in perception), 201
Classification paradigm, 214
Clustering of features, 215-219
  of stimulus patterns, 219
Clusters, winner-take-all, 215-217
CM-NETtalk, 336-338
Context in perception (CIP), 201
Coalitions of nodes, stable, 99, 168
Coarse coding, 41-44, 220
Coarse-fine coding, 41-44
  example of, 44
  tuning of, 45
Code learning, example of in ART, 266-269
  stabilization of, 260-262
Cognitive sciences, theories of, 14

Combinatorial implosion, 176
Communication between connectionist modules, 302-303
Communication problems of holographic models, 352
Compact representation position, 345-350
Competition and cooperation, 176
Competition, intermodality ART, 255
Competitive learning, 205, 215-222, 243-283
  architecture for, 215-217
  correlated teaching inputs in, 232-233
  detecting different levels of structure in, 232
  experimental results of, 222
  feature discovery by, 247-248
  formal analysis, 220-222
  formal analysis, examples of, 226-228
  geometric interpretation of, 218
  learning role for, 217
  letter similarity effects in, 232
  stability of, 222-223
  use in discovering horizontal and vertical lines, 234-239
  use of multilayered systems for, 206
  formalization of, 274-279
  vs. interactive activation, 251
Comprehension errors in natural language, 183
Compression, bandwidth, 329
Computation rates, estimates of, neural, 343
Computation with limited precision, 41
  number of units required for, 30
Computational temperature, 289-290
Computational units, self-scaling, 256-257
Computer simulation of CID, 75
Computer vision, 2, 155
Computers, massively parallel, 5-6, 14, 200, 329, 340
Computers, parallel, 3
Computing units, neuron-like, 19-28
Concept representation, biological, 311
Concepts, connectionist representation of, 341-363
Conceptual hierarchy, 158
Conditioned reflexes, encoding of, 351
Conference on Mechanization of Thought Processes, 211
Configurations, minimum energy, 288-290
Conjunctive connection strategy, 32-34
  connections, 21-22
  connections, examples of network using, 34

# SUBJECT INDEX

Connecting nodes, three-way, 5
Connection Machine, 200, 329-340
Connection strategy, conjunctive, 32-34
Connectionism and interactive activation, 94
Connectionist networks
  scaling of, 3, 339
  implementation of recursion in, 92-93
  learning in, 4
  models, hierarchical, 97-117
    AI and, 343
    controlling physical motion, application of, 54-57
    history of, 1-3
    local vs. distributed, 8, 66
    motor control of the eye, application of, 52
    object recognition, application of, 47-52
    serial sequential programming in, 93-94
    statistical mechanics, application to, 285
    symbol manipulation and, 287-288
  systems, natural language, 4
  bottom-up processing in, 97
  modules, communicating information between, 302-303
  programmable vs. hardwired, 64-67
  representation of concepts, 341-363
  scaling of, 339
  timing in, 7-8
  top-down processing in, 97-117
Connections
  conjunctive, 21-22
  number required, 40
  symmetric, 287-288
  top-down and bottom-up, 48
  vs. symbols, 16
Consistent interpretations, use of decay in finding, 167
Constraint propagation, 2
Constraint-satisfaction searches, 285-307
Constraints, combining of, 341
  linguistic, 107
Content-addressable memory, 309-327
Context, example of connectionist model for, 196-199
Context, models of, 195
Context-free grammar, 189
Contrast-enhancement in short-term memory, 259-260
Control, 211

Controlling physical motion, application of connectionist models to, 54-57
Convergence of networks, 35
Convergence rate for gradient descent algorithms, 327
  steepest descent algorithms, 327
  neural computation, 26
Converting time to space, 56
Cooperation, competition and, 176
Cooperative computational theories, 15
Correlated memories, discriminating between, 322-324
Correlated teaching inputs in competitive learning, 232-233
Correlation-based learning schemes, 356
Cortical neurons, number of input fibers to, 354
Cray-2, 329-330
Credit-assignment problem, 291, 357
Critical feature patterns, 254, 256-257
Cross-talk, 18, 352-353
  and interference, 80-85
  in neural encodings, 349
  spatial coherence, use to avoid, 46

## D

Death of neurons, 347-348
Decay, effects on grandmother memory, 321
  role in reuse of network modules, 198
  use of in finding consistent interpretations, 167
Decision making, 211
  in neural models, 35
Dendritic spikes, 21
Detectors for features and properties of images, 48
Detector, regularity, 214
Detectors, position specific letter, 230
Deterministic parsers, 191
Developmental consequences of the learning rule and training procedure, 123
Developmental effects in the LW model, explanation of, 136
Developmental neural model of visual word perception, 119-143
Digital spreading activation, 184
Direct access to learned codes in ART, 257
Direct simulation of logical processes, 211
Discovery of features, 205
  by competitive learning, 247-248
Discrete states, neural models with, 24

Dishabituation, 25
Disjunctive firing conditions, 21
Distributed and punctate models, arguments for and against, 348
Distributed associative maps, 267
Distributed holographic models, encoding propositions in, 354
Distributed models, comparison of local and, 344
Distributed neural activity, 348
Distributed representation for CID, 87–89, 91
Distributed representation vs local, 301–303
Distributed vs. grandmother memory, 321
Distributed vs. local connectionist models, 8
Distributed vs. local representations, 85–86
Dynamic connections, 201
Dynamic activity of cell assemblies, 351

E

Editing, prearticulatory, 106
Effect(s),
  of pseudowords on letter perception, 111
  of sequential redundancy on LW model, 141–143
Emergent property(ies)
  of network interactions, 244–246
  of neural networks, 356
  logic as an, 4
Encoder problems, use of Boltzmann Machines for, 295–301
Encoding(s) of,
  conditioned reflexes, 351
  propositions in distributed holographic models, 354
  spatial frequency, 348
Energy configurations, minimum, 288–290
Energy landscape, 302, 313–314
Energy minima, global, 289–290
Environment, modeling of, 292–294
Equilibrium states in competitive learning, 221
Error back propagation algorithm, 309
Error correction, 3, 310
Error rate, relationship to learning, 318
Error tolerance, 310–311, 322–325
Errors in comprehension, modeling of, 191–194
Errors, misordering, 108
  speech, 103
  substitution, 108

Event shape diagrams, 194
Events, segmentation of, 6
Evidence, accumulation of, 166
Evidential vs. logical reasoning, 3–4
Exemplars, recognition of prototypes and, 258
Expectations
  as exemplars or as prototypes, 270–271
  learned, 261
Experiment on word bias effects in letter feature perception, 142
Experimental results of competitive learning, 222
Expert systems, 2

F

Fault-tolerance, 3, 5–6, 351
Feature correlations and orthographic information in the LW model, 123–124
Feature detectors, 206
Feature discovery, 205
  by competitive learning, 247–248
Feature spaces, hierarchy of, 158
Feature-phoneme feedback, 108
Feature-to-phoneme feedback during production, evidence for, 215
Features, clustering of, 215–219
Feed forward networks, 309–327
  capacity vs. Hopfield networks, 319
Feedback, feature to phoneme, evidence for, 115
  feature-phoneme, 108
  phoneme-to-morpheme, 103–106
Feedback system as a "lexical editor," 104–106
Fermi function, 312
Figure/ground perception, 6
Fine-grained parallel computers, 329–340
Finite automata, classical, 22
Firing conditions, disjunctive, 21
Firing, neural models of, 22
Focus concepts, 4
Forced learning, 209
Forgetfulness, 309–327
Forgetting, 320
Formalisms, mathematical importance to network models, 244
Fragment masking, visual letter, LW experiment on, 136
Frame problem, 2
Function, logistic, 331

# SUBJECT INDEX

Function, objective, 294
Function, sigmoid, 331
Functional decomposition of networks, 40
Fundamental network modules, 245

## G

Gain control,
  in ART, attentional, 255
  attentional, 263-264
Garden path sentences, 183, 191
  preventing backtracking in, 191
Gated dipole field, 262
Generalization, 351
  and hierarchy, 356
  in phrase structure grammars (GPSG), 189
Generative model for phonology, 102
Geometric interpretation of competitive learning, 218
Global energy function, minimizing, 311
Global energy minima, 289-290
Globally consistent interpretations in origami world, 165
Grammar, semantic, 183
Gradient descent algorithms, rate of convergence of, 327
Grandmother cells, 9, 309-327, 343
  evocation of, 314
  relearning after loss of, 321
Grandmother memory, 309-327
  effects of decay on, 321
  vs. distributed memory, 321

## H

HEARSAY II speech understanding system, 184, 201
HOPE, 191
Hardwired vs. programmable connectionist networks, 64
Harmony function for the BSB model, 127
Hebb/Hebbian,
  cell assemblies, 38
  learning, 207
  learning rule, 124, 352
  learning rule, shortcomings of, 357
Hetero-associative memory, 310, 314-319
Heuristic search, 1
Hidden units, 267, 292, 311
Hierarchical connectionist models, positive feedback in, 97-117
Hierarchical dependencies in the LW model, 124

Hierarchy and generalization, 356
  of feature spaces, 158
  of parameter spaces, 158
  category-based, on example of, 346-347
Higher-level langauges and intermediate levels of computation, 18
Holograms, optical, 351
Holographic memory models, 343-344, 350
  comparison of, punctate and, 344
  models, advantages of, 351
  representational schemes, problems with, 352-354
  vs. punctate models, 344
  communication problems of, 352
  spin-glasses as, 349
Homunculus, 190
Hopfield network, 310, 338
  capacity vs. feed-forward networks, 319
Horizontal and vertical lines, discovery by competitive learning, 234-239
Hough transform, 48, 157, 163
Hypercube corner, 127
Hyperspheres, mapping patterns onto, 217-218
Hypotheses, simultaneous evolution of, 15-16

## I

IA (interactive activation), 120, 145-148
IPM (information processing model), 13-17
Identification of perceptual features, 97
Ill-formed input, 183
Ill-formed syntactic structure, 191
Images, detecting features and properties of, 48
Implementation of network learning, 329-340
Indexing problem, 49
Information processing models (IPM), 13-17
Inhibition, 21
  lateral, 25-28, 98
  mutual, 25-28
Inhibitory connections in origami world, 164
Innate organization, 6
Input vectors, noisy, 294
Integrating semantic and lexical strategies, 191
Integration of knowledge sources, 183-184
  modules, 5
  syntax, semantics and context, 181-204

Interacting between attentional and orienting systems, 254
Interactive activation (IA) model, 96, 120, 243–248
  and connectionism, 94
  and crosstalk in, 80–85
  competitive learning vs., 251
  results of simulations, 78–79
  word perception in, 145–148
  word recognition in, 63–96
Intermodality competition in ART, 255
Interpreted symbol system in brains, 18
Intrusion errors in word recognition, 63–96
Invariance to translation and rotation, 159
Invariant object properties, 49
Iterative refinement, 158

## K

Knowledge acquisition, 2
Knowledge representation in semantic networks, 329
Knowledge store in CID, 73–76

## L

LISP, 18
LSP project, 183
LTM (long-term memory), 259–263
LW (letter-in-word),
  mixed-case effects experiment, 131–135
  visual letter fragment masking experiment, 136
LW (letter in word) model,
  explanation of development effects in, 136
  of orthographic and case effects in, 136
  computer simulation details for the training procedure, 129
  development of orthographic information in, 123–125
  feature correlations and orthographic information in the, 123–124
  hierarchy of dependencies in, 124
  learning algorithm in, 127–129
  learning rule, 123
  network model, 120–125
  training procedure, 123
Language production system, variability in input to, 107
Language, applications of connectionist networks to, 97–117
  human, 97
  parallel network models of, 100

Language (cont.)
  psychological evidence for positive feedback in, 109
Large number of input fibers to cortical neurons, 354
Lateral inhibition, 25–28, 98
  in origami world, 164
  in natural language processing, 184–186
Law of mass action, 351
Laws of thermodynamics, 211
Leaky learning model, 229
Learned codes in ART, direct access to, 257
Learned expectations, 261
Learning, 309–327, 357–359
  algorithm for Boltzmann Machines, 285–307
  algorithms, 285–307
  back propagation, 6
  biological, 311
  connectionist, 4
  derivation of, 304–306
  forced, 209
  from experience, 6
  Hebbian, 207
  history of research in, 206–213
  in origami world, 178
  in the LW model, 127–129
  leaky, 229
  learning rule, LW model, 123
  methods for stabilizing, 250
  of stimulus sets that are not linearly separable, 205
  prewired structure in, 357
  relationship to error rate, 318
  role for competitive rule in, 217
  schemes, correlation-based, 356
  self-stabilized, 249–251
  spontaneous, 209
  temporally unstable, 248
  triggering of by atypical features, 121
  triggering of by mismatched features, 256
  words and letters, 228
Letter and word levels, assumption of, 252
Letter detectors, position specific, 230
Letter feature perception, experiment on word bias effects in, 142
Letter perception, effect of pseudowords on, 111
Letter recognition LW model, accuracy testing procedure for, 130
  psychological phenomena in, 100

Letter similarity effects in competitive learning, 232
Letters and words, 228
Letter-in-word (LW) model, 120–154
Letter-within-word perception, Adam's model of, 144–145
Level interactions, 246–247
Levels of structure in competitive learning, 232
Lexical access, 182, 186
  modular account of, 110
Lexical bias effect, 103
  in slips, 109
  in speech errors, 110
  evidence of effect on slips, 112
  feedback model account of, 111–113
  single-level account of, 111–113
  time dependence of, 112
Lexical context phenomena, 100
Lexical decisions, semantic priming in, 110
Lexical editor, feedback system as a, 104–106
Lexical neighborhood model, 114–115
Lexical strategies, integrating semantic and, 191
Limited precision computation, 41
Line drawings, understanding of, 155
Linear threshold elements, 20
Linearly separability, learning stimulus sets that lack, 205
Linguistic constraints, 107
Links, multiplicative, 70
Local and distributed models, comparison of, 344
Local energy minima, 288
  use of noise to escape from, 289–290
Local vs. distribute representations, 8, 66, 85–86, 301–303
Logic, 1, 3–4
  as an emergent property, 4
  relationship to statistics, 210–211
Logical processes, direct simulation of, 211
Logistic function, 331
Long term memory (LTM), 259–263
  LTM traces, 248–249
Loosely-coupled systems, 24

## M

Map, retinotopic, 48
Marker passing, 2
  algorithms, 184

Masking field, 264–267
Masking stimulus, relation to word superiority effect, 136
Mass action, law of, 351
Massively parallel computers, 5–6, 14
Matching, best vs. exact, 3
Maximum-flow algorithm, 336
Mechanics, statistical, application to connectionist modeling, 285
Memory,
  associative, 4–5
  auto-associative, 310, 314–319
  content-addressable, 309–327
  hetero-associative, 310, 314–319
  neural substrate of, 341
Memory capacity of networks, 310
Memory search in ART, self-adjusting, 257
Mental images, 195
Microfeatures, 5, 194–199, 201, 352–353
Microworlds, 4
Minima, local, noise used to escape from, 289–290
Minimal attachment, 183, 201
Minimization using Newton's method, 313
Minimizing global energy function, 311
Minimum energy configurations, 288–290
Minsky's Neural Net Machine, 207
Misordering errors, 108
Missing information
  in origami world, 171–174
  rate of convergence with, 172
Mixed-case effects, LW experiment on, 131–135, 139–140
Modeling an environment, 292–294
Models,
  biologically plausible, 17–19
  information processing, 13–17
  phonological encoding, 101–103
  vergence control, 54–56
Modifiers, network using, example of, 33
Modular account of lexical access, 110
Modulation of attentional vigilance, 258–259
Modules in CID, programmable, 71
  in ART, fundamental, 245
  integration of, 5
Morphemes, 100–103
Motor control of the eye, application of connectionist models to, 52
Motor performance, skilled, 54
Motor sequence, 54

378 SUBJECT INDEX

Multi-layered networks, 331
Multi-layered system, use in competitive learning, 206
Multi-level interactive explanations of psycholinguistics phenomena, 109–110
Multiplicative links, 70
Mutual inhibition (see also lateral inhibition), 25–28
Mutually reinforcing nodes, 99

**N**

Natural and artificial intelligence, 8
Natural language,
  comprehension errors in, 183
  connectionist systems for, 4
  interpretation of, 181–204
Natural language processing, examples of connectionist systems for, 186
  lateral inhibition in, 184–186
  spreading activation in, 184–186
  representation schemes for, 2
Nearly decomposable knowledge systems, 182
Necker Cube, 29, 183
NETL, 200
NETtalk, 267, 273, 329–330
  speed on various machines, 336–338
Network(s),
  adaptability of, 249
  coarse-fine, tuning of, 45
  convergence of, 35
  feed-forward, 309–327
  functional decomposition, 40
  Hopfield, 310, 338
  learning, implementation of, 329–340
  memory capacity of, 310
  multi-layered, 331
  plasticity of, 249
  programmable CID, 68–69
  reduplicated vs. sequential, 66–67
  regulated, 35–39
  regulated, example of, 37
  stable states of, 35
  unit/value, 28–37
  using conjunctive connections, example of, 34
  using modifiers, example of, 33
  winner-take-all (WTA), 37, 99
Neural activity and perception, 345–348
Neural computation rates, estimates of, 343

Neural computation, convergence rate of, 26
Neural computing elements, speed of, 13–14
Neural doctrine, Barlow's, 345
Neural encodings, cross-talk in, 349
Neural model of visual word perception, developmental, 119–143
Neural models of firing, 22
Neural models, decision-making in, 35
Neural nets on digital computers, simulation of original idea for, 208
Neural network model, letter-in-word (LW), 120–125
Neural networks, emergent properties of, 356
Neural reorganization studies, 354–355
Neural representation of letters and words, 146–148
Neural substrate of memory, 341
Neuron-like computing units, 19–28
Neurons,
  death of, 347–348
  spatio-temporal summation in, 21
  speed of, 20
Neuroscience, 2
Neuroscience, theoretical, 341
Newton's method, minimization using, 313
Noise,
  graceful degradation in presence of, 156
  threshold, 166–167
  tolerance, 309–327
  tolerance in origami world, 170
  use to escape from local minima, 289–290
Noisy input, coping with, 100
Noisy input vectors, 294
Non-Hebbian associative law, 280
Non-monotonic logics, 2
Nongrammatical text, 183
Normalization factor, use of, 166
Nuclear magnetic resonance, 2

**O**

Object recognition, application of connectionist models to, 47–52
Objective function, 294
Occlusion, 156
Occlusion in origami world, 171–174
Optical computers, 339
Optical holograms, 351
Organization, innate, 6
Orienting subsystems, interactions between, attentional and, 262

## SUBJECT INDEX

Orienting systems, interacting attentional and, 254
Origami,
  objects, visual recognition of, 155
  polyhedia, representation of, 162
Origami world, 155
  bottom-up connections in, 163
  examples of solutions in, 168-177
  globally consistent interpretations in, 165
  hierarchy, 161
  inhibitory connections in, 164
  lateral inhibitory connections in, 164
  learning new objects, 178
  missing information in, 171-174
  occlusion in, 171-174
  recognizing novel objects in, 178
  top-down connections in, 164
Orthographic,
  and case effects in the LW model, explanation of, 136
  information in the LW model, development of, 123-125
  information in the LW model, feature correlations and, 123-124

## P

Parallel computers, 3, 5-6, 14, 329-340
Parallel models of language production, 100
  of typing, 100
Parallel parsing, 181-204
Parameter spaces, hierarchy of, 158
Parity, detection with perceptrons, 212-213
Parsing, word expert, 184
  parallel, 181-204
Pattern associator paradigm, 214
  classifiers, 206
  completion, 214, 352
  correction, 311
Patterns, simultaneous processing of, 63-96
Perception model, word, 228
Perception of letters, 110
  developmental neural model of visual word, 119-143
  figure/ground, 6
  neural activity and, 345, 348
  printing words, 15, 97-117
Perceptrons, 1, 206, 208
  C', 210
  convergence theorem, 291
  effects of Minsky and Papert's book on, 212-213

Perceptrons (cont.)
  learning theorem, 208
  limits of, 213-215
  random, 212
  relationship with artificial intelligence, 210
  research, 210-212
  two layer, 211
Perceptual features identification of, 97
Perceptual invariants in the BSB model, 142
Perceptual rivalry, 28
Phoneme-morpheme feedback, 103-106, 109
Phoneme-morpheme feedback, problems with, 105
Phonemes, 100-103
Phonological encoding, model of, 101-103
Phonological production, 100
Phonological similarity, activation based on, 114
Phonology, generative model for, 102
Physical symbol system, 1
PLANES project, 183
PLANNER, 2
Plasticity of networks, 249
Pontifical cell, 343
Position specific letter detectors, 230
Position specific bigram detectors, 141
Positive feedback, 26
  in hierarchical connectionist models, 97-117
  word level to letter level, 97
Potential, models of action, 21
Prearticulatory editing, 106
Precision, computation with limited, 41
Predicate calculus, 18
Preference semantics, 196
Prewired structure in learning, role of, 357
Primal sketch, 162
Priming, 28, 186
  in art, attentional, 255
  of networks, 101
  attentional, 263
Principle of early closure, 183
Processing, levels and interactions, 246-247
Processing, bottom-up and top-down, 120
Production systems, 18
Production of words, 97-117
Production, phonological, 100
Programmable CID networks, 68-69
  connectionist mechanisms, 64

Programmable CID networks (cont.)
  vs. hardwired connectionist networks, 64–67
  modules in CID, 71
Programming style, von Neumann, 199
PROLOG, 2
Pronunciation of text, 330–331
Properties, invariant object, 49
Propositions encoding in distributed holographic models, 354
Protocol analysis, 1
Prototypes, 4, 254
Prototypes and exemplars, recognition of, 258
Psycholinguistic phenomena,
  multi-level interactive explanations of, 109–110
  single-level structural explanations of, 109–110
Psychological evidence for positive feedback in language production, 109
Psychological phenomena in letter recognition, 100
Punctate and distributed models, arguments for and against, 348
Punctate and holographic models, comparison of, 344
Pure feature representations, 356

## Q

Qualification problem, 2
Quantitative shape recovery of visual objects, 156, 159

## R

Random perceptron, 212
Rate of convergence with missing information, 172
Rate of speaking, 103
Reaction time LW model, testing procedure, 130
Real-time network models, importance of, 244–246
Reasoning, evidential vs. logical, 3–4
Recognition categories, 248
Recognition prototypes and exemplars, 258
Recognition, speech, 329
Recognizing novel objects in origami world, 178
Recovering shape using skewed symmetry, 159

Recruiting compact representation for new concepts, 358
Recursion, 3
  connectionist implementation of, 92–93
READER system, 201
Reading, partial theory of, 15
Redundancy information, sequential, 122
  spatial, 122
Reduplicated vs. sequential networks, 66–67
Refinement, iterative, 158
Regularity detector, 214
Regulated networks, 35–39
Relaxation processes, 158
Relaxing constraints in parsing, 183
Relearning after grandmother cell loss, 321
Repeated phoneme effect, 106–109
  interactive feedback account of, 111
  single-level accoun tof, 110
Representation of concepts in a Boltzmann Machine, 301–303
  concepts, connectionist, 341–363
  letters and words, neural, 146–148
  pure feature, 356
  schemas, natural language, 2
Representing a category-based hierarchy, example of, 346–347
Representing words as state vectors, 126
Retinotopic map, 48
Right association, principle of, 183
Role of prewired structure in learning, 357
Rotation, invariance to, 159
Router operations, 332

## S

Saturation, modeling of, 24–25
Sausage Machine, 191
Scaling up connectionist architectures, 3, 339
Scan operations, 332
Schemas, 194
Search, elimination of need for, 176
Segmentation of events, 6
Selectional restrictions, 194
Self adjusting memory search in aRT, 257
Self evolution, 6
Self occlusion, 174
Self organization, 250, 256
Self scaling computational units, 256–257
Self stabilized learning, 249–251
Semantic and lexical strategies, integrating, 191

# SUBJECT INDEX

Semantic garden path sentences, 183, 192-195
Semantic grammar, 183
Semantic memory, 15
Semantic networks, knowledge representation in, 329
Semantic priming, 185
Sequential redundancy information, 122
Sequential redundancy on LW model, effects of, 141-143
Sequential vs. reduplicated networks, 66-67
Serial sequential programming in connnectionist models, 93-94
Shape recovery, quantitative, 159
Shape, skewed symmetry, use in recovering, 159
Short-term memory (STM), 259-263
  contrast-enhancement in, 259-260
  reset and search, 262
Sigma-pi units, 338
Sigmoid function, 331
SIMD computers, 332
Simulated annealing, 285-387
Simulation of training procedure, LW models, 129
Simulation of neural nets on digital computers, 208
Simultaneous evolution of hypotheses, 15-16
Simultaneous processing of patterns, 63-96
Single-level structural explanations of psycholinguistic phenomena, 109-110
Skewed faces, representation of, 162
Skewed symmetry, 159
  use in recovering shape, 159
Skilled motor performance, 54
Slips of the tongue, 97, 102
  evidence for lexical bias in, 112
  frequency asymmetries in, 105
  lexical bias in, 109
Sound similarity effect, 115
Spatial coherence, use to avoid cross-talk, 46
Spatial frequency encoding, 348
Spatial redundancy information, 122
Spatio-temporal summation in neurons, 21
Speech acts, 107
Speech errors, 103
Speech lexical bias in, 110
Speech model, 273-274
Speech rate, 103
Speech recognition, 329
Speed of computing elements, 13-14
Speed of neurons, 20
Speed of NET talk on various machines, 336-338
SPICE circuit simulator, 336
Spikes, dendritic, 21
Spin-glasses as holographic memory models, 349
Spontaneous learning, 209
Spreading activation, 101
  in natural language processing, 184-186
  theories, 15
Stability of competitive learning, 222-223
Stability-plasticity dilemma, 249-251, 262
Stabilization of code learning, 260-262
Stable coalitions of nodes, 38-39, 99, 168, 186
Stable code learning in ART, 264
  of sparse patterns, theorem, 280
  formalization of, 279
Stable states of networks, 35
State vectors, representing words as, 126
Statistical mechanics, application to connectionist modeling, 285
Statistics, relationship to logic, 210
STM (see short-term memory), 259-263
  reset and search, 262
Steepest descent algorithms, rate of convergence, 327
Steepest descent, limitations of, 313
Stimulus patterns, clustering of, 219
  structure of, 219
Structural preferences, 191, 201
Subjective contour, 28
Substitution errors, 108
Summary of articles in this book, 9
Symbol manipulation and connectionist models, 287-288
Symbols system in brains, interpreted, 18
Symbols vs. connections, 16
Symmetric connections, 287-288
Symmetric weights, systems with, 303
Symmetry, skewed, 159
Syntactic structure, integrating with word selection, 107
Syntax, autonomy of, 189
Systems with symmetric weights, 303
Systems, loosely-coupled, 24
Syntax, autonomy of, 184

## T

Temperature, computational, 289–290, 312–313
Template matching, top-down, 260–262
Temporarily unstable learning, 248
Testing procedure, letter recognition to LW model, 130
Text, pronunciation of, 330–331
Theorem, perceptron convergence, 291
Theoretical neuroscience, 341
Theory of reading, 15
Thermal equilibrium, 290–291
Thermodynamics, laws of, 211
Three-way connecting nodes, 5
Threshold elements, linear, 20
Threshold, noise, 166–167
Time dependence of lexical bias, 112
Time to space conversion, 56
Timing in connectionist networks, 7–8
Tolerance to errors, 310–311, 322–325
Top-down and bottom-up
  connections, 48
  mechanisms, 254–255
  processing, 120
Top-down connections in origami world, 164
Top-down processing in connectionist networks, 97–117
Top-down template matching, 260–262
Toy AI problems, 4
Training procedure, LW model, 123
  computer simulation of, 129
Transgraphemic information, 122
Translation, invariance to, 159
Trihedral objects, 159
Tuning of coarse-fine networks, 45
Turing universality, 14
Two layer perceptrons, 211
Two thirds rule, 263–264
Typing, parallel network models of, 100

## U

Ungrammatical inputs, 183
Unification, 4
Unit/value networks, 28–37
Unit/value organization, evidence for, 28
Units,
  disjunctive normal form, 22
  hidden, 292, 311
  numbers required for computation, 30
  visible, 292

## V

Variability of input to language production system, 107
Variable binding, 3
Variable mappings, 30
Varying threshold learning methods, 230
VAX 750, 314
VAX 780, 330, 337
Vector coding, example of, 126
Vergence control, model of, 54–56
Vigilance, modulation of attentional, 258–259
Virtual processors (VP's), 335–336
Visible units, 292
Vision, animal, 156
Vision, computer, 2, 155
Visual attention, role of, 177
Visual letter fragment masking, LW experiment on, 136
Visual object, quantitative shape recovery, 156
Visual recognition of origami objects, 155
Visual word perception, developmental neural model of, 119–143
VLSI technology, 330
von Neumann Machine, 199–200
von Neumann programming style, 199
VP's (virtual processing), 335–336

## W

Wait and see parser, 182
Weak constraints, 287
Weight decay, 321
Weight transport, 271, 273
Weight, need for limits on growth, 209–210
Wildrow-Hoff algorithm, 331
Winner-take-all (WTA) networks, 35–38, 59, 99, 168
  clusters, 216
Word bias effects in letter feature perception, 142
Word detection, 231–232
Word expert parsing, 184, 186
Word level to letter level, positive feedback in, 97
Word perception, 97–117
  interactive activation model of, 145–148
  intrusion errors in, 63–96
  model, 228
  models, criticism of, 247
  of printed, 15

Word perception (cont.)
 production of, 97-117
 recognition, interactive activation model of, 63-96
 visual, developmental neural model of, 119-143
Word selection, influence on from syntactic structure, 107
Word superiority effect, 15, 105, 100, 110, 342
 dependence on masking stimulus, 136
Words and letters, learning, 228